Supply Chain Strategies and the Engineer-to-Order Approach

Richard Addo-Tenkorang
University of Vaasa, Finland

Jussi Kantola
University of Vaasa, Finland

Petri Helo
University of Vaasa, Finland

Ahm Shamsuzzoha
Sultan Qaboos University, Oman

A volume in the Advances in Logistics, Operations, and Management Science (ALOMS) Book Series

BUSINESS SCIENCE Reference
An Imprint of IGI Global

Published in the United States of America by
Business Science Reference (an imprint of IGI Global)
701 E. Chocolate Avenue
Hershey PA, USA 17033
Tel: 717-533-8845
Fax: 717-533-8661
E-mail: cust@igi-global.com
Web site: http://www.igi-global.com

Library of Congress Cataloging-in-Publication Data

Names: Addo-Tenkorang, Richard, 1978- editor.
Title: Supply chain strategies and the engineer-to-order approach / Richard Addo-Tenkorang, Jussi Kantola, Petri Helo and Ahm Shamsuzzoha, editors.
Description: Hershey, PA : Business Science Reference, [2016] | Includes bibliographical references and index.
Identifiers: LCCN 2016002535| ISBN 9781522500216 (hardcover : alk. paper) | ISBN 9781522500223 (ebook : alk. paper)
Subjects: LCSH: Business logistics.
Classification: LCC HD38.5 .S8962436 2016 | DDC 658.5/038--dc23 LC record available at http://lccn.loc.gov/2016002535

This book is published in the IGI Global book series Advances in Logistics, Operations, and Management Science (ALOMS) (ISSN: 2327-350X; eISSN: 2327-3518)

British Cataloguing in Publication Data
A Cataloguing in Publication record for this book is available from the British Library.

All work contributed to this book is new, previously-unpublished material. The views expressed in this book are those of the authors, but not necessarily of the publisher.

For electronic access to this publication, please contact: eresources@igi-global.com.

Advances in Logistics, Operations, and Management Science (ALOMS) Book Series

John Wang
Montclair State University, USA

ISSN: 2327-350X
EISSN: 2327-3518

Mission

Operations research and management science continue to influence business processes, administration, and management information systems, particularly in covering the application methods for decision-making processes. New case studies and applications on management science, operations management, social sciences, and other behavioral sciences have been incorporated into business and organizations real-world objectives.

The **Advances in Logistics, Operations, and Management Science** (ALOMS) Book Series provides a collection of reference publications on the current trends, applications, theories, and practices in the management science field. Providing relevant and current research, this series and its individual publications would be useful for academics, researchers, scholars, and practitioners interested in improving decision making models and business functions.

Coverage

- Networks
- Marketing engineering
- Risk Management
- Decision analysis and decision support
- Production management
- Services management
- Operations Management
- Computing and information technologies
- Finance
- Information management

IGI Global is currently accepting manuscripts for publication within this series. To submit a proposal for a volume in this series, please contact our Acquisition Editors at Acquisitions@igi-global.com or visit: http://www.igi-global.com/publish/.

The Advances in Logistics, Operations, and Management Science (ALOMS) Book Series (ISSN 2327-350X) is published by IGI Global, 701 E. Chocolate Avenue, Hershey, PA 17033-1240, USA, www.igi-global.com. This series is composed of titles available for purchase individually; each title is edited to be contextually exclusive from any other title within the series. For pricing and ordering information please visit http://www.igi-global.com/book-series/advances-logistics-operations-management-science/37170. Postmaster: Send all address changes to above address.

Titles in this Series

For a list of additional titles in this series, please visit: www.igi-global.com

Optimal Inventory Control and Management Techniques
Mandeep Mittal (Amity School of Engineering and Technology, India) and Nita H. Shah (Gujarat University, India)
Business Science Reference • copyright 2016 • 406pp • H/C (ISBN: 9781466698888) • US $170.00 (our price)

Handbook of Research on Strategic Supply Chain Management in the Retail Industry
Narasimha Kamath (JDA Software, India) and Swapnil Saurav (JDA Software, India)
Business Science Reference • copyright 2016 • 429pp • H/C (ISBN: 9781466698949) • US $235.00 (our price)

Handbook of Research on Military, Aeronautical, and Maritime Logistics and Operations
Alberto Ochoa-Zezzatti (Juarez City University, Mexico) Jöns Sánchez (Consejo Nacional De Ciencie Y Tecnologia (CONACYT), Mexico) Miguel Gastón Cedillo-Campos (Transportation Systems and Logistics National Laboratory, Mexican Institute of Transportation, Mexico) and Margain de Lourdes (Polytechnic University of Aguascalientes, Mexico)
Business Science Reference • copyright 2016 • 587pp • H/C (ISBN: 9781466697799) • US $310.00 (our price)

Innovative Solutions for Implementing Global Supply Chains in Emerging Markets
Ashish Dwivedi (University of Hull Business School, UK)
Business Science Reference • copyright 2016 • 313pp • H/C (ISBN: 9781466697959) • US $195.00 (our price)

Managerial Strategies and Practice in the Asian Business Sector
U Zeyar Myo Aung (Tourism Promotion and International Relations Department, Myanmar) and Patricia Ordoñez de Pablos (The University of Oviedo, Spain)
Business Science Reference • copyright 2016 • 395pp • H/C (ISBN: 9781466697584) • US $195.00 (our price)

Designing and Implementing Global Supply Chain Management
Sudhanshu Joshi (School of Management, Doon University, India) and Rohit Joshi (Indian Institute of Management IIM, Shillong, India)
Business Science Reference • copyright 2016 • 342pp • H/C (ISBN: 9781466697201) • US $195.00 (our price)

Innovative Management Education Pedagogies for Preparing Next-Generation Leaders
Shalini Rahul Tiwari (Institute of Management Technology, India) and Lubna Nafees (Institute of Management Technology, India)
Business Science Reference • copyright 2016 • 313pp • H/C (ISBN: 9781466696914) • US $195.00 (our price)

IGI GLOBAL
DISSEMINATOR OF KNOWLEDGE
www.igi-global.com

701 E. Chocolate Ave., Hershey, PA 17033
Order online at www.igi-global.com or call 717-533-8845 x100
To place a standing order for titles released in this series, contact: cust@igi-global.com
Mon-Fri 8:00 am - 5:00 pm (est) or fax 24 hours a day 717-533-8661

Table of Contents

Detailed Table of Contents

managing the various phases of the supply chain. It was also found that the use of collaborative models such as Public Private Partnership (PPP) can prove rewarding when stakeholders involved in a project have an understanding of the model and its dynamics. This chapter contributes to the use of PPP collaborative models in construction project delivery in South Africa.

Following Nelson's primary typology of enterprises based on a characterization of their technical change process (bulk commodities; complex systems producers; and chemical products producers) this chapter presents a case study on the innovation diffusion in the European tile industry based on the analysis of its manufacturing process and supply chain. The producers of ceramic tiles, the machinery producers and the frits and glazes producers fit each of Nelson's categories. Further, the European ceramic tile industry is clustered in two southern European regions: Emilia Romagna (Sassuolo – Italy) and Valencia Region (Castelló de la Plana – Spain) with a different configuration of the type of predominant companies within the cluster but also regarding surrounding clusters. This relative specialization, the interactions between surrounding clusters, and the strong interlink supplier-producer between both ceramic tile clusters make the study of the innovation diffusion in the European ceramic tile industry supply chain an interesting case.

This chapter presents a systematic methodology for modeling and control of supply chain networks, especially focusing on formal representation and control synthesis aspects. A supply chain is represented as a discrete event dynamic system. The use of high-level Petri nets is proposed to formulate event related rules commonly seen in supply chains and analyze cause-effect relationships between events. Petri nets have been successfully introduced as an effective tool for describing control specifications and realizing the control in automated supply chain processes. The extended net representation of the productive task flows can provide more synthetic specifications for consistent management and control of supply chain systems by a top-down refinement methodology. Software implementation is described to simulate and control real supply chain processes.

Each company is original from the point of production processes. Application of standard enterprise information system (SAP, proAlpha, Baan, etc.) need difficult adaptation, especially for conditions of small or medium enterprises (SMEs) and the price of that system is relatively high. Therefore, the proposed capacity planning model is much more suitable for the conditions, requires and demands of SMEs. Specifics of production processes of RS Ltd. are: - combinations of discrete and continuous processes, - this small company is connected to a supply chain (KANBAN) with its mother company, which defines the level of some chosen products in the expedition warehouse by the end of a month. For the above mentioned reason it had to be designed the new original production logistic system.

Section 3
Resilience and Risk within Engineer-to-Order Supply-Chain Management Businesses

Global business communities are facing tremendous challenges from market places with respect to reduce cost and offer true customized products or services to the end customers. To cope such challenges companies are nowadays considering forming a business network with the objective to achieve several business benefits. However, to execute such business network is not risks free but always facing some problems for its continuation successfully. In such situation, it is necessary to formulate risk mitigation plan and strengthen the resilience within business network. The objective of this article is therefore to identifying and sharing risks within the collaborative business network and proposing necessary mitigating plan and resilience for it. In this research, a framework is also highlighted that provides a structural approach for identifying and assessing potential risks and resilience in business networks and their possible impacts on different levels of collaboration. The study is concluded with future research directions.

Section 4
Strategies in Supply-Chain Management and Engineering-to-Order Processes

Supply chain management involves efficiently integrating suppliers, manufacturers, warehouses, stores, and customers. To survive in a highly competitive business environment, manufacturing, resource planning and scheduling and distribution operations are the significant drivers that need to be optimized within supply chain management. In this chapter, we design the supply chain system considering dual demand management strategies simultaneously for the same company, both engineer-to-order (ETO) and make-to-order (MTO). This research has been inspired from the window manufacturer which manufactures and distributes vinyl windows to meet new construction and replacement/remodeling sector demand in the State of Ohio. The company manufacturers windows based on make-to-order strategy for new construction projects and at the same time builds replacement windows to individual customer specifications in very small quantities to be used for replacement in homes. In this study, a total of 174 individual customers and six big contractors are considered throughout the State of Ohio. This paper proposes to separate products based on demand management strategy and develop different supply chain networks for each group. The idea is to design bigger facilities for high volume (make-to-order products) as transportation cost per unit is reduced due to economies of scale whereas to place smaller and more facilities for low

Editorial Advisory Board

List of Reviewers

Foreword

The Engineer-to-Order (ETO) approach in manufacturing supply chain offers flexibility and responsiveness to address the changing customer or market requirements. By the virtue of ETO, it enables the integration of supply chain activities and therefore, the treatment of ETO and supply chain strategies is novel and appropriate. The integration eventually would lead to efficient and effectiveness along the supply-chain operations.

The book treats the subject well highlighting the importance of ETO in developing a well-integrated supply chain in order to be flexible and responsive, but at the same time to be cost efficient. Furthermore, it emphasizes the role of an information management system in order to facilitate a better communication along the supply chain and in turn improved performance in ETO companies. Supply chain strategies and the Engineer-To-Order are complementary to each other in developing agile manufacturing systems.

The book has been tailored to researchers, industrial R&Ds and academia. There are 11 chapters in this book. Chapter 1 presents the role of information systems in ETO supply chains. Chapter 2 describes the improvement of the communication system between the company and customers. A communication network is analyzed for achieving sustainable supply chain in Chapter 3. Chapter 4 demonstrates the importance of collaborative modelling in improving the construction supply chain. The innovation diffusion within the Europena ceramic tile supply-chain network is presented in Chapter 5. Chapter 6 articulates a systematic methodology for modelling and synthesis of supply chain networks using Petri's nets.

The application of production capacity planning and scheduling models in terms of supply-chain management is presented in Chapter 7. Chapter 8 deals with management risk and resilience within a collaborative business network. Chapter 9 elaborated the supply-chain design approaches for dual demand management strategies. The study reveals that dual demand management strategies are designed simultaneously for both engineer-to-order and make-to-order business scenarios. Hence, this study is motivated from by the case of a window manufacturer; this manufactures and distributes vinyl windows to meet new construction and replacement/remodelling sector demand. Furthermore, this study proposes to separate products based on demand management strategy and develop different supply chain networks among others. Therefore, the basic contribution to this research approach is to design bigger facilities for high volume (make-to-order products) as transportation cost per unit is reduced due to economies of scale. Thus, placing smaller and more facilities for low volume engineer-to-order products to be closer to the clients where it may not be feasible to carry only few products over long distances. The hybrid supply chain strategies in wind business are described in Chapter 10. The final chapter presents the physical and digital integration strategies of electronic supply chains and their applicability to ETO supply chains.

The book outlines the critical success factors, optimized processes, efficient and effective enterprise systems enablers and applications. This book will be useful for professionals and academics in the field of industrial engineering and management, supply-chain management and value-chain management. It introduces innovating new processes, novel operational architectures, models, and framework strategies.

This book is a valuable reference material for the professionals and the academic scholars or researchers in global industrial-project and operation's executives, global-industrial engineers and enterprise supply-chain management professionals, global industrial-research and development professionals, and global industrial-enterprise manufacturing consumers. This book is a great addition to the body of manufacturing supply chain.

Angappa Gunasekaran
University of Massachusetts Dartmouth, USA

***Angappa Gunasekaran** is a Professor of Operations Management in the Department of Decision and Information Sciences at the Charlton College of Business, University of Massachusetts Dartmouth (USA).*

Preface

Supply and logistics chain management is a growing concern in today's competitive business environment. Without efficient and effective supply and logistics chain it is almost impossible to achieve competitive business environment. The success of a supply chain also often very much depends on the product development strategies such as Engineer-to-Order, Assemble-to-Order, Make-to-Order, etc. The focus of ETO product development strategy is highly focused on customized product design and development, where customers can directly participate in the design process to get their chosen products. It is therefore, critical to concentrate supply-chain management especially focused on customized product's development and delivery. The characteristics of companies that are adopting ETO strategy can be described in terms of their markets, products and the internal processes. All such features can be categorized in the context of current trends in supply-chain management. The products as developed through the ETO projects are usually complex in nature and there exists uncertainties of markets. These uncertainties can be reduced through integrating marketing and supply chain with other business processes design and manufacturing.

There is a growing concern among ETO companies in all sectors to find ways to reduce costs, shorten product development times and manage risk. The transactions between companies in supply chains are characterized to reduce costs, risk and lead-times associated with business processes. The business process as approach to describing the ETO production scenarios helps to identify the nature of relationships between customers and suppliers. The ETO production scenario is particularly complex, which is involved in many different types of supplier relationships. The importance of classifying supply chain's relationships according to particular products and markets has been widely addressed in the supply-chain literature. The supply-chain management for ETO systems suggests to improve the interface between product development activities and the supply chain. To improve overall supply chain strategy it is necessary to deal with the information management system that can help manufacturers in a project supply chain reduce their demand uncertainty. Especially, this information management system may help manufacturers to reduce the stock of finished goods. Implementation of information database may explore the potential of utilizing web and product identification technologies to overcome logistical challenges of a project supply chain. Information management system also can help companies from using an identity-based system for tracing, tracking and control of project deliveries.

The work presented in this book focus towards the manufacturing companies producing customized products following ETO production scenario. From this book, the companies can be benefited to revisit their existing ETO strategies and find out the limitations or bottlenecks of their existing ETO strategy. Not only the manufacturing organizations but the researchers in the field of supply chain and logistics sectors, especially dealing with ETO scenarios will also be benefited from this book through getting an up-to-date research works in the field. Academicians and researchers can consult different chapters of this book to get an insight knowledge and expertise in the field of ETO based supply-chain management strategy and lead their current research towards an innovative pathway.

There are 11 chapters of this book. Each chapter contains standalone information to be suitable for managing efficient supply chain strategy in an organization or within a collaborative business network. These chapters' highlights will be very much useful to the business and research communities to identify the needs and suitability to execute efficient logistics and supply-chain management. The brief summary from each of the chapters is highlighted at the following paragraphs.

Chapter 1 outlines the benefits of information systems that can be used as the enabler to manage an ETO supply chain. It describes the ETO business strategy which is gaining more and more popularity within market segments. This strategy promotes customisation through offering custom-tailored product. In ETO, strategies are often having to rely on the combined design capabilities of the entire value chain in order to satisfy customer demands. Because of extended relationship between customers and business partners within the value chain, it becomes very important to maintain such a relationship and store the valuable information or data within the database systems. The information systems can be leveraged to streamline the communication process and improve data exchange between the members of the value chain. This chapter will examine how information systems can support the ETO supply-chain management and identify some of the issues involved. This chapter also articulates the scalability and knowledge management within distributed and reconfigurable supply-chain management.

Chapter 2 describes the improvement of the communication system between internal public and customers of an industrial establishment. This improved communication process supports internal processes such as recruitment, selection, motivation and development of an industrial establishment along supply-chain management. In this chapter, the authors analyse the communication channel that is to be to maintain the talented employees within a company and be able to attract new customers while maintaining the old ones and satisfy their needs. This relationship management between business stakeholders helps industrial communities to face the continual challenges from the external environment, such as globalisation, liberalization, new technologies, etc. It also supports to open new business opportunities through improved organizational performance by enhanced human factor and higher customer satisfactions. The enhanced human factor is analysed describing issues like: communication, recruiting, attracting, retaining, motivating, developing employees under the impact of new technologies. This strategy supports performance-generating factor in obtaining a competitive advantage along supply chain, human factor and business competitiveness. The up-to-date communication channels such as E-Communication, E-Supply Chain, E-Recruitment, E-Development is also discussed within the scope of this chapter.

Chapter 3 highlights a communication network that is analysed for achieving sustainable supply chain. This sustainable supply chain ensures competitive advantage for the manufacturing firms. An ETO product development strategy is adopted in order to explain this communication network. This proposed communication channel enables to the implementation of efficient and effective product-development support processes in the quest of achieving shorter product development lead-times and higher return on investments (ROIs). Engineer-To-Order (ETO) product capacity projects such as large electric machine, huge centrifugal pumps, diesel/natural fuel power plant engines, steam turbine, boiler, ship, etc., have challenges concerning their long product-development lead-times. The challenges as confronted by the enterprises can be effectively and efficiently managed through the communication network or flow of technical communication among the main stakeholders for complex / new product-development.

Chapter 4 explains the collaborative modelling that can be applied to improve supply-chain management, especially within the construction supply chain. It is believed that the presented collaborative modelling contributes to the improvement of global construction supply chain network. This chapter especially reports on some of the challenges faced by the construction industries in order to maintain their supply network successfully. One case example as taken from the department of environmental affairs (DEA) headquarters in South Africa is outlined in this chapter. Qualitative interviews were conducted within five professional bodies, which were part of the construction supply chain team that completed the DEA project in South Africa. Research findings show various challenges within the construction supply chain. These challenges necessitate the use of collaborative means of managing the various phases of the supply chain. The research outcomes as elaborated within this chapter also found that the use of collaborative models such as Public-Private Partnership (PPP) can prove rewarding when stakeholders involved in a project have an understanding of the model and its dynamics. This chapter contributes to the use of PPP collaborative models in construction project delivery in South Africa.

Chapter 5 presents the innovation diffusion within the European ceramic tile industry supply chain network. This research work follows Nelson's primary typology of enterprises based on a characterization of their technical change process (bulk commodities; complex systems producers; and chemical product's producers). This chapter highlights a case study on the innovation diffusion in the European tile industry based on the analysis of its manufacturing process and supply chain. The producers of ceramic tiles, the machinery producers and the frits and glazes' producers fit each of Nelson's categories. Furthermore, the European ceramic tile industry is clustered in two southern European regions: Emilia Romagna (Sassuolo – Italy) and Valencia Region (Castelló de la Plana – Spain) with a different configuration of the type of predominant companies within the cluster but also regarding surrounding clusters. This relative specialization, the interactions between surrounding clusters, and the strong interlink supplier-producer between both ceramic tiles clusters make the study of the innovation diffusion in the European ceramic tile industry supply chain an interesting case.

Chapter 6 articulates a systematic methodology for modelling and synthesis of supply chain networks using high-level Petri's nets. A supply chain is represented as a discrete event dynamic system. Petri nets have been successfully introduced as an effective tool for describing control specifications and realizing the control in automated supply chain processes. The use of high-level Petri's nets is proposed to formulate event-related rules commonly seen in supply chains and analyse cause-effect relationships between events. The extended net representation of the productive task flows can provide more synthetic specifications for consistent management and control of supply chain systems by a top-down refinement methodology. Software implementation is described to simulate and control real supply chain processes.

Chapter 7 highlights the production capacity planning and scheduling models in terms of supply-chain management. It explains that an application of a standard enterprise information system standard enterprise information system standard enterprise information system standard enterprise information system (SAP, proAlpha, Baan, etc.) need difficult adaptation, especially for conditions of small or medium enterprises (SMEs), and the price of that system is relatively high. Therefore, the proposed capacity planning model is much more suitable for the conditions, requires and demands of SMEs. A case example is highlighted in this chapter where the specifications of production processes of the case company are: combination of discrete and continuous processes is connected to a supply chain (KANBAN) with its mother company. This connection defines the level of some chosen products in the expedition warehouse by the end of a month. For the above-mentioned reason, it had to be designed the new original production logistic system.

Chapter 8 is concerned with the management of risk and resilience within a collaborative business network. It defines the general risks as collaborative business partners are faced during executing their business network. Without proper resilience and mitigation plans the benefits of the business network cannot be achieved. This chapter also discussed the benefits of forming the business network within companies as well as the challenges as faced, they faced within the network. To execute such a business network not risks free but always facing some problems for its continuation successfully. In such situation, it is necessary to formulate risk mitigation plan and strengthens the resilience within a business network. The objective of this chapter is therefore, to identify and sharing risks within the collaborative business network and proposing necessary mitigating plan and resilience for it. In this research, a framework is also highlighted that provides a structural approach for identifying and assessing potential risks and resilience in business networks and their possible impacts on different levels of collaboration.

Chapter 9 deals with the supply chain design approaches for dual demand management strategies. This dual demand management strategies at=re designed simultaneously for both engineer-to-order and make-to-order business scenarios. This research has been inspired from the window manufacturer, which manufactures and distributes vinyl windows to meet new construction and replacement/remodelling sector demand. One case example is presented where the case company manufactures windows based on make-to-order strategy for new construction projects and at the same time builds replacement windows to individual customer specifications in very small quantities to be used for replacement in homes. This study proposes to separate products based on demand management strategy and develop different supply chain networks for each group. The basic idea is to design bigger facilities for high volume (make-to-order products) as transportation cost per unit is reduced due to economies of scale whereas to place smaller and more facilities for low volume engineer-to-order products to be closer to the clients where it may not feasible to carry a only few products over long distances.

Chapter 10 is described the hybrid supply chain strategies in wind business. The objective of this chapter is to increase the understanding of possible supply chain models and their fit and the most effective configuration in a typically engineer-to-order sector, namely the wind business. The findings from this research work argue the pertinence of a Hybrid Supply Chain model combining elements of the Lean, Agile and Leagile paradigms upstream and the Project Management paradigm downstream in the case of a Wind-Turbine Manufacturer. From the research outcomes, it is found that depending on the complexity of the market and supply diversity, multiple decoupling points emerge. Through the presentation of the Wind-Energy Supply Chain case, it is found that for effective management of a global supply chain characterized by a high number of references and multiple technologies, it is necessary to mix different manufacturing paradigms in order to provide a quick, agile and competitive response to the customer.

Chapter 11 is identified physical and digital integration strategies of electronic device supply chains and their applicability to ETO supply chains. The growth in the manufacture and distribution of electronic devices presents a source of continuing innovation. Electronic devices are products that integrate physical forms (i.e. hardware) and virtual forms (e.g. software) to deliver value to customers. These forms are very different from a product design and supply chain perspective, but nevertheless, they need to work closely together in order to create value for the customers. For electronic device manufacturers. It is important that processes are in place to facilitate the seamless integration of both forms throughout the engineering, production, distribution and support stages of the product life-cycle. This chapter examines the role of physical and virtual supply chain innovation strategies in electronic device supply chains by exploring the commonalities and differences between the design, manufacturing, and distribution models of digital and physical elements. It also explores to what extent such as strategies that could be employed for engineer-to-order (ETO) supply chains.

The supply-chain strategy and ETO product development scenario have emerged as a major supply chain structure and is set to become an increasing importance, as more customised products are demanded across a range of industries. This book has presented a comprehensive research works the ETO synthesising literature relating to supply-chain management. It explains the ETO based supply chain, which has been characterised as having completely customised production dimension. Various strategies for the ETO sector have been presented in this book. Relationships between the supply-chain network, ETO production approach and information management have been explored in this book. By analysing these different strategies as highlighted in this book will be useful to provide a base for thinking about ETO based supply chain systems. The presented works will be beneficial to developing a comprehensive definition of ETO supply chains and developing tailored strategies for the ETO supply chain and logistics network.

Acknowledgment

The editors would like to acknowledge the help of all the people involved in this project and, more specifically, to the authors and reviewers that took part in the review process. Without their support, this book would not have become a reality.

First, the editors would like to thank each one of the authors for their contributions. Our sincere gratitude goes to the chapter's authors who contributed their time and expertise to this book.

Second, the editors wish to acknowledge the valuable contributions of the reviewers regarding the improvement of quality, coherence, and content presentation of chapters. Most of the authors also served as referees; we highly appreciate their double task.

Finally, the editors will like to sincerely acknowledge our IGI Global book project development representative and the entire editorial management team. Without their timely advice and professional guidance this project would not have been able to manage the tight planned editing project schedule. Above all we thank God almighty for making it a success.

Richard Addo-Tenkorang
University of Vaasa, Finland

Jussi Kantola
University of Vaasa, Finland

Petri Helo
University of Vaasa, Finland

Ahm Shamsuzzoha
Sultan Qaboos University, Oman

Section 1

Information Communication Systems in Supply-Chain Management Engineering-to-Order

Chapter 1
Information Systems as the Key Enabler in Engineer-to-Order Supply Chain Management

Arun Nambiar
California State University, USA

ABSTRACT

Engineer-to-Order (ETO) environments are gaining more and more popularity these days with customers demanding custom-designed products to meet their specific needs. ETO enterprises are often having to rely on the combined design capabilities of the entire value chain in order to satisfy customer requirements. Due to the increased level of interaction with customers and between partners in the value chain, it becomes imperative to have an effective means of communication and data storage. Information systems can be leveraged to streamline the communication process and improve data exchange between the members of the value chain. This chapter will examine how information systems can be the key enabler in ETO supply chain management and identify some of the issues involved. The chapter will conclude with suggestions on future direction for research in this area.

INTRODUCTION

Engineer-to-Order environments are increasingly becoming more commonplace these days with the advances in design and manufacturing making it more and more feasible and practical. However, it is not sufficient to have good design software and manufacturing capabilities to be able to succeed in this environment. An integrated information management system is indispensable especially in today's highly dispersed supply chains and uncertain markets in which these companies seek to thrive. The requirements of an information management system in an engineer-to-order environment are vastly different from that of a make-to-stock environment where information about the product variety and manufacturing processes are known ahead of time. In an engineer-to-order system, the product mix is governed by cus-

DOI: 10.4018/978-1-5225-0021-6.ch001

tomer choice and hence not all information is available ahead of time. Most of the information processing happens on a real-time basis once the order is received. Besides handling the nitty-gritty details of order management, such information systems also help with knowledge management, production planning, performance measurement, and traceability.

In order to be able to succeed in an engineer to order environment, it is imperative to provide customers with numerous choices for the product. Keeping track of these numerous product variants can be an onerous task in itself in any organization. Moreover, these product variants have to be efficiently translated to manufacturing for quick and speedy implementation. Knowledge management allows companies to leverage the resident knowledge about product variants and their translations to generate automated systems that would efficiently transfer customer design specifications to manufacturing instructions.

Efficient production planning is a key enabler for an agile manufacturing system that responds quickly to changing customer demands. Information systems allow dynamic scheduling based on the inflow of customer orders thus helping manage the plethora of disparate product mixes that are a hallmark of engineer-to-order environments. Decision support systems that leverage the underlying knowledge management systems can be employed to assist in the process.

Performance measurement is critical in such elaborate and complex systems. In order to be effective, performance management systems have to be tailored specifically to the company's environment and its long-term goals. In an engineer-to-order environment, it is important to have specific key performance indicators in order to track innovation and its rate of transfer from engineering to manufacturing. Besides these, it is also important to monitor the efficiency and reliability of manufacturing logistics in fulfilling highly customized orders, which is the mainstay of engineer-to-order environments.

Information systems also help bridge the gap between engineer-to-order and mass customization, which aims to improve the organization's competitiveness in a customer-centric market. While engineer-to-order focuses on tailor-made products for specific customers, mass customization aims to expand this to the larger population by providing enough choices so that most customers get their desired product features from the available product mix.

With the growing popularity of sustainable practices in manufacturing, life cycle management is of paramount importance in improving the company's sustainability quotient. This becomes more complicated in an engineer-to-order environment due to the sheer variety of customer-centric products and the complexity of manufacturing logistics. This is further compounded by the disparate systems employed by partners in the supply chain. Technologies such as RFID help with improving traceability but require an underlying information system to manage all the data collected. Information systems need to be able to communicate with one another and be seamlessly integrated to facilitate efficient exchange of data to improve traceability throughout the life cycle of each product.

Thus, it can be seen that information systems has a vital role to play in the success of an engineer-to-order environment. In this chapter, we will review the different types of information systems currently being used and identify some of the salient features of these systems. We will also identify benchmark features that are indispensable for a fully integrated information system that facilitates seamless exchange of data across all of the organization's partners while monitoring the performance of the entire supply chain. Specifically, we will examine how information systems act as the key enabler in implementing manufacturing paradigms such as lean and agile principles, quick response manufacturing and mass customization with a particular interest in sustainability.

BACKGROUND

The advent of globalization has resulted in unprecedented growth and development in many parts of the world resulting in an increased demand for goods and services. This combined with the rapid growth in population has undoubtedly put a tremendous strain on the limited resources available. Hence, there is a growing awareness about the need to ensure that this rapid pace of growth and development is sustainable. The essence of sustainability is captured in the definition included in the Brundtland Report (UN 1987) as "the development that meets the needs of the present generation without compromising the ability of future generations to meet their own needs". This report was a watershed moment in sustainable development leading to significant efforts to conserve resources while promoting progress for the betterment of the society.

The principles of sustainable development focus on the entire lifecycle of the product or service and strive to ensure that the process is streamlined and seamlessly integrated. A holistic systems approach (Pham et al., 2008, Rusinko, 2008) helps to integrate all activities in the value chain with a central purpose of achieving objectives in the most efficient manner. Often times, a sustainable approach is combined with other paradigms (Nambiar, 2009a, Nambiar, 2010a) such as lean principles and agile manufacturing (Nambiar, 2009b).

In today's digital age, information holds the key to success for organizations world over. An effective and efficient information management system (Nambiar, 2009b; Nambiar, 2010b) is indispensable for the company's survival and growth.

INFORMATION SYSTEMS IN MANUFACTURING

Information technology and the devices that enable information technology have been game changers for numerous industries leading to cataclysmic improvements in the way business is carried out. One such industry that has benefited significantly from this information technology revolution is manufacturing. The manufacturing industry has come a long way since Henry Ford (Ford & Crowther 1922) pioneered the concept of mass production for his Model T automobile about which he famously said "...customer can have a car painted any color that he wants so long as it is black." This quote is a reflection of the times where customers were happy to have access to cheap products and were not particular about their individual preferences. Today however, another manufacturing paradigm called *mass customization* is gaining increased traction among manufacturers (Nambiar 2009a). The objective here is to produce highly customized products catering to customer preferences at prices comparable to mass production prices. Information system is one of the pillars on which the edifice of mass customization stands and is critical for its success. Numerous researchers Vrechopoulos (2004), Karpowitz et al. (2008), Dean et al. (2008), Shin et al. (2008)) have focused on the effective use of information systems in mass customization.

Another such paradigm on the same lines is the Engineer-to-order (ETO) paradigm where products are engineered to the customers' specific needs and requirements. This category has been defined (Porter et al. 1999) as a system where a standard product may be customized by adding modifications through direct involvement of the customer. This model has been reported (Grabenstetter & Usher, 2013) to be used by more than a quarter of manufacturing firms in North America. These tend to focus mainly on capital goods with tenders being called for contracts to be awarded. There are two main types of activities (Bertrand & Munslag, 1993) in this environment - physical and nonphysical. The non-physical group of activities focuses on the design and plans while the physical group of activities concern with the actual manufacture of components and assembly of the product. The core competency of an ETO company is often the non-physical group of activities such as its ability to leverage the functionality afforded by individual components to create a custom-designed product that integrates these individual components to achieve tasks that meet the customers' requirements. It is has been shown by Pandit and Zhu (2009) and Little et al. (2000) that this non-physical group of activities account for a significant portion of the total product lead-time which is really critical in today's time-based competitive market. Thus, in an ETO environment, the non-physical activities assume more significance than in traditional environments due to its paramount importance as a core competency and due to the high level of interaction between customers, parent company and suppliers involved at this stage. McGovern et al. (1999) illustrate the complexity of this interaction in a data-flow diagram, which underscores the need to make use of the benefits afforded by information systems to streamline the process.

Towards the middle of the twentieth century, multinational corporations from the United States began venturing offshore in search of cheap labor and access to much wanted natural resources to cater to the growing needs of the population. This led to the wave of new era globalization resulting in highly dispersed operations. For example, a company might source its raw materials from one country, refine it in another, manufacture parts in yet another country and finally assemble the product in a different country and ship it to markets across the globe. This has compounded the problems in supply chain management necessitating drastic changes in product design, development, manufacturing and assembly.

Globalization has also increased customer awareness and changed their perspectives about products. Customers have now begun to expect frequent product improvements resulting in shorter product life-cycles. Agile manufacturing (Nambiar 2009b) principles help companies to reduce design-to-market times thus providing them with the competitive advantage of being first to introduce a new product into the market. In the beginning, customized products could be charged a premium and customers were willing to pay for products that cater to their specific needs. However, the advent of globalization brought more players into the already crowded marketplace thus leading to a price battle. Companies are striving hard to reduce their operating costs so that they can continue to make profits even though the prices are near mass production levels. This led to the popularity of the lean paradigm (Nambiar 2010a), which sets forth five principles aimed at eliminating inefficiencies and thus reducing costs.

Computers have been used in manufacturing since the 1950s (Cooper 1957, Gravel et al 1957). Computer methods have been used to assist with production planning and scheduling since the late seventies (Hammond & Kocchar 1978, Kocchar 1978, Heward 1979) and eighties (Kocchar 1981, Kimber 1988). Computers have been used to generate optimal solutions to mathematical models that take into account the different parameters influencing the production plan. This may be achieved through perturbation analysis (Mehra et al 1994), and simulation-based experiments (Tzafestas et al 1997, Hicks and Braiden 2000, Hussein et al 2009, Monostori et al 2010). Simulation can be used to evaluate the robustness of production plans (Yan & Kou 2009) by comparing multiple alternatives thus affording the opportunity to tweak the production plan to better handle uncertainties. A more recent concept is cloud-based manufacturing (Husejnagic & Sluga, 2015) where, similar to cloud-based computing, manufacturing and design facilities are harnessed on a one-off basis to design and deliver products to customers.

It has been shown (Porter et al. (1999)) that manufacturing companies invest a great deal of capital in information systems as part of their information technology infrastructure. However, often times, companies have diverse requirements and most of the systems available tend to be geared towards the pre-defined HVLV (high-volume-low-variety) or LVLV (low-variety, low-volume) product mixes. Some of the more popular systems include Materials Resource Planning (MRP) and its extension, Manufacturing Resource Planning (MRP II). There are also other software systems geared towards production planning in the aforementioned environments.

INFORMATION SYSTEMS IN ETO SUPPLY CHAIN MANAGEMENT

The ETO supply chain management is vastly different from traditional make-to-order or assemble-to-order supply chain environments. In the traditional systems, the product mixes are predefined, the components that go into the products are pre-determined and the production schedules are often pre-planned. However, the scenario is quite different in the ETO supply chain. Here, the customer determines the type of product needed, which in turn determines the components, required. The production schedule is determined based on the lead-time for the components. Moreover, the increased level of interaction between customers and the company necessitates an information system that has the capability to efficiently translate the product requirements into design specifications and effectively communicate these to all partners in the supply chain. There are two major stages of interaction (McGovern et al 1999) between the customers, parent company and suppliers. The first stage is during the call for tenders. At this point, the parent company has to interact with customers to determine the needs and communicate with suppliers to determine costs for the components and the expected lead times. The second stage is after the tender has been accepted and contract has been signed. At this point, more communication ensues between the customer, parent company and suppliers to iron out the detailed plan for production. It has been shown (Bertrand & Munslag, 1993; Little et al., 2000) that traditional information management systems such as MRP and MRP II often come up inadequate to manage the supply chain in an ETO environment. The areas in which information systems have a critical role to play in the engineer-to-order environment include knowledge management, supply chain management, and lifecycle management.

Knowledge Management

Knowledge Management (KM) has been defined (Yang 2010) as a system to “harvest the tacit knowledge residing in individuals and make it a firm asset, rather than to only leave it in the heads of the particular individuals”. Creating an effective knowledge management system involves four steps (Ni et al. 2010) - knowledge acquisition, knowledge accumulation and storage, knowledge exchange, and finally knowledge application. A robust information system is critical (Jiang et al. 2009) to building a knowledge management system that helps its users to respond quickly to customer demands.

A crucial feature integral to Engineer-to-Order environments is the product design. Information systems can be leveraged to speed up the design process to help with quick turnovers. Product design generators (Roach et al. 2003) have been developed to assist with this process. This tool automates the design process by plugging the various design modules in response to the user preference and includes all pertinent information required for its manufacture. Decision support systems (DSS) that aid with the decision-making utilize KM systems to harness the captured knowledge, to learn from previous experiences and to make intelligent predictions about future outcomes. Expert systems is another technology that again leverage KM systems to assist with short-term decision-making. However, most of the DSS and Expert Systems mainly focus on downstream activities such as production, quality control, etc. There has been limited application of these technologies in the design phase (Reyes & Raisinghani, 2002). The authors underscore the importance of capturing resident knowledge and converting it into intellectual capital in order to be harnessed for innovation to obtain competitive advantage.

KM systems have been shown (Yang 2010) to have significant positive impact on the company's performance when implemented right. However, Nambiar (2013) identify some of the challenges companies face in implementing effective KM systems. These include encouraging collaboration and cooperation among supply chain partners, leveraging Internet and Web 2.0 technologies, and developing distributed systems.

Supply Chain Management

In the engineer-to-order environment, due to the high involvement of customer in the design process, the requirements of the supply chain management are vastly different than that of a traditional high-volume manufacturing environment. There is a general lack of emphasis on research (Hicks et al, 2000) into supply chain management strategies for the engineer-to-order environment. This might be primarily because of the low volumes that is so characteristic of this environment. However, due to the increased interaction with customers, the role of information systems becomes more protracted. In an engineer-to-order environment, customers need to be made aware of the options available to them regarding their customization choices. On the other hand, companies functioning in this environment need to become more aware of market trends and need to understand the pulse of the customer, the choices they make, the rationale behind it and the factors that influence it. This calls for an efficient customer relationship management system.

Vertical integration is an important factor in engineer-to-order companies. There are companies that do everything in-house and there are companies that are engaged only in the design of the product. A vertically integrated company has greater flexibility in terms of product configurations and have the advantage of being able to respond quickly to customer demands. However, they are often limited in the range of products they can provide. Hence, in recent years, companies in the engineer-to-order environment are focusing on their core competencies and outsourcing other routine activities (Hicks et al. 2001). This increases the level of interaction with suppliers and vendors. Unlike in traditional manufacturing environments, since the products are rarely standardized, there is need for increased communication between partners in a supply chain. It is really important that the product requirements are transferred accurately to all partners of the supply chain right from when the customer requests specific features which need to be accurately translated into product specifications which in turn have to be communicated to all the suppliers of components that go into making the product.

Information sharing along the supply chain is really critical for the success of ETO companies. The Japanese supply chain management system has been popularized around the world (Womack et al 1999, Womack & Jones 2003) and attempts have been made to implement these successful strategies in the Western world. This close partnership with suppliers among Japanese companies began after World War II with companies forming what are known as "keiretsu's" - a network of closely knit group of suppliers. An integral part of the success is information sharing both during research and development and during production. Due to the long-term relationship between suppliers and parent companies, both parties were willing to share information and make changes in their operations to suit the partners. A recent survey (Kawai et al. 2013) of Japanese companies found that companies that have shared information with suppliers often reap benefits such as improvement in quality and profits and reduction in costs.

This becomes all the more pertinent in the ETO supply chain environment because of the peculiar nature of this environment. ETO companies typically function in the capital goods market and capital goods typically contain components that range from low-end standard components to high-end, highly sophisticated components. With stiff time-based competition (McGovern et al 1999), companies often rely on the suppliers' capability to design and manufacture such specialized components while focusing mainly on the design of the product and its assembly. Suppliers have to be involved very early on in the process to be able to effectively deliver the components required for assembly.

Life Cycle Management

In today's world of increased emphasis on material and energy conservation and with companies striving to improve their "green image", life cycle management is of paramount importance. Life cycle assessment (LCA) focuses on the environmental impact of the product right from its raw material stage to its recycling or reuse stage with LCA forming the basis. ISO 14000 series of standards focus on environment management with special emphasis on tools to "manage environmental responsibilities" (ISO, 2015a). ISO 14040 (ISO, 2015b) is a set of standards that focuses on methodology for applying LCA. Life cycle management (LCM) focuses on putting environment and sustainability on the forefront of every thought process related to the design, manufacture, distribution, use, and recycling of the product. It encourages companies to seek collaboration with all partners in the supply chain with the aim of "preserving the environment for future generations to come" as stated in the Brundtland Report (UN, 1987). Companies often find it challenging to implement LCM and the underlying LCA due to a variety of reasons. Some of the reasons might be

- Lack of understanding of how various processes impact the environment.
- Unavailability of sufficient data to perform meaningful analysis.
- Lack of cooperation among supply chain partners.

Fullana et al (2011) suggest companies to adopt the "three-thirds principle" where companies spend one-third of the project resources allocated for LCA for seeking consensus and understanding the process of LCA, one-third on educating the customer about the benefits of the project, and the final one-third on the actual LCA study. Chen et al (2009) develop an ontology-based system to capture product knowledge spread across the supply chain. It can be seen that information systems is critical for an effective implementation of LCM. With proper information systems, data can be collected about the various processes that take place right from design till delivery of the product to the customer. When these information system packages can interact with one another, information across the entire supply chain can be analyzed for potential improvements with the aim of reducing environmental impact.

SOLUTIONS AND RECOMMENDATIONS

Based on review of the existing literature related to information systems in ETO, some of the shortcomings of existing approaches or solutions have been identified below. While not all systems are deficient in all systems and there is rarely a one-size-fits-all solution, the following seem relevant in today's world and more so in the future.

Distributed and Responsive Design

The very essence of Engineer-to-order firms is that the products are designed specifically to meet customer needs. This calls for increased communication between customers and product design team at the ETO enterprise. As seen earlier, ETO enterprises are focusing more on their core competencies such as design and outsourcing more routine functions, the supply chain becomes more complex. Moreover, in today's world of stiff competition, companies are striving to hold on to as many customers as possible through innovative and creative products and services. This necessitates harnessing the design capabilities of suppliers and vendors as well. All of these factors call for a mechanism for ETO enterprises and partners in the supply chain including customers to be able to freely collaborate and cooperate to develop new designs. This is a huge ask in today's world of geographically distributed suppliers and customers. Technology can be leveraged to handle this requirement. Information technology has made great strides in the recent decades and Internet has permeated even the most remote corners of the world. Computers around the globe have been connected together and their processing abilities have been harnessed to solve computational tasks previously considered unsolvable. Thus, technology can be put to good use to build distributed and connected information systems that allow partners in the ETO supply chain to work together towards developing designs that meet or exceed customer requirements.

Due to the rapid pace of change in technologies used, and the volatile nature of customer demand in the ETO industry, it is important for such a distributed system to be responsive to changes as well. Drastic changes in customer requirements might necessitate collaborations with new suppliers and vendors. Changes in customer needs might also require complete overhauling of the product range as well. Due to the diverse nature of the global supply chain, suppliers and vendors might be located in different countries with different federal requirements to adhere to. For example, take the accounting practices. In the United States, the standard used for financial reporting is GAAP. Many countries around the world are moving towards a global standard known as International Financial Reporting System (IFRS). Now, for these financial systems to communicate with each other, it is important for these systems to understand each other.

Interoperability

With advances in information system models and newer products purporting to do things better and more efficiently being introduced into the market quite frequently, it becomes imperative that these systems are able to communicate with each other. A company may be upgrading from its legacy systems to a newer and more powerful information system. It would still have copious amounts of useful data within its legacy system. In order to be able to fully leverage that data while employing the power of newer systems, it is important that the newer system be able to interact seamlessly with such legacy systems. Moreover, as mentioned earlier, there is rarely a product that does everything a company wants from its information systems. Companies with the help of its information technology personnel often stitch together a collection of systems into one big scaffolding to achieve its diverse information system needs. Thus, it becomes all the more critical that these systems communicate not just with legacy systems but with each other as well.

Moreover, the increased involvement of the customer and the suppliers adds to the complexity of the interactions that information systems has to be capable of handling in the ETO environment. With growing number of suppliers and their geographic diversity, information systems have to be leveraged to effect reliable communication. Researchers (Caron & Fiore (1995), Pandit & Zhu (2009)) have underscored the need for effective information exchange in an ETO environment. Thus, it becomes vital for the varied systems to be able to seamlessly integrate into one large system.

In order to facilitate seamlessly interaction between varied systems, it is essential that the systems be able to understand the underlying system of data storage. There are two main mechanisms of achieving this:

1. **Data Exchanger:** In this approach, each system can adopt its own format for data storage. If a system wants to use data from another system, it would do so through the data exchanger (Nambiar et al., 2009c), which would handle the task of converting data from one format to another. This is cumbersome from software development standpoint since separate data exchangers need to be developed to interoperate with difference systems. Moreover, if a particular vendor decides to change its data format, new data exchangers would have to be developed in order to communicate with that system. Hence, it is not future-proof. However, this would be the easier approach than what is described below purely from the implementation perspective.
2. **Standard Data Format:** In this approach, all system developers agree upon a uniform data format for storing persistent data. This could be along the lines of an eXtensible Markup Language (XML) and could be achieved by developing an industry specific markup language. While this is the ideal approach since it makes it easy for systems to communicate with each other and makes it independent of future variations of the system since all changes would adhere to the agreed upon markup language. Chen and Chen (2009) describe the use of XML in supply chain for simulation purposes and describe how XML can be used to represent information within the supply chain.

Reconfigurability, Adaptability, and Scalability

Porter et al (1999) mention that the one of the primary reasons why smaller companies lag behind in adopting newer information system products is the lack of reconfigurability of such systems. Even though companies in the engineer-to-order environment share similar characteristics, there still exist differences that place diverse demands on the information system, which necessitate the ability to configure such systems to meet the specific needs and requirements of a particular company. Moreover, as the company grows through increased sales and product diversification, the requirements of the information systems also change thus making it imperative for such systems to be adaptable and scalable.

Performance Measurement

Performance measurement is really critical to measure the success of an organization. Kleijnen and Smits (2003) examine performance measurement in traditional supply chain environments. Though a lot of research has been done in the area of performance measurement, it has been shown (Folan and Browne, 2005) that research in performance measurement mechanisms for an extended enterprise such that of an ETO environment is one area that is lagging behind that of traditional performance measurement systems. A balanced scorecard approach (Kaplan & Norton, 1992; Kleijnen & Smits, 2003) that takes into account all the factors that influence performance in an ETO supply chain could help companies identify their strengths and weaknesses. Kaplan and Norton (1996) provide a 4-stage framework for applying the balanced scorecard as a management strategy with the aim of improving communication and business planning. Bradley (1996) proposes a tri-axis cube strategy for performance measurement with design coordination, cost and quality, and the manufacturing environment forming the three axes. Strategic Performance Indicators (SPI) can be developed that focus on all three axes while keeping in mind the company's critical success factors and customer requirements. Customer Enquiry Management (CEM) is another system specific to the ETO environment that helps with performance measurement. CEM has been defined (Zorini et al. 2008) as the process that focuses on all aspects of order management right from when the customer calls for a bid to the product being delivered. The authors emphasize the need for greater integration between value chain partners to have a positive impact on the performance of the enterprise as a whole.

Gunasekaran et al (2001) identify some key performance metrics for a typical supply chain. However, in an environment such as that of ETO where there is increased reliance on suppliers even during the design process, a performance measurement system has to cross the boundaries of traditional enterprise to morph into a performance management system (Smith & Goddard, 2002, Folan & Browne, 2005) that focuses on intra-organizational performance. Bresnan (1996) underscores the need for increased research into the interorganizational relations and their impact on performance. However, Folan and Browne (2005) caution about the possibility of disparate performance measurement systems being developed leading to incompatibilities. This is where again the power of information systems can be leveraged to develop systems that can talk to each other thus enabling the construction of integrated performance management system that monitors the health of all firms linked to a typical ETO enterprise.

FUTURE RESEARCH DIRECTIONS

Based on the analysis, it is evident that ETO enterprises have very specific information system requirements. The power of technology needs to be tapped into in order to design efficient information systems that allow all players in the ETO supply chain to freely interact with one another for mutual benefit. It is also important to develop standards for data to facilitate easy exchange of data across diverse systems. Harnessing Web 2.0 technologies and using XML-based data formats will help solve some of the potential problems identified so far such as interoperability and responsiveness. Cloud-based systems that work on the service-oriented architecture (Wang et al. 2006, can be implemented to handle the distributed nature of the ETO environment. In order to be scalable and adaptable, it is important for such systems to be modular so that the necessary modules can be coupled together to meet the requirements as the situation might warrant. Another important research direction is in the area of performance measurement of ETO supply chains. As previously discussed, intra-organizational performance measurement is critical for the ETO environment. Data exchange mechanisms and standard data representation formats can enable the individual performance measurement systems to be integrated into a performance management system as outlined in Folan and Browne (2005).

CONCLUSION

Engineer-to-order (ETO) enterprises are engaged in the business of providing highly customized products engineered specifically to the needs of the customer and function in a highly volatile market with rapidly fluctuating demands (Hicks & McGovern, 2009). With more and more ETO firms relying on the combined design capabilities of its value chain, there is an increased level of interaction between its partners. With much of the lead-time in the ETO business spent on designing the product, it becomes imperative that efforts be made to reduce the design time so that products can be delivered to customers at the right time and in the right place. Information systems can be leveraged to facilitate seamless communication and data exchange between various partners in the value chain all the way from customers to suppliers and distributors thus enabling ETO businesses to thrive successfully in today's global markets characterized by time-based competition, location-independent partners, and customer-centric strategies.

REFERENCES

Agouridas, V., Allen, M., McKay, A., Pennington, A., & Holland, S. (2001). Using Information Structures to Gain Competitive Advantage. *Proceedings of the 2001 PICMET - Portland International Conference on Management of Engineering and Technology, Portland, OR* (Vol. 2, pp. 681–692). doi:10.1109/PICMET.2001.952416

Bertrand, J. W. M., & Munslag, D. R. (1993). Production Control in Engineer-to-Order firms. *International Journal of Production Economics, 30/31*, 3–22. doi:10.1016/0925-5273(93)90077-X

Bradley, P. (1996). *A Performance Measurement Approach to the Re-engineering of Manufacturing Enterprises* [PhD Thesis]. CIMRU, NUI Galway, Ireland.

Bresnen, M. (1996). An Organizational Perspective on Changing Buyer-Supplier Relation: A Critical Review of the evidence. *Organization, 3*(1), 121–146. doi:10.1177/135050849631007

Caron, F., & Fiore, A. (1995). 'Engineer to Order' companies: How to integrate manufacturing and innovative processes. *International Journal of Project Management, 13*(5), 313–319. doi:10.1016/0263-7863(95)00023-J

Chen, Y.-J., & Chen, Y.-M. (2009). An XML-based Modular System Analysis and Design for Supply Chain Simulation. *Robotics and Computer-integrated Manufacturing, 25*(2), 289–302. doi:10.1016/j.rcim.2007.12.005

Chen, Y.-J., Chen, Y.-M., & Chu, H.-C. (2009). Development of a mechanism for ontology-based product lifecycle knowledge integration. *Expert Systems with Applications, 36*(2), 2759–2779. doi:10.1016/j.eswa.2008.01.049

Cooper, A. H. (1957). Integration of computers with factory processes. *Journal of the British Institution of Radio Engineers, 17*(8), 431–441. doi:10.1049/jbire.1957.0044

Dean, P.R., Tu, Y.L., Xue, D. (2008). A Framework for Generating Product Production Information for Mass Customization. *International Journal of Advanced Manufacturing Technology*, 38(11-12), 1244 - 1259.

Folan, P., & Browne, J. (2005). A Review of Performance Measurement: Towards Performance Management. *Computers in Industry, 56*(7), 663–680. doi:10.1016/j.compind.2005.03.001

Fullana i Palmer, P., Puig, R., Bala, A., Baquero, G., Riba, J., & Raugei, M. (2011). From Life Cycle Assessment to Life Cycle Management. *Journal of Industrial Ecology, 15*(3), 458–475. doi:10.1111/j.1530-9290.2011.00338.x

Gosling, J., & Naim, M. M. (2009). Engineer-to-order Supply Chain Management: A literature review and research agenda. *International Journal of Production Economics, 122*(2), 741–754. doi:10.1016/j.ijpe.2009.07.002

Grabenstetter, D., & Usher, J. M. (2013). Defining Job Complexity for the Engineer-to-Order Environment. In A Krishnamurthy, & W.K.V Chan (Eds.), *Proceedings of the 2013 Annual Industrial and Systems Engineering Research Conference* (pp. 2404 – 2413).

Gravel, J.-P., & Kavanagh, T.F. (1957). Computers -- the Key to Modern Manufacturing Scheduling. *IRE Transactions on Industrial Electronics*, 4, 90-93. doi: .10.1109/IRE-IE.1957.5007790

Gunasekaran, A., Patel, C., & Tirtiroglu, E. (2001). Performance Measures and Metrics in a Supply Chain Environment. *International Journal of Production and Operations Management*, *21*(1-2), 71–87. doi:10.1108/01443570110358468

Hicks, C., & Braiden, P. M. (2000). Computer-aided Production Management Issues in the Engineer-to-Order Production of Complex Capital Goods explored using a simulation approach. *International Journal of Production Research, 38*(18), 4783–4710. doi:10.1080/00207540010001019

Hicks, C., & McGovern, T. (2009). Product Life Cycle Management in Engineer to Order Industries. *International Journal of Technology Management*, *48*(2), 153–167. doi:10.1504/IJTM.2009.024913

Hicks, C., McGovern, T., & Earl, C. F. (2000). Supply Chain Management: A Strategic Issues in Engineer to Order Manufacturing. *International Journal of Production Economics*, *65*(2), 179–190. doi:10.1016/S0925-5273(99)00026-2

Hicks, C., McGovern, T., & Earl, C. F. (2001). A Typology of UK Engineer-to-Order Companies. *International Journal of Logistics: Research and Applications*, *4*(1), 43–56. doi:10.1080/13675560110038068

Husejnagic, D., & Sluga, A. (2015). A Conceptual Framework for a Ubiquitous Autonomous Work System in the Engineer-to-Order Environment. *International Journal of Advanced Manufacturing Technology*, *78*(9-12), 1971–1988. doi:10.1007/s00170-015-6798-7

ISO. (2015a). ISO 14000 - Environmental Management. Retrieved from http://www.iso.org/iso/home/standards/management-standards/iso14000.htm

ISO. (2015b). ISO 14040:2006 - Environmental Management: Life Cycle Assessment - principles and framework. Retrieved from http://www.iso.org/iso/catalogue_detail?csnumber=37456

Jiang, H., Liu, C., & Cui, Z. (2009). Research on Knowledge Management System in Enterprise. *Proceedings of the 2009 International Conference on Computational Intelligence and Software Engineering* (pp. 1-4). doi:10.1109/CISE.2009.5363935

Kaplan, R., & Norton, D. (1992). The Balanced Scorecard - measures that drive performance. *Harvard Business Review*, *70*(1), 71–79. PMID:10119714

Kaplan, R., & Norton, D. (1996). Using the Balanced Scorecard as a Strategic Management System. *Harvard Business Review*, *74*(1), 75–85.

Karpowitz, D. J., Cox, J. J., Humpherys, J. C., & Warnick, S. C. (2008). A Dynamic Workflow framework for Mass Customization using Web Service and Autonomous Agent Techniques. *Journal of Intelligent Manufacturing*, *19*(5), 537–552. doi:10.1007/s10845-008-0136-y

Kawai, T., Sakaguchi, J., & Shimizu, N. (2013). Transition of buyer-supplier relationships in Japan: Empirical Evidence from manufacturing companies. *Journal of Accounting and Organizational Change*, *9*(4), 427–447. doi:10.1108/JAOC-07-2011-0033

Kleijnen, J. P. C., & Smits, M. T. (2003). Performance Metrics in Supply Chain Management. *The Journal of the Operational Research Society*, *54*(5), 507–514. doi:10.1057/palgrave.jors.2601539

Little, D., Rollins, R., Peck, M., & Porter, J. K. (2000). Integrated Planning and Scheduling in the engineer-to-order sector. *International Journal of Computer Integrated Manufacturing*, *13*(6), 545–554. doi:10.1080/09511920050195977

McGovern, T., Hicks, C., and Earl, C.F. (1999). Modelling Supply Chain Management Process in Engineer-to-Order Companies. *International Journal of Logistics Research and Applications*, 2(2), 147-159.

Nambiar, A. N. (2009a, July 1-3). Mass Customization - Where do we go from here? *Proceedings of the 2009 World Congress on Engineering*, London, U.K (*Vol I).*

Nambiar, A. N. (2009b, July 6-9). Agile Manufacturing: A Taxonomic Framework for Research. *Proceedings of the 2009 International Conferences on Computers and Industrial Engineering* (pp. 684-689). doi:10.1109/ICCIE.2009.5223676

Nambiar, A. N. (2010a, March 17-19). Modern Manufacturing Paradigms - A Comparison. *Proceedings of the 2010 International Multiconference of Engineers and Computer Scientists*, Hong Kong (Vol. 3).

Nambiar, A. N. (2010b, January 9-10). Challenges in Sustainable Manufacturing. *Proceedings of the 2010 International Conference on Industrial Engineering and Operations Management*, Dhaka, Bangladesh.

Nambiar, A. N. (2013). A Current Perspective of Knowledge Management in a Global Economy. *International Journal of Soft Computing and Software Engineering*, *3*(3), 236–239.

Nambiar, A. N., Judd, R. P., & Greenblatt, J. (2009c). Data Exchange among different modules in a manufacturing or service framework. *International Journal of Product Development*, *8*(2), 122–140. doi:10.1504/IJPD.2009.024184

Ni, G., Wang, W., Wang, J., Zong, Z., & Xie, M. (2010). Research on the Knowledge Management System of the Vicarious Management Corporation. *Proceedings of the 2010 International Conference on Information Science and Management Engineering* (pp. 62–67). doi:10.1109/ISME.2010.128

Pandit, A., & Zhu, Y. (2009). An Ontology-based Approach to support Decision-making for the design of ETO (Engineer-to-order) products. *Automation in Construction*, *16*(6), 759–770. doi:10.1016/j.autcon.2007.02.003

Pham, D. T., Pham, P. T. N., & Thomas, A. (2008). Integrated production machines and systems - beyond lean manufacturing. *Journal of Manufacturing Technology Management, 19*(6), 695–711. doi:10.1108/17410380810888094

Porter, K., Little, D., Peck, M., & Rollins, R. (1999). Manufacturing Classifications: Relationships with production control systems. *Integrated Manufacturing Systems, 10*(4), 189–199. doi:10.1108/09576069910280431

Reyes, P., & Raisinghani, M. S. (2002). Integrating Information Technologies and Knowledge-based Systems: A Theoretical Approach in Action for Enhancements in Production and Inventory Control. *Knowledge and Process Management, 9*(4), 256–263. doi:10.1002/kpm.153

Roach, G. M. (2003). *The Product Design Generator - A Next Generation Approach to Detailed Design.* Provo, UT: Brigham Young University.

Rusinko, C. A. (2008). Towards more sustainable management systems: Through life cycle management and integration. *Journal of Cleaner Production, 16*(10), 1071–1080. doi:10.1016/j.jclepro.2007.06.006

Shin, Y., An, S.-H., Cho, H.-H., Kim, G.-H., & Kang, K.-I. (2008). Application of Information Technology for Mass Customization in the housing construction industry in Korea. *Automation in Construction, 17*(7), 831–838. doi:10.1016/j.autcon.2008.02.010

Smith, P., & Goddard, M. (2002). Performance Management and Operational Research: A Marriage made in Heaven? *The Journal of the Operational Research Society, 53*(3), 247–255. doi:10.1057/palgrave.jors.2601279

Report on the World Commission on Environment and Development: Our Common Future. (1987). *UN.* Retrieved from http://www.un-documents.net/our-common-future.pdf

Verchopoulos, A. P. (2004). Mass Customization Challenges in Internet Retailing through Information Management. *International Journal of Information Management, 24*(1), 59–71. doi:10.1016/j.ijinfomgt.2003.09.013

Wang, Z., Zhang, D., & Xu, X. (2006). Service-Oriented Infrastructure for Collaborative Product Design in ETO Enterprises. *Proceedings of the 10th International Conference on Computer Supported Cooperative Work in Design, Nanjing, China* (pp. 1–6). doi:10.1007/11686699_1

Womack, J. P., & Jones, D. T. (2003). *Lean Thinking: Banish Waste and Create Wealth in Your Corporation.* New York, NY: Simon & Schuster.

Womack, J. P., Jones, D. T., & Roos, D. (1999). *The Machine that Changed the World.* New York, NY: Rawson Associates.

Yang, J. (2010). The Knowledge Management Strategy and its Effect on Firm Performance: A Contingency Analysis. *International Journal of Production Economics, 125*(2), 215–223. doi:10.1016/j.ijpe.2010.03.012

Zorzini, M., Hendry, L., Stevenson, M., & Pozzetti, A. (2008). Customer Inquiry Management and Product Customization: An Empirical multi-case study analysis in the Italian capital goods sector. *International Journal of Operations & Production Management, 28*(12), 1186–1218. doi:10.1108/01443570810919369

Chapter 2
Improving Communication with Internal Public and Customers of an Industrial Company: A Major Challenge along Supply Chain

Florea Nicoleta Valentina
Valahia University of Targoviste, Romania

Tanasescu Irina Antoaneta
Valahia University of Targoviste, Romania

ABSTRACT

The present chapter deals with an important topic, namely communication with its internal public and its customers by improving internal processes like recruitment, selection, motivation and development of an industrial company along supply chain management. We analyse this relationship taking into account that having talented employees the company will be able to attract new customers, maintain the old ones and satisfy their needs, that organizations must face the continual changes from the external environment, and that factors like globalization, liberalization, new technologies, and market opening have a major impact on organizational performance, on the human factor and on customers. As astrategic performance-generating factor in obtaining competitive advantage along supply chain, human factor is analysed describing issues like: communication, recruiting, attracting, retaining, motivating, developing employees under the impact of new technologies in order to obtain performance along supply chain.

DOI: 10.4018/978-1-5225-0021-6.ch002

INTRODUCTION

The general purpose of this chapter is to show the important role of communication with its internal public and its customers of companies from industry area, taking into account the continuous change of environment, of supply chain management, of markets, of competition and of human capital behaviour. So, business communicators become important for many reasons: to communicate effectively, to act strategically, to make effective and ethical decisions, to use new technology in order to meet business goals and customers needs, to cultivate trust and credibility among employees and other stakeholders, to embrace diversity, to build strong relations and strategic alliances along supply chain, and to implement best practices.

The main objectives of this chapter are:

- To show that communication with customers suppose developing a strong supply chain relationship based on attracting and developing the best employees in order to know the customers' needs and to satisfy them better, faster, and cheaper.
- To show that attracting and maintaining high-quality employees, the industrial organizations will obtain those unique technical skills, economical competences, behaviours and attitudes, using the competition principles and being aware of the industrial consumer's profile.
- To show that motivating the employees organizations is offering a clear vision, training for achieving individual and organizational performances, developing their team work spirit.
- To show that improving internal communication will be avoided redundant flows, encouraged creativity, harboured conviviality, decreased the number of conflicts and implemented some efficient plans for monitoring and measuring the causes, for rewarding employees' positive behaviours and creating a positive image in the internal and external environment of the industrial organization.
- To show that using new internal practices and new technologies the industrial organizations will obtain performance and satisfy customers' needs.
- To show that using simulation and modelling technics organizations will may predict risks and benefits for employees, for customers and also for organization.

Also, we shall analyze several tendencies present in the organizational culture, in the area of industrial communication on the labour market, which influence the future evolution of the development of communication management and the human capital contribution to obtaining performance and creativity.

BACKGROUND

A major challenge in supply chain management is developing and maintaining information and communication systems. This means more than implementing information technology, means: making strategic plans, customer and supplier evaluation, developing joint objectives or project review (Lambert, 2008,

p. 242). Once a supply chain management process is implemented, there is the need to continuously improve it (Lambert, 2008, p. 253). An important aspect of implementing supply chain management is the formation of appropriate linkages between members of supply chain. The main challenge is to find effective methods for developing tha right type of relationships, alliances and partnerships. In an environment characterized by scare resources, increased competition, higher customer expectations, and faster rates of change, managers must stregthten supply chain intehration and provide sustainable competitive advantage. Partnerships provide new skills, experience, knowledge, information, and may decrease costs

To obtain performance organizations must discover, analyse and satisfy customers needs, and to do so along supply chain it must implement best internal practices. John Gattorna said "Supply Chain is a combination of processes, functions, activities and relationships between products, services and information and financial transactions move within and between organizations."

Supply Chain involves moving of all these processes from the producer to the final consumer, and all members of the organization are involved in this to become real. One of the significant paradigms of modern business management is that individual businesses no longer compete as autonomous entities, but with supply chains. In this competitive environment, success will depend on the organization's management ability to integrate the complicated network of the organization into business relationsheeps (Drucker, 2003).

Training, experience, knowledge, capabilities, skills and attitudes of people will depend on the success or failure of the organization in its relations with suppliers and creating value through innovation, communication, information, co-revolution, retention and winning new customers. If organization' customers are enthusiastic, than the level of involvement by an organization is high to meet the expressed needs of customers and then appaer a long-term relationship based on trust and loyalty and it surprise customer by anticipating their needs and wishes.

Consumers develop a strong emotional connection with the offer, which in turn becomes part of their lives. Customer enthusiasm can not be achieved without the total involvement of the entire organization. Employees of the organization must not only confront consumers, but it should satisfy their needs and the organization offers solutions for them. Organization with customers satisfied and loyal has impact on employee motivation.

Companies with good reputation can attract the people best trained in the labor market. Quality people, operating in a motivated environment, are valuable goods and a necessary condition for generating value for consumers. A manager with a clear strategic vision, able to motivate people, plays a key role in consumer and employee enthusiasm. Integration of an organization business processes along the supply chain involves well-trained and performant people. On them will depend the success or failure of the organization.

Organizations in the Supply chain management face the problem that concerns the education of the staff. The general objectives of the job are achieving sales target, development of customer relationship and the roles that a category may play are: attracting new customers; increasing level of transactions; help the organization create a well known product; generate cash and generate gross margins further.

ISSUES, CONTROVERSIES, CONSEQUENCES

The Impact of Globalization on the Internal Processes

In the modern era, globalization has been linked with free-market concept, which is characterized by global competition, privatization, deregulation, and by self-correcting mechanisms that avoid any necessity for state intervention (Truss, Mankin & Kelliher, 2004, p. 12).

Globalization has two simultaneous faces (Vaseth, 2002, p. 1):

The rewarding face of globalization made the trade and investment liberalisation increase: the competition, productivity, innovation, minimizing waste and using new technologies, company earnings, economic growth, prosperity, creation of jobs, transfer of know-how, of new ideas, of best practices, introduced new knowledge, welfare, technical innovation in order to improve the quality of people' lives (OECD, p. 41, 51). Markets disrespect borderlines, goods, services, people, money, ideas and cultural artefacts within local, national and especially global markets, but respect their own internal regulations. The governance of global markets therefore is beyond the boundaries of any individual nation-state (Vaseth, 2002, p. 1). Regulation is not separated from the free market; it is in fact an integral part of the free market (McCleskey, 2010, p. 20). Technological revolution brought on by computers is tying the world together as never before. Now, its information available 24/7, the Internet allows even small businesses to have global ambitions, and it places pressure on nations to relax regulations and cut taxes that make it more difficult for their businesses to compete (Veseth, 2002, p. 20). Size is no longer a determinant of whether an organization can be described as global, facilitating collaboration between large numbers of employees, being called as talent-intensive mega-institutions (Lowell & Joyce, 2007). Open markets have reasonable prices, are more abundant and are far freer. For-profit firms are the vehicles of production. They decide what to produce and how, on the basis of their assessments about what is most profitable (Harcourt, 2011, p. 8).

The uncomfortable face of globalization made the countries, organizations and employees and also the customers to be affected by their competitors' improvements in efficiency. Despite the fact that open markets produce overall gains, some segments experienced income losses, job losses, raising tariffs and prices, lowering quality (OECD, 1998, p. 51), structure changes, new systems development, and workers confronted with different stress situations (Mott, 2004, p. 53). The environment is more volatile than ever, due to different challenges: continuous change, deregulation, fierce competition, digitalization, open communication, and globalization. In order to be better, faster, quicker, and cheaper, any organization must engage in competition, continuous change, and innovation (Hamel, 2007, p. 9-11). Competition is no longer just company-against-company, but also network-against-network, which means that organizations with access to the best networks can outperform competitors (Hanna, 2010, p. 35).

Globalization brought a "perceptual revolution, an effort to produce new ideas":

- Increasing technical ability to manage and systematize globalization.
- Changing other systems and forced them to adapt and to survive.

- Traditional structures have lost ground to alternative structural principles.
- Flexibility is critical for sustaining innovation in global markets (Sparrow, Brewster & Harris, 2004).
- Hierarchies are flatter, but they have not disappeared.
- Organizations had to make structural changes in order to keep pace with global trends.
- Increase mergers, acquisitions, and alliances using size and brand name.
- Technology and communications have helped the world to become flatter as people make connections across organizational and national boundaries (Truss, Mankin & Kelliher, 2004, p. 12).
- Human resources field has shaped and reshaped itself in response to the changing demands, helping to give birth to the modern practices.

A survey made on 200 large multinational US companies, reveald that human resources has a different role today, and human resources brings value-added for the company (Kaufman, Beaumont. & Helfgott, 2003, p. 403). Organizations, due to changing conditions and evolution, ask rapid responses and strive for profits in the face of increased competition, talent shortages and regulations. "Human resources management is perceived by a few personnel managers as being an old wine in new bottles" (Armstrong, 2006, p. 51). The difference between human resources management and industrial resources management would be that the first treat people as key-resources and strategic part of organization management. Using Internet, new ways of working and competing have been opened up through e-commerce (Truss, Mankin & Kelliher, 2004, p. 11). The impact of e-commerce in this century will be as significant as the impact of mass production methods in the last century (Sparrow, Brewster & Harris, 2004).

The Internet revolution has changed the human resources practices thoroughly, including the recruiting and selection ones (Stegăroiu & Florea, 2013, p. 110) and, particularly, the human capital development ones, by the use of the new e-mentoring and e-coaching and also and e-learning methods. One of the conditions of the modern business is that organizations and human capital should quickly adapt to the continually changing economic environment (Florea, 2014, p. 139). The greatest challenge facing human resource professionals working in an international capacity is how to operate effectively at both a global and a local level; to have the ability to think globally but act locally. To be successful, human resource professionals must understand global trends and the issues that matter most to stakeholders, In order to create value in the eyes of investors, customers, line managers and employees, must align human resources practices with the stakeholders' needs (Ulrich & Brockbank, 2005). The global recession has highlighted the importance of their role in obtaining performance; that is why the retention of the key employees is critical in most organizations (Truss, Mankin & Kelliher, 2004, 37).

The Impact of New Technologies on Attracting Qualified Employees

Over the past twenty years, along with the opening of new markets, the world economy has known a work revolution. The intangibles (workforce capabilities and its unique knowledge of production, of markets, of relationships with stakeholders, brand value and reputation) took the place of the other organization resources as the strategic asset, which is driven by human talent (Cheese, Thomas & Craig,

2008). Converging economic, societal and technological forces have all elevated the importance of talent for 21st century organizations. Even if talent is important for an organization' success, few organizations manage it strategically (Van Der Sluis &Van De Bunt-Kokhuis, 2009, p. 2).

Talent has become a precious resource fought over by competitors in a global war for talent. Finding, selecting, and judging good quality online content is an increasingly important competence in today's complex world. The talented employees must be used to all kinds of filters in order to be selected by a performance company. Communication by publicity is the clearest method of attracting candidates. Advertisement is the most important and most popular source for finding manpower. Advertisement is the most fragile contact between the future candidates and potential employers (Pattanyak, 2005, p. 57). Advertisement is the most efficient method to attract candidates. The question is where the advertisement should be used, and whether the recruitment should be done by the organization or by recruitment agents (Armstrong, 2006, p. 96). The publishing of recruitment advertising may prove a costly activity, especially when properly elaborated (Martin & Jackson, 2008, p. 166). That is why the industrial organizations use the ads-metry (Patriche, 1994, p. 237) a technique used in order to determine the effects of promotion actions on evolution of the amount of sales. The human factor is recognized as one of the key factors by which organizational success is achieved. Based on customer need anticipation, an industrial organization can develop specific plans for recruiting, selecting and retaining those employees who satisfy these needs at the highest level. Only by anticipating and retaining those employees fit for the respective jobs, an organization can be successful in a global, dynamic and changing competition environment (Sims, 2002, p. 2). The recruitment process re-groups the set of activities aimed at analyzing individuals and competences best fitting the exigencies of the vacant job (Cornet & Warlaud, 2008, p. 65). Recruitment represents a set of actions carried out by an organization to attract candidates owning the necessary skills for the immediate or future filling up of a vacant job (Lakhdar, 2001, p. 225).

We are entering an era focused everywhere in the direction of talent, resources, markets, and networks. There are no longer any boundaries, the world is flat (Friedman, 2005) and careers are boundaryless (Athus & Rousseau, 1996). The recruitment processes have evolved increasingly more lately, and organizations have realized that their success depends on their ability to attract, develop and retain skilled personnel. Once, after the recruitment and selection processes, organizations were able to select well-trained candidates, but nowadays it is very hard to find a good candidate (Holbeche, 2001, p. 170). The recruitment process is the best known dimension of the human resource management (Benchemam & Galindo, 2009, p. 49).

The traditional Japonese employment practices are characterized by close relationships in the value chain, long-term employment security and seniority based wages. The conservative attitude is under pression from a progressive young generation, a deceleration of the economy and an aging society. A rapid acceleration in the development of skills such as English language, cross-cultural communication, diverse thinking and international leadership is required. In some industrial companies, the vacancies are not even advertised, they are known through networking, personal relationships, and internship (van den Ende, 2013, p. 24). For example, in the Japonese organizations is used a tradition, that of lifetimeemployment, which offers social and financial security for employees until the retirement. That because they invest much in training of new recruits. This technique is based not on a contract but on mutual trust and loyalty. Then again, it is unlikly that employees would like to switch companies due to seniority-based wages: promotions, better consideration on working preferences, higher salary, and less demanding work requirements (van den Ende, 2013, p. 21). Large companies prefer to recruit fresh graduates from a selection of the best institutions, the practise being called shinotsu (new graduate) and taking place in March or October at the end of university semester/year. The candidates are pressured to graduate until 24 years, just to have a better career in large companies.

Another example, is the one from France, where are made pedagogical partnerships, which are a solid source of recruitment for organizations (Bournois, Point, Rojot & Scaringella, 2007, p. 118). Bouygues Construction organyzes each year a national game, Defi Bouygues Construction, involving 16 large schools of engineering and commerce from Europe. The objective is to attract talent and Bouygues can show its reference states as employer. In 2006, Renault has put in place the foundations for some teaching courses in France with Japanese, Koreans and Brazilians students. France, also make appeal to cooptation centres in order to attract candidates. For example, Bouygues Construction use this method, offering, the employee' who braught a new collaborator, shopping vouchers and an amount which varies depending on the level of cooptation. Another way to attract the talented candidates is employment of young diplomats. Technip Company recruited 116 employees of which 42 professionals under the age of 25 years and Renault about 75% of young graduates with less than three years experience. The open and public advertisments are most effective recruiting strategy reflects US individualistic cultural values (Cullen & Parboteeah, 2013, p. 501). An important recruitment practice in the US is to avoid any form of discrimination language in job advertisments. Most companies tend to recruit using newspaper and employment agencies. In fact, inequality is very prevalent in Mexico and in many Latin American countries, where in jobs ads are used age or gender criteria of selection (Cullen & Parboteeah, 2013, p. 502).

Starting with 1990, Canada began to compile anonymous CVs, and recruitment should be based on competence and not based on "a priori" criterions (Lakhdar, 2001, p. 290). Law since spring 2006 made compulsory in organizations with more than 50 employees in France, but not published and not implemented (Marchal & Rieucau, 2010, p. 15). Even when concluding selection, CV must become anonymous, thus allowing its application to another post, even in the same organization (Mullenders A., 2009, p. 91). It was implemented for the first time by AXA from France in January 2005; L'Oreal it took action too, removing mailing address from the candidate's CV form; ACCOR has implemented a project that mobilizes all managers to fight against discrimination, an adaptation program for employees at local culture and also cultural diversity present in the organization (Bournois, Point, Rojot & Scaringella, 2007, p. 119). Today's globalized world faces an increasing shortage of talent. Many organizations invest in their talented potential, which now is different in age, gender, language, culture, and technological and digital competencies (Van Der Sluis, Van De Bunt-Kokhuis, 2009, p. 2). Digital competencies (hardware, software, and communication) are now among the major skills for future talents. Surveys show that between 70 and 90% of large organizations now use e-recruiting systems, and it is anticipated that over 95% of organizations plan to use them in the near future (Guental & Stone, 2005).

The online job market is growing faster than the conventional recruitment market, growing from 82% of the cases in 2004 (Czerny, 2004) at approximately 100% in 2006 (Krishnamurthy, 2006, p. 63) and its usability has increased its traffic up to 100% especially in large organizations (Rankin, 2009, p. 220). Organizations have reduced recruitment costs by 85% using online recruitment ads (Bach, 2005).

Nike, with e-recruitment, obtained an average time to fill job positions reduced from 62 to 42 days and the recruitment costs dropped by 54% (Bondarouk, 2009). While t-recruitment is characterized as a sequential process, e-recruitment is characterized as a continuous and online process. E-recruitment refers to marketing and promotional efforts aimed at recruiting prospective employees by electronic means (Doran, 2001). Some indicators show that e-recruitment strategy is the integration and use of Internet technology to improve the efficiency of this process: 71% of employers are using sites for employments, 92% prefer Internet for the rapid access, 89% appreciate that there are specific criteria to search for a vacancy (Bournois, Point, Rojot &Scaringella, 2007).

Other important advantages of using e-recruitment: winning time, lower costs; relatively simple technique for all, increasing productivity (Bournois, Point, Rojot &Scaringella, 2007), quick response time, broad range of applicants, more educated applicants (Bondarouk, 2009), worldwide accessibility at their own convenience (24/7) (Galanaki, 2002).

The Impact of Employees' Motivation on Satisfying Consumers and Obtaining Performance

Communication can contribute to motivating the employees, building confidence, creating a common identity or can increase personal involvement; it offers individuals a way of expressing their feelings, of sharing hopes and dreams and of celebrating achievements. Communication is the way by which individuals and groups understand the organization, what it is and what it represents.

Flexibility, creativity and innovativeness are needed to act and react in a continuously changing business environment. These dynamics make organization to motivate talented workforce focused on self-governance, skills development, ethical behaviour, and professionalism. Professionalism is a product of the motivation to learn, the awareness of the need to learn, the need for development of knowledge, skills, values, and attitudes (Truss, Mankin & Kelliher, 2004, p. 36). Given the strategic importance of talent and the ability to attract, retain and motivate it, nurturing and developing talent has to become everyone's responsibility in every organization. In a talent-driven economy, organizations need to understand, acknowledge and live that responsibility. These changes in the type and content of work are why the word "talent" has come to be used so much now in business, as opposed to the general term "labour". The traditional use of talent refers to a special gift; in business, it refers to various attributes of people (knowledge, experience, skills, behaviours, career anchors, ambitions, norms, personality, and values) (Van Der Sluis, Van De Bunt-Kokhuis, 2009, p. 13). So, talent is those candidates that any organization would want to acquire, retain, motivate and develop in order to achieve its objectives and obtain performance. Motivation is becoming more and more important as a process nowadays, due to the intensification of competition at the workplace and to the unemployment rate increase (Hertel, 2003, p. 3). By motivating the organizations, it is ensured that the most creative and performance-achieving employees shall remain in the organization, identifying themselves with it and contributing to the accomplishment of its goals.

Motivation consists of three elements (Saemann, 2009, p. 1): effort - a more motivated person will put more efforts in his/her activity, desire - an employee desires to achieve its goals if he is highly motivated, and affect - the employee enjoys learning and doing its tasks if he/she is motivated enough.

Motivation and retention of the best employees can be influenced by communication practices (Phillips & Edwards, 2008). Whatever the existing motivation may be - financial or non-financial - in organizations there must be an efficient communication process between manager and employee. A pertinent, clear and concise communication shall motivate the employee in obtaining individual and implicitly organizational performances.

Industrial sector companies value external customer satisfaction more than the service sector companies (Mayrhofer, Brewster & Morley, 2007, p. 406). High-value employees are finding it harder than ever to balance the demands of their personal and professional lives. Organizations understand that their success depends largely on these employees, and for this reason they must develop effective strategies in order to attract, motivate, and retain the best employees.

The Impact of Internal Communication on Relationships with Customers along Supply Chain

Communication dominates modern life and it is a foundation stone on which any relationship is based. Good communication requires engagement, promotion, and time and as much listening as speaking, and if it is done well it can positively improve the participants' experience (Simpson, Robinson, Fletcher & Wilson, 2005, p. 12). Communication is important because it performs three vital functions: it unites people, it spreads knowledge and it leaves no room for assumptions and secrecy (Akay, 2006, p. 88). Managers of large companies report that 14% of their work/week is lost for reasons of inefficient communication. At the workplace, an inefficient communication leads to occurrence of errors, increase of costs, loss of customers, decrease work satisfaction and performance (Wood, 2010, p. 7). Communication represents the vital essence necessary to business development and helps helps accomplish the organizational functions (Kumar, 2010, p. 8). Whatever the goals people may have for developing their career, they must develop their communication abilities, to reach performance, professional success and satisfy customers' needs.

So, communication is essential, solving many problems and improving many processes:

Communication with Customers

Communication with customers represents the most important part of a business. No business can survive without customers. That is why, it is very important that the organization should be able to communicate with them correctly and own abilities of listening, of answering the complaints, of customer relation management and of customer attraction and satisfaction (Gopal N., 2009).

The goal of a retail business is to have loyal customers (Tang & Lim, 2012).

To obtain this, enterprises must:

- Use new technology which is now a fundamental competitive tool along supply chain.
- Use databases which can provide data about consumers behaviors.
- Use e-communication with customers because computers have become a must in any competitive business (Cant, 2005, p. 17) and an ability to attract and retain customers (Menne, 2008, p. 1).

- Create and offer an efficient message taking into account the market trends and customers' preferences (Krizan, Merrier & Logan, 2011, p. 2).

Communication with industrial customers can be realized by different means of communication: brochures, technical information, prospects, catalogues of productive enterprises, industrial sales force, notes, technical data offering a high degree og technicity and utility for buyers (Patriche, 1994, p. 233), telephones, public relations, service, radio and television messages, newspapers and magazines, sale/purchase points, posters, displays, all of them being used to reach the intended goals (Kumar, 2010, p. 12). Due to the existence of many competitors on the market and the technicity of organizations, it is important that the organization should efficiently communicate with the customers and organize market researches and studies (Gopal, 2009) since only in this way it will be able to communicate more efficiently and sell better based on information, persuasion, image, trust, facilities, notoriety (Patriche, 1994, p. 237), as a communication method in industrial marketing. Industrial organizations are aware that attracting skilled, creative employees, motivated to satisfy customers is essential for differentiating a company from its competitors (Bhattacharya, 2009, p. 396). Organizations must be aware that they must attract talented employees, especially in sensitive departments, such as production, purchasing, selling, marketing, in order to increase their share market. The most talented people will want to work in those organizations they regard as good employees (Payne, Frow, 2013, p. 139).

Communication with Candidates during Recruitment and Selection Processes

The human resource specialist needs to be able not only to use the technology, but also to master the art of creating pertinent ads, a good ad being one of the main keys to a good recruitment process, especially for industrial companies. Communication with the outside world represents an indicator allowing an analysis of the organizational health, strategy evolution, employment policy and human resource management. Other organizational health symptoms are: recruitment volume, nature of labour contracts, career shaping and evolution, selection and interview methods (Cadin, Guerin & Pigeyre, 2009, p. 293).

Communication with Employees

For instance, top managers use 75 up to 95% of their time for communication. Another study shows that the organizational employees use between 30 and 50% of their time to write (Kumar, 2010, p. 10). By means of communication, people develop continual personal relations, to face their continuously changing needs and this becomes efficiency when the desired reaction or answer, are obtained from the person who has received the message (Kumar, 2010, p. 13). Communication is important in itself, to develop and maintain the personal, professional and cultural relations (Wood, 2010, p. 7). Employees are very important for organizations. Organizations need to communicate with their employees to cooperate, to improve sociability, to stimulate innovation and creativity, to strengthen employee morale and their identification with the organization and to ensure that employees know how to accomplish their own, specialized tasks (Cornelissen, 2011, p. 163

Every manager must develop solid communication skills in three important areas if he/she wants to be successful (Sears, 2007, p. 37): active listening and giving clear information, and getting clear and unbiased information. Giving clear information to subordinates means obtaining compliance, performance, and results. More and more, communication is a tool for gaining collaboration, not just compliance, and for inviting talented individuals into colleague relationships. Managers dedicate 75% of their time at work to communicate (Mintzberg, 1980, from Svetlik I. & Ilic, 2006, p. 213). Listening consumes more than reading, writing and speaking together represents a very important interpersonal ability (Lehman & DuFrane, 2011, p. 51). The organization's ability to communicate affects not only its own success, but also the success of each employee, including their promotion (Lehman & DuFrane, 2011, p. 8). Organizational communication represents a zone of major interest and it is important in maintaining relations, in supporting the morale, especially as they spend long hours together every day (Wood, 2010, p. 14). Managerial communication is a complex process and is influenced by a number of factors (Lehman & DuFrane, 2011, p. 16): legislative and ethics factors: national and international laws, ethics code, diversity, technological factors: security issues, information accuracy, databases, and teamwork environment: confidence, roles, tasks, goals, expectations, leadership.

In a competition-based environment, efficient internal communication becomes an important element of the organizational strategy, while the management mobilizes the staff, raises its self-esteem, motivates it, orients it towards certain priorities (Agarwal, 2010, p. 10). Internal communication occurs between the employees of one company who work to attain its operational goals and plans (Kumar, 2010, p. 11).

METHODOLOGY

In our research we started from the fact that:

- The primary focus of an organization must be its employees, because it is the foundation for customer relation management system (Lamb, 2012) along supply chain, and because between the organization and customers there are one-time events named touch-points, having potential for innovation (Matthews, 2009, p. 83).
- A good internal process depends on a good communication (Nykiel, 2007, p,262).

So, we want to demonstrate that there is a direct, a linear and a positive connection between communication process and internal processes, in our case between recruitment and development processes. Analysing these processes through the prism of using new technologies, we analyse the e-coomunication, e-recruitment and e-development processes using official data from European level -EU(28) (www.eurostat.com). We realyze that we can use a mathematical model to show this connection, and is reffering to linear regression.

This is a parametric method, which predict the value of a dependent variable Y, based on an independent variable X (Darbyshire & Hampton, 2014). Is among the most popular data-driven models for their easy application and very well-known techniques (Araghinejad, 2013, p. 48).

The formula for this model is:

$$y = b_0 + b_1 \times x_1 + b_2 \times x_2$$

where:

Y = Enterprises sharing electronically information on sales or on purchases with the software used for any internal function (named by us e-communication).
X1 = Individuals using internet looking for a job/send an application (named e-recruitment).
X2 = Individuals using internet for looking for information about education and training (named e-development).

Thus, we want to demonstrate that e-communication process used at enterprises level is depending on individual e-skills used to improve internal processes as recruitment and development (using new technologies). The strength between the dependent variable (y) and the independent variables (X1 and X2) can be determined by calculating the coefficient of determination (denoted by R^2), which can take the values between 0 and 1 (Dixon, 2011, p. 129; Walish, Michael & Benyoun, 2014, p. 99):

0 The dependent variable cannot be predicted from the independent variable. A lower value implies that the dependent variables have some variance unaccounted for by the model.
1 There is no error in the relationship between the dependent and independent variable.

- **Sample size:** The reasearch was made using data and information from the oficial site of www. eurostat.com.
- **Tools for data collection:** Observation and data analysis.
- **Tools for analysis:** Modeling and simulation using linear regression function, Excel program.

Research Objectives

The article it is analysing the influence could have the individuals using internet looking for a job/send an application (X1) and individuals using internet for looking for information about education and training (X2) on enterprises sharing electronically information on sales or on purchases with the software used for any internal function (Y), using regression function.

Data Presentation and Analysis

The research was made on oficial data from www.eurostat.com, taking into account the variables described above. We used a range of data on the evolution of the three indicators of results for the period (2008-2013) (Table 1).

We used Excel- Data Analysis, and we obtained the following results (Figure 1):

Table 1. Data recorded at EU level (EU28) 2008-2013

Year	E-Communication (y)	E-Recruitment (x_1)	E-Development (x_2)
2008	41	13	20
2009	40	15	23
2010	40	15	23
2011	40	17	29
2013	44 (from 2012)	17	31

Source- www.eurostat.com

Using the linear regression and Excel we obtained:

$$y = 53,021 - 2,6304 \times x_1 + 1,1304 \times x_2$$

Multiple R is the multiple coefficient of correlation between X_1, X_2 and Y. Between the value X_1, X_2 and y recorded in the analyzed organization in the period 2008-2013 there is a direct and strong connection due to the amount obtained by Multiple R (0.90).

R^2 means that 81,88% of the variance in the dependent variable Y is predicted or explained by the independent variables X_1 and X_2. The strength of the linear relationship between variables is measured by the simple correlation coefficient (Araghinejad, 2013, p. 51), so the R^2 show a value of 0,8188, which is closer to 1 indicating that the most of the variance in the dependent variables originate in the variation of the regressors, as propagated through the model. Adjusted R Square is a coefficient of determination corrected with the freedom degree having the same signification as R^2.

Figure 1. Determine the linear parameters using the regression function from Excel

SUMMARY OUTPUT

Regression Statistics	
Multiple R	0,904898105
R Square	0,81884058
Adjusted R Square	0,637681159
Standard Error	1,04257207
Observations	5

ANOVA

	df	SS	MS	F	Significance F
Regression	2	9,826086957	4,913043	4,52	0,1811594
Residual	2	2,173913043	1,086957		
Total	4	12			

	Coefficients	Standard Error	t Stat	P-value	Lower 95%	Upper 95%	Lower 95,0%	Upper 95,0%
Intercept	53,02173913	7,935527065	6,681565	0,021674	18,877922	87,16555632	18,877922	87,1655563
e-recruitment (X1)	-2,630434783	1,119090233	-2,35051	0,143136	-7,445491	2,184621862	-7,4454914	2,18462186
e-development (X2)	1,130434783	0,40670189	2,779517	0,108731	-0,619462	2,880331779	-0,6194622	2,88033178

RESIDUAL OUTPUT

Observation	Predicted e-communication (Y)	Residuals	Standard Residuals
1	41,43478261	-0,43478261	-0,58977
2	39,56521739	0,434782609	0,589768
3	39,56521739	0,434782609	0,589768
4	41,08695652	-1,08695652	-1,47442
5	43,34782609	0,652173913	0,884652

PROBABILITY OUTPUT

Percentile	e-communication (Y)
10	40
30	40
50	40
70	41
90	44

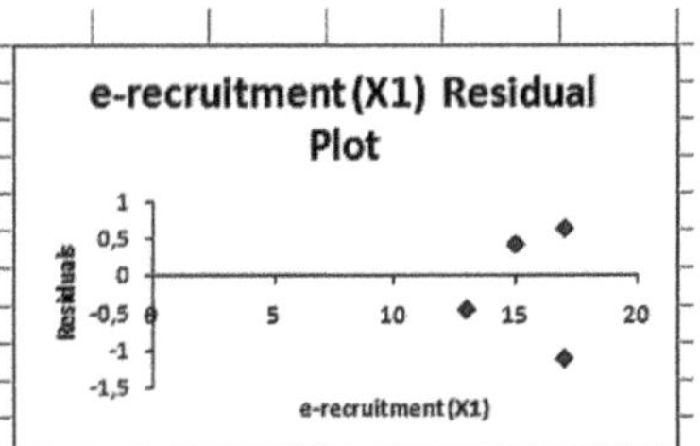

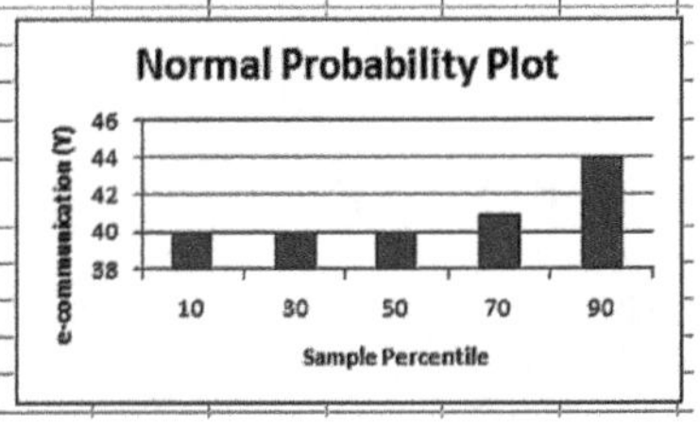

Standard Error is the standard error which shows the average of the values observed deviate from the theoretical values y_i of the regression line, Y_i (in this case with ±1,042).

Observations are n, the number of observations, n=5.

ANOVA reprezents the analysis table of tha variance. For the variance due to factor x, Regression, Residual due to the others unregistred factors, and Total due to the all factors:

- df- degrees freedom: k – number of explivative variables x (simple regression k=1), n-k-1 for residuals (5-1-1=3 degree freedom) and n-1 for total variance (5-1=4); The sum df for Regression and Residual is equal with df / Total: $k + (n - k - 1) = n - 1$.
- Intercept is for the free term (constant) of the model and has the value (0).
- Coefficients- contain the estimated values of coefficient b.
- Lower 95%, Upper 95% – upper and lower limits of the confidence interval for the parameter. The limits for 0.05 are calculated automatically, regardless of the initialization procedure Regression. It could be said that the linear model parameters are within the following ranges:

$18,87<b_0<87,16$

$-7,44< b_1 < 2,18$

$-0,61< b_2 < 2,88$

Analysis of model quality is presented in Figure 2.

Figure 2. Residual Plot for variable X1

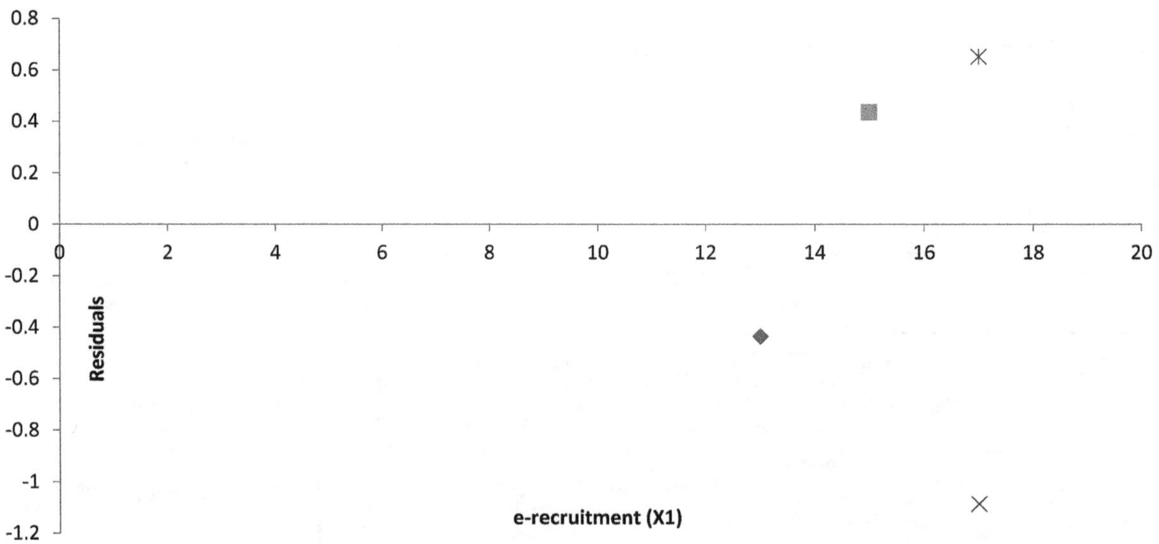

After the form of the points position, results that there is no correlation between the independent variables and residuals, thus, the model was well choosed.

From the simulation model results that X1 and X2 asre very important factors in the evolution of Y. To increase with a monetary unit the X1 will get an decrease of 2,63 monetary units of Y and and to increase with a monetary unit of X2 will get an increase of 1,13 monetary units of Y.

We note that the value of 53,021 of free term is not very high, which allows us to conclude that the factors were not taken into account in the model construction having a material impact on the evolution of the e-communication process. The positive value of the free term reveals that the variables were not included in the econometric model has a positive effect on the evolution of communication.

We propose to increase with 10% the data regarding e-development in 2014 (because we observe the trend is increasing with almost 10% annually) and maintain the data for e-recruitment, due to tha fact that it has increased its usability up to 100% in 2007. Thus, we want to simulate Y, depending on the forcasted analyzed variable (Table 2).

$$y_{2014} = 53,021 - 2,6304 \times 17 + 1,1304 \times 34 = 46,73$$

$$y_{2015} = 53,021 - 2,6304 \times 17 + 1,1304 \times 37 = 50,121$$

$$y_{2016} = 53,021 - 2,6304 \times 17 + 1,1304 \times 41 = 54,641$$

The graphical representation is made below (Figure 3).

As we may see, applicating simulation and modeling, the process of e-communication at enterprises level is depending on e-recruitment and e-development of individuals who work in this enterprises. Using statistical analysis from Excel, we showed that the e-communication process is depending the two analyzed variable used in this model: erercuitment and e-development and are very important factor in the evolution of e-communication. Having better and talented employees, the organizations may improve the other processes including the communication with employees and customers aloong supply chain.

RESULTS, DISCUSSIONS AND RECOMMENDATIONS

Now, after 2000, the enterprises are in the fifth stage of the supply chain management evolution, that of e-SCM (Ross, 2002, p. 6). This means that using internet along supply chain, enterprises and its

Table 2. Forecasted evolution of Y for 2016 depending on new X2

Year	Y (forcasteed)	X1 (maintained)	X2 (+10% annually)
2014	?	17	34
2015	?	17	37
2016	?	17	41

Figure 3. Evolution of e-communication forcasted for 2014-2016

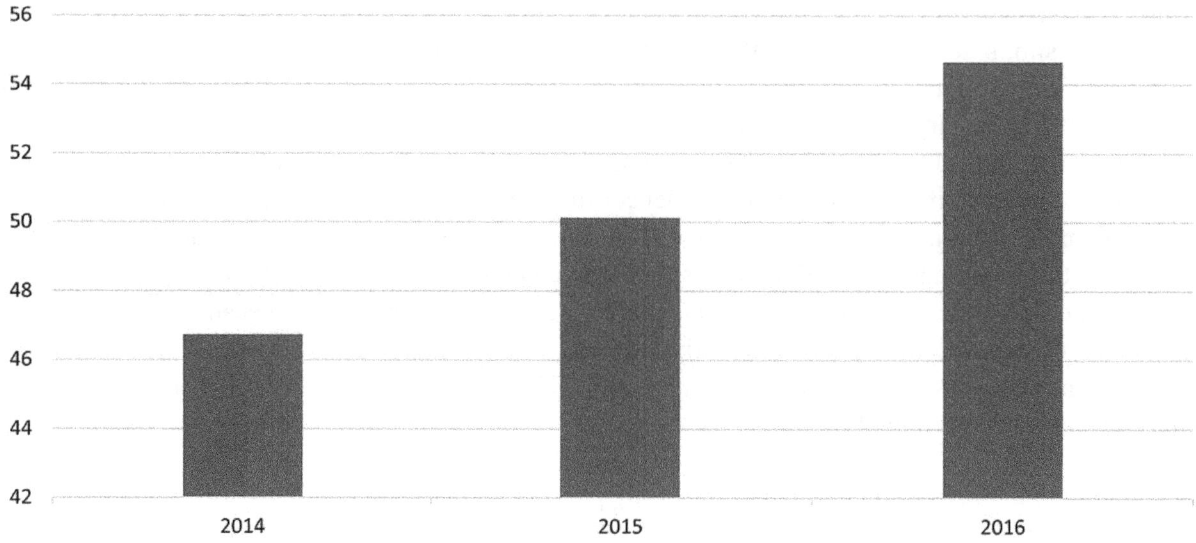

employees will improve relationships with customers and suppliers, will reduce costs, will improve internal processes (like communication, recruitment and develpment), will share new information, will gain new skills and competencies (like employees' e-skills and management agility), will share rapidly databases, will used them to simulate and model business and will obtain supply chain synchronization. Increasing business relations management along supply chain has become Supply Chain Management (Lambert, 2004).

So, improving communication, and human resources processes any industrial company may obtain performance through: using simulation and modelling, using integrated communication to improve relationships with suppliers and customers along supply chain and new technology in order to cut costs (in our case e-development methods like e-coaching and e-mentoring), improve internal and external processes, improve access to knowledge, experience, and skills, and improve time management.

To improve performance of an industrial organization along supply chain, we propose solutions like:

- *Using integrated communication to improve relations with employees and customers*

This concept represents the mix of all tools, approaches and communication resources to maximize the impact on clients and to conduct to minimal costs and maximum profits. Integrated communication, used before conceptualization, was rapidly extended due to technological boom and the many advantages in regarding the need and desire of consumer. As response to mutations from different areas, integrated communication lead to pozitive relations building, and not only on external markets, but also with its internal public, which is extremelly more important, because it is favourise loialty and commitment (Tanasescu, Lupu & Petrescu, 2014, p. 41). Cornelissen & Lock (2000, p. 7) sustain that integrated marketing communication is not more then a fashion in management and could become at least a managerial fad. The key of developing a integrated communication plan is in respecting the following eight principles (Schultz & Schultz, 2004, p. 49-66): company must be focused on clients, company must make its planning from outside, company must be focused on total experience of clients, company must align the clients' objectives with its own, company must objectively establish the clients behaviors, company must treat its clients as precious assets, company must streamline its functional activities, company must orient its marketing communication towards the same destination. Thus, any organization, in order to win the confidence of employees and clients, must build a favourable image and brand, through an integrated effort, a coherent approach who must consist of complex and flexible communicational programs. Only integrated communication may bring clarity and may help finding a balance between the global coherence and local differences (Tanasecu, 2014, p. 49).

- *Using new technologies to improve e-communication along supply chain*

New technologies have led to increased consumer expectations and a changing society, and enable people to stay in touch with one another across any distance and any moment (Simpson, Robinson, Fletcher & Wilson, 2005, p. 6, 12). People spend more and more time online (Greiffenstern, 2010, p. 1). The Internet is the last method of electronic communication of the 20th century; it is a decentralized network, providing information internationally and changing ideas and information (Janoschka, 2004, p. 43). The Internet appears as a privileged support of today's communication, facilitating the integrated presentation of image, of writing and of the voice. As we can note, the number of people which are using the Internet and on-line communication is increasing, becoming the most important form of communication. It is said that it is closer to face-to-face communication than all the other forms of communication (Schachtebeck, 2005, p. 3). The main aim of online communication is that people may communicate with anybody connected to the Internet, from anywhere (Bakshi & Godse, 2009, p. 499). Since the appearance of Internet communication, a new form of communication has developed - communication in the new era of the cyber-space (Jain, 2011, p. 241). E-communication was adopted by businesses throughout the world, transforming their relationships with stakeholders. In this era of new technology, employees working in a new digital workplace need new digital competencies to face these new challenges (Van Der Sluis L., Van De Bunt-Kokhuis, 2009, p. 36). Beside face-to-face communication, managers use emails, videoconferencing or enterprise software to communicate with the employees (Cornelissen, 2011, p. 165). IBM, for example, offers more than 5,000 audio and video podcasts "episodes" to employees, in order to watch or listen to them whenever they have time.

- *Using new technologies to improve e-development programs (e-mentoring, e-coaching)*

Human resource processes, including development is on-line: e-mentoring, e-coaching. Large companies create e-coaching/e-mentoring programs on the Internet, which helps to improve communication, development and employee training, providing information and knowledge at a very low price (Law, 2007, p. 12). They provide immediate knowledge to specific questions: for example in occupational safety, and help employees in making complex decisions. The program selects a command to send or to receive questions/responses to/from the virtual consultants (Horton, 2001, p. 58). Many have realized the potential development of e-mentoring and built an online platform to connect better people with able body of experts, the aim being collaboration between many organizations (Burri & Cottier, 2012,

p. 446). The platform offers: professional management, regular updating of learning resources, continuous improvement platform, feedback to participants, and facilitates communication. E-mentoring is a relatively new and still in research (Headlam-Wells, Gosland & Craig, 2006, p. 273).

- *Using new technologies to improve e-recruitment process*

In today's increasingly electronic world, the Internet is an important tool for applicants and for organizations to interact in an e-business and e-commerce era. Technology has allowed organizations to send monthly emails, newsletters and updated jobs right to the target candidates. E-recruitment refers to marketing and promotional efforts aimed at recruiting prospective college students by electronic means (web, sms, e-mail) (Doran, 2001). E-recruitment has become an indisputable tool and the main competitor for mass-media (Bournois, Point, Rojot & Scaringella, 2007). Recruitment can happen today 24/7 from everywhere, anytime, anyway, at any moment, reducing the costs and increasing the efficiency of this process. The messages transmitted by the top managers have an important impact on the employees' attitudes and behaviours. Consequently, communication must be a process based on trust, openness, understanding, listening, transparency, feedback, and empathy in order to maintain the best candidates and face the challenge from the external environment. Many software companies (like SAP, Oracle, PeopleSoft) offer online program which help manufacturers, distributors and supply-chain participants with ERP systems (Bidgoli, 2004, p. 712). Internet is now a fundamental part of the recruitment process (Armstrong, 2007) and will definitely continue to play a very prominent role as the world becomes more digitized (Bondarouk, 2009). Globalizations and use of Internet has revolutionized the world and implicitly the human resource processes. Recruitment, which is using Internet, bring for organizations important advantages.

FUTURE RESEARCH DIRECTIONS

Our research aims a clear understanding of the analyzed processes by taking into account the new technologies proposed above. An ever-growing interest in industrial organizations is to bring better practices of communication and human resource, in order to predict risks, obtain performance, and satisfy customers along supply chain.

Based on these objectives, we show the implications brought by this chapter:

Theoretical Implications

This chapter focused on the study of literature in the field of communication and human resources management. The chapter was based on the analysis of the concepts like globalization, informatization, communication, recruitment, selection, sustainable development, non-discrimination, ethics, equality, fairness, analysis and monitoring of costs. We also have analyzed the best practices associated with these processes, the effectiveness of communication, and the performance of organizations, employees, and processes like communication, recruitment, selection, and development under the impact of new technologies.

Methodological (Practical) Implications

Managerial Implications by Improving Strategies

The evidence brought in here is suggesting future trends in:

- Developing effective communication strategies, in order to improve all the organization' processes and implicitly the human resource processes using simulation and modeling.
- Implementing good practices of recruitment, selection, development and communication, taking into account factors such as globalization, internationalization, liberalization of markets, and using new technologies.
- Developing certains styles of communication in order to obtain performance and increase organization' productivity, and reducing conflicts.
- Sending almost all the employees to training based-communication in order to improve relationships between colleagues and also the external image among stakeholders.

Due to eventual risks, costs of these processes, and time management, industrial companies may develop new strategies, new procedures and new politics.The structure, the culture, its specificity of products, customers and markets and the size of industrial companies influence the future of human resource and communication techniques, and the quantity and the quality of data used by them.

Employees' Implications by Improving Processes

Developing effective processes of communication, recruitment, selection, and employees' development will lead to new research directions as:

- Knowing communication and human resource strategies at organization level.
- Implication of employees and organizations in actions that make easy the implementation of-communication and human resource strategies, by using strong principles of ethics, equity, not-discrimination, social risponsibility, and sustainable development.
- Developing team skills by intensifying training programs.
- Investing in people development, without considering simple costs.
- Calculating performance indicators and implicitly the costs of these processes in order to have a clear vision of their measuring and obtain performance.
- Implementing analysis in order to make these processes more effective, by using benchmarking, scorecard, some analysis matrix.
- Developing clear and written recruitment and selection procedures and make them known to all the employees.
- Using new technologies, managers can develop new programs of recruitment, selection, development, motivation and communication, having new expectations, and may bring new talented staff at the right place and time, and at the right cost.

Using e-recruitment, e-selection, and e-communication in industrial companies and all other organizations it may be obtain many advantages, such as:

- Winning time, lower costs, relatively simple and usable technique for all, growing productivity compared to analyzing written and submitted resumes.
- Reaching large numbers of persons, including those in international labour markets, streamlining burdensome administrative processes, enabling the organization to evaluate the success of its strategy.
- Speeding up the process, cutting out intermediaries, ethicality of computer-based processes, improved accessibility to information.
- Quick response time, broad range of applicants, more educated applicants.
- Worldwide accessibility at their own convenience, a shorter cycle and lower cost.

Supply Chain Implications by Improving Relationships with Customers and its Sustenability

Supply chain can be an information flow between users and receiving party (Borgstom & Hertz, from Delfmann, 2007, p. 39). Information exchange seems to be dependent on a change in procedures in order to drive efficiency improvements in the supply chain.

To obtain sustainable supply chain success, longer-term relations (in our case with candidates, with employees and with customers) are necessary, beneficial to all parties.

The advantages and benefices of using e-tools along supply chain:

- Save resources like time and money.
- Improve people development and careers.
- Improve relationships, collaboration, and learning in the supply chain.
- Inform more rapidly the customers.
- Attract new customers due to the attraction of new talented candidates.
- Develop appearance of strategic alliances.
- Developing new performance procedures and rules.
- Gaining new knowledge, experiences, and skills.
- Develop new working processes.
- Develop new IT systems and processes.
- Obtain economic and social gains.
- Synchronization of internal processes such as: communication, recruitment, selection, motivation and development with supply chain.
- Align internal processes to the best internal processes of other companies present in the supply chain.

CONCLUSION

Our objective was to explore important concepts like communication, and internal processes under the impact of new technologies and supply chain management from three perspectives:

1. A benefits perspective- Implementing a supply chain suppose investing time and resources. To assure that a supply chain was effective, managers must assess if benefits was achieved: communication with customers was improved, number of new customers has increased, new knowledge of new employees was gained, profit of the company has grown.
2. A relational perspective: between organizations with its internal public and customers, in order to satisfy their needs and to obtain performance.
3. A technological perspective: which consists in using new tools to improve these relationships: e-communication, e-recruitment, e-selection and e-development.

Competitive advantage in tomorrow's environment will go to those enterprises that can consistently anticipate and implement customer-winning supply chain competencies (Ross, 2002, p. 1). An efficient internal communication contributes to increaseing the employees' satisfaction at the workplace, their morale, productivity, commitment, confidence and learning, improving the communicational climate and public relations, and increasing quality and profit. In an industrial automation field, where almost all the processes are under the impact of new technologies, the processes of communication and of human resources are in the same situation.

E-communication, e-recruitment and e-develoment become important parts of organizations which are ambassadors of their effectiveness in the external environment, having the role of a healthy indicator for stakeholders. Its positive image from inside will be refelected into the image from outside through brand confidence, loyalty consumers, good relationships with distributors and a good implication and strategic alliences with competitors. Such a process will achieve internal-external integration and performance on long term. Inside the company, it strengthens the employees' trust in the organizational vision and mission, connects the employees to the reality of the respective business, stimulates the company's development process, facilitates the changes necessary to progress and contributes to the employees' behavioural change.

Integrated communication, implicitly e-communication can help industrial organizations achieve its objective and also can offer an easy way to communicate with its customers and its employees. E-communication, offers many advantages, such as: it can spread the mission, the objectives, the vision, and values throughout the organization, announce changes in structure, procedures, legislation, help integrate the new employees, offer special advises, ask questions, offer responses, remaind employees their tasks, their responsibilities, encourage employees, motivate them, provide information to customers, gather information about the employees' efficiency, provide databases, improve interrelations inside and outside the organization.

It was shown that there is a direct link between communication strategy, organization productivity and employee satisfaction (Kusluvan, 2003, p. 395). The most effective organizations are those that have the most effective communication across the whole organization and along supply chain. Human resources, being the most important asset in an organization, the organization must effectively communicate with its capital, must attract the right people, develop, retain and motivate it, in order to satisfy customers and obtain performance. Attracting the right person to the right place at the right moment and cost, communicating effectively with these persons, they will be more implicated, more loyal and will try to satisfy the customers' needs and obtain performance.

REFERENCES

Agarwal, M. (2010). *Business Communication*. India.

Satyendra, R., & Akay, V. (2006). *Business Is the People and People Are the Business. Humanizing the Corporation*. Pine Lake Road, USA: iUniverse.

Armstrong, M. (2006). *A Handbook of Human Resources Management Practice*. USA: Kogan Page.

Arthus, M. B., & Rousseau, D. M. (1996). *The Boundaryless Career: A New Employment Principle for a New Organizational Era*. NY, USA: Oxford University Press.

Bach, S. (2005). *Managing Human Resources: Personnel Management in Transition*. USA: Blackwell Pub.

Bakshi, U. A., & Godse, A. P. (2009). *Communication Engineering*. India: Technical Pub. Pune.

Benchemam, F., & Galindo, G. (2009). *Gestion des resources humaines*. Paris: Gualino.

Bhattacharya, C. (2009). *Services marketing*. New Delhi: Excel Book.

Bidgoli, H. (2004). *The Internet Encyclopedia* (Vol. 1). USA: John Wiley&Sons. doi:10.1002/047148296X

Bondarouk, T. (2009). *Handbook of Research on E-Transformation and Human Resources Management*. Netherlands: Idea Group Inc.

Bournois, F., Point, S., Rojot, J., & Scaringella, J. L. (2007). *RH - les meilleures pratiques. Eyrolles*. Paris: Editions d'Organisation.

Burri, M., & Cottier, T. (2012). *Trade governance in the digital age*. NY, USA: Cambridge University Press. doi:10.1017/CBO9781139136716

Cadin, L., Guerin, F., & Pigeyre, F. (2009). Gestion des ressources humaines. Paris: Ed.

Dunod, C. M. C. (2005). *Introduction to retailing*. Cape Town, South Africa: Juta&Co. Ltd.

Cheese, P., Thomas, R., & Craig, C. (2008). *The Talent Powered Organizations*. London: Kogan Page.

Cornelissen, J. (2011). *Corporate Communication: A Guide to Theory and Practice*. CA, USA: Sage Pub. Ltd.

Cornelissen, J. P., & Lock, A. R. (2000). Theoretical Concept or Management Fashion? Examining the Significance of IMC. *Journal of Advertising Research*, *40*(5), 7–15. doi:10.2501/JAR-40-5-7-15

Cornet, A., & Warlaud, Ph. (2008). *Gestion de ressources humaines et gestion de la differentité*. Paris: Dunod.

Cullen, J., & Parboteeah, P. (2013). *Multinational management*. Cengage Learning.

Darbyshire, P. (2014). *Hampton D., Hedge fund modelling and analysis using MATLAB*. John Wiley & Sons.

Dixon, K. R. (2011). *Modelling and simulation in Ecotoxicology with applications in MATLAB and Simulink*. CRC Press. doi:10.1201/b11089

Doran, A. (2001). *E-Work Arhitect: How HR Leads the Way Using the Internet*. Austin, Texas, USA: IHRIM Press Book.

Drucker, P. (2003). Management's new paradigms.

Florea, N. V. (2014). Training, Coaching, Mentoring. Methods and Models. Bucuresti: Ed.

Beck, C. H., & Friedman, T. L. (2005). *The World Is Flat. A Brief History of the 21st Century*. NY: Farra, Straus and Giroux.

Galanaki, E. (2002). The Decision to Recruit Online. *Career Development International*, *7*(4), 243–251. doi:10.1108/13620430210431325

Coronas, T. T. (2008). *Encyclopedia of HR Information Systems: Challenges in E-HRM*. Hershey, PA, USA: IGI Global.

Gopal, N. (2009). *Business Communication*. India: New Age International.

Greiffenstern, S. (2010). *The Influence of Computers, the Internet and Computer-Mediated Communication on Everyday English*. Berlin, Germany: Bibliografische Inf.

Guental, H. G., & Stone, D. L. (2005). The Brave New World of eHR. HRM in the Digital Age. SF, USA: Jossey-Bass a Wiley Imprint.

Hamel, G. (2007). *The Future of Management, Harvard Business Press, Hanna, N.K., (2010). Enabling Enterprise Transformation*. USA: Springer.

Harcourt, B. (2011). *The Illusion of Free Markets*. Harvard University Press.

Headlam-Wells, J., Gosland J. & Craig, J. (2006). Beyond the organization: the design and management of e-mentoring systems. *International Journal of Information Management*, 26, 272-285.

Hertel, F. (2003). *Motivation in the Workplace*. GRIN Verlag.

Holbeche, L. (2001). Aligning Human Resources and Business Strategy. *Business Economics (Cleveland, Ohio)*.

Horton, W. K. (2001). *Leading e-learning. USA: ASTD*

Jain, A.K., (2011). Professional Communication Skills. ND, India: S.Chand&Co.Ltd.

Janoshka, A. (2004). *Web Advertising: New forms of communication on the Internet.* USA: John Benjamin Pub. doi:10.1075/pbns.131

Kaufman, B. E., Beaumont, R. A., & Helfgott, R. B. (2003). *Industrial Relations to Human Resources and Beyond: The Evolving Process of Employee Relations Management.* M.E. Sharpe.

Krizan, A. C., Merrier, P., & Logan, J. P. (2011). *Business Communication.* Cengage Learning.

Kumar, R. (2010). *Basic Business Communication.* ND, India: Excel Books.

Kusluvan, S. (2003). *Managing Employee Attitudes and Behaviors in the Tourism and Hospitality Industry.* NY, USA: Nova Science Pub.

Lakhdar, S. (2001). *Gestion de ressources humaines.* Canada: Université de Boeck.

Law, H. (2007). *The psychology of coaching, mentoring and learning.* England: J. Wiley & Sons.

Lamb, C. W. (2012). *Marketing.* Cengage Learning.

Lambert, D.M. (2008). *Supply Chain Management: Processes, Partnerships, Performance.* Supply Chain Management Inst.

Lehman, C. M., & DuFrene, D. D. (2011). *Business Communication.* USA: Cengage Learning.

Lowell, B., & Joyce, C. (2007). *Mobilizing Minds.* McGraw Hill.

Marchal, E., & Riencau, G. (2010). *Le recrutement.* Paris: Edition La Decouverte.

Martin, M., & Jackson, T. (2008). *Practica de personal.* Bucuresti: Editura CODECS.

Matthews, J. R. (2009). *The customer-focused library.* ABC-CLIO.

Mayrhofer, W., Brewster, C., & Morley, M. (2007). *HRM in Europe.* Taylor&Francis.

McCleskey, S. (2010). *When Free Markets Fail: Saving the Market When It Can't Save Itself.* John Wiley&Sons.

Menne, P. (2008). *Potential of Geo-Marketing-Tools for the development of advance online-marketing business models* [Master thesis].

Mott, W. H. (2004). *Globalization. People, Perspectives, and Progress.* USA: Greenwood Pub.Group.

Mullenders, A. (2009). *E-DRH. Outil de géstion innovant.* Bruxelles, Belgique: Edition de Boeck Universite.

Nykiel, R. A. (2007). *Handbook of Marketing Research Methodologies for Hospitality and Tourism.* Routledge.

OECD. (1998). *Open Markets Matter. The Benefits of Trade and Investment Liberalisation*. OECD Pub.

Pattanayak, B. (2005). *Human Resources Management*. New Delhi, India: Prentice Hall.

Payne, A., & Frow, P. (2013). *Strategic Customer Management. Integrating Relationship Marketing and CRM*. UK: Cambridge University Press. doi:10.1017/CBO9781139057417

Phillips, J. J., & Edward, S. L. (2008). *Managing Talent Retention. An ROI Approach*. JohnWiley&Sons.

Rankin, N. (2009). *IRS best practice in HR handbook*. Routledge.

Ross, D. F. (2002). *Introduction to e-Supply Chain Management: Engaging Technology to Build Market-Winning Business Partnerships*. NY: CRC Press. doi:10.1201/9781420025415

Saemann, B. (2009). *Motivation in Second Language Acquisition*. Auflage.

Schachtebeck, T. (2005). *Internet chat communication. A tightrope walk between oral communication and written communication. Seminar paper*. Berlin: Grin.

Schultz, D. E., & Schultz, H. (2004). *The Next Generation. Five Steps for Delivering Value and Measuring Returns Using Marketing Communication* (pp. 49–66). McGraw Hill.

Sears, W. H. (2007). *The Front Line Guide to Communicating with Employees*. USA: HRD Press.

Simpson, L., Robinson, P., Fletcher, M., & Wilson, R. (2005). *E-Communication Skills*. Oxon, UK: Radcliffe Pub.Ltd.

Sims, R. R. (2002). *Organizational Success through Efective Human Resources Managemen*. USA: Greenwood Pub. Inc.Group.

Stegaroiu, I. & Florea, N.V., (2013). *Tehnici şi instrumente de recrutare şi selecţie*. Bucuresti: C.H.Beck.

Sparrow, P., Brewster, C., & Harris, H. (2004). *Globalizing HRM*. London: Routledge.

Svetlik, I., & Ilic, B. (2006). *HRM's Contribution to Hard Work*. Bern: European Acd.Pub.

Tanasescu, D. (2008). *Relaţii publice şi de protocol în organizaţii*. Târgovişte: Editura Bibliotheca.

Tanasescu, D., Lupu, M.-M., & Petrescu, M. (2014). *Comunicarea integrata*. Targoviste: Editura Bibliotheca.

Tang, A., & Lim, S. (2010). *Retail operations. How to run your own store*. Jurong, Singapore: Prentice Hall.

Truss, C., Mankin, D., & Kelliher, C. (2012). *Strategic HRM*. Oxford: Oxford University Press.

Ulrich, D., & Brockbank, W. (2005). *The HR Value Proposition*. Harvard Business School Press.

Van Der Sluis, L., & Van De Bunt-Kokhuis S. (Eds.), (2009). *Competing for talent*. Assen.

van den Ende, R. (2013). *A guide to an internship in Japan for industrial design engineers*. Lulu Enterprises Inc.

Veseth, M. (2002). *The Rise of the Global Economy*. Taylor&Francis.

Wood, J. T. (2010). *Communication in Our Lives*. MA, USA: Cengage Learning.

Walish, P., Michael, E. L. R., & Benyoun, M. D. (2014). *MATLAB for neuroscientists: an introduction to scientific computing in MATLAB*. Academic Press. Retrieved from www.internetworldstats.com

Chapter 3
Engineer-To-Order Product Development:
A Communication Network Analysis for Supply-Chain's Sustainable Competitive Advantage

Richard Addo-Tenkorang
University of Vaasa, Finland

Petri Helo
University of Vaasa, Finland

Jussi Kantola
University of Vaasa, Finland

ABSTRACT

Industrial manufacturers' complex product-development activities have seen various advancement and improvement approaches over the past decades. In order to enable the implementation of efficient and effective product-development support processes in the quest of achieving shorter product development lead-times and higher return on investments (ROIs). Engineer-To-Order (ETO) product capacity projects, including large electric machine, huge centrifugal pumps, Diesel/Natural fuel power plant engines, steam turbine, boiler, ship, etc., have challenges concerning their long product-development lead-times. The challenges confronting these enterprises industrial Original Equipment Manufacturers (OEMs) are enormous with one of the major ones being the effective and efficient network or flow of technical communication among the main stakeholders for complex / new product-development. Moreover, with all the industrial manufacturing complex product-development process improvements, in terms of complex engineering design and delivery, there are still a lot more variances to be addressed on the 'better, faster and cheaper' paradigm. Furthermore, attention is needed on efficient information exchange systems as well as effective operational communication in their complex product-development processes for a sustainable competitive advantage. Therefore, this paper presents a proposed optimum conceptual information technology systems' architecture towards enhancing an industrial sustainable competitive advantage: By employing social network theory (SNT) analysis to advise on a strategic and effective communication network for industrial supply-chain (SC) sustainable competitive advantage.

DOI: 10.4018/978-1-5225-0021-6.ch003

INTRODUCTION

Engineer to order as popularly known as ETO; (ETO) is a manufacturing process whereby finished goods are built to unique customer product specifications. Engineer to order products may require a specific set of item numbers, bills of material (BOM), and routings; these are usually complex with long lead times. ETO is a product-development process, which starts with a product specification and finishes with an engineering design as its deliverable, it rarely sometimes includes a bit of practical manufacturing processes. It is usually limited to an engineering design process which involves the tasks of engineering analysis, concept design, architectural design, detailed design and manufacturing process design. These are all technical communication network processes of the complex / new product to be developed, which needs proper effective and efficient coronation for sustainable competitive advantage. Therefore, in order to sustain competitive advantage and operational survival; industrial organizations have been compelled to implement new strategies, based on collaboration with their SC partners and an advanced utilization of information technologies (IT) systems and Internet-based services (Geunes, Pardalos, & Romeijn, 2002). According to Musa, et al., (2014) end-to-end supply-chain product visibility (i.e., product tracking and tracing) has been exploited as a means of product support security, process control and optimisation in many industrial sectors. Including huge complex products such as jet engines - aviation, ship power engines - marine, automobiles, etc., (Maier, et al., (2008); Hsu, and Wallace, (2007); Addo-Tenkorang, and Eyob, (2012)). Chandra, and Grabis, (2007) Identified some key triggers for designing and implementing SC with regard to effectiveness, efficiency, flexibility and responsiveness. This key triggers include; introduction of new/complex product(s), upgrades for existing product(s); introduction of new or improvement in an existing product development support process (es). In addition are, allocation of new or re-allocation of the existing resource(s); selection of new supplier(s), de-selection of existing ones; changes in demand patterns for complex product(s) manufactured; changes in lead-times for product and/ or product support process life cycle; and changes in commitments among the SC network partners, etc.

The challenges confronted by industrial SC networks are severe. Some of the real-life pressures include growing customer demands, inflexible global competition, and the need to improve new/complex product development's time-to-market (Murman, Rebentisch, & Walton, (2000); Molina, Aca, & Wright, (2005); De Brentani, Kleinschmidt, & Salomo, (2010)). All these challenges can be effectively dealt with, if the information exchange flow as well as communication network among them is effective and efficient. Hence, integration approached programs, applications and processes involving all the SC stakeholders in complex / new product-development *(i.e. customers, suppliers, marketers, accountants, design engineers, production engineers, manufacturing engineers, etc.)* during the capacity management, enhances the ETO supply chain. Thus, this integration process turns to shortening the delivery lead-time for customer specification (Swafford et al., 2006). Moreover, customer relationship effects in ETO process integration in the supply-chain, increases understanding in customer demand, managing complaints and improving customer satisfaction which critical responsive element through the ETO value chain (Li et al., 2006; Holweg, 2005). According to Zhou and Benton (2007) quality information, sharing on a common and assessable platform among stakeholders enhances ETO supply chain practices such as just-in-time (JIT) and lean lead-time delivery. Thus, the capability to collect and disseminate various data or information stakeholders across ETO supply-chain network concurrently is imminent in order enable internal manufacturing strategies to respond effectively to the needs of customers.

This paper analysis and elaborates on the implication of technical communication network in engineer-to-order product-development as a sustainable supply-chain management perspective. An original scientific social network theory (SNT) analysis research to analyse industrial organization's SC technical communication network for a complex product-development in a case study approach. The rest of this paper is structured as follows: background review; methodology; results & findings of the analysis and finally, concludes and recommends.

BACKGROUND REVIEW

In recent years, industrial organizations' ability to implement the requisite information technology systems to enhance their operational activities faces numerous challenges, including system interface as well as change management problems with employees' (multi-disciplinary teams') collaboration issues, information flow and exchange as well as communication network issues, etc. However, in the past industries in almost all markets have been facing a boosting level of competitiveness. There are many reasons for this, but most of them can be followed to some of the following principal trends: shortening product life cycles, globalisation of the market, rapid technological changes, environmental issues, and higher complexity of products, customers demanding products with more features, higher quality, lower cost, and demand for more and more customized products. Concurrent Engineering (CE) is an integrated product development approach that emphasises the response to customer expectations by producing better, cost-effective and much faster product-development process (Addo-Tenkorang, 2011). CE also supports multidisciplinary team values of cooperation and trust. Thus, sharing and exchanging required knowledge and information in a manner that will enhance decision-making processes. Furthermore, the CE principle also emphasises on simultaneous consideration during the design stage and all the other Product Life Cycle (PLC) aspects of the product development (Addo-Tenkorang, & Eyob, 2012). Product Life Cycle is the progress of a product from its concept through to its development and introduction to the market and then its withdrawal from the market or final disposal. One of the most salient means to reduce development time is through the use of "concurrent engineering." According to the Institute for Defence Analysis (IDA), concurrent engineering is the systematic approach to the integrated concurrent design of products and related processes, including manufacture and support processes (Winner, Pennell, Bertend, & Slusarczuk, 1988).

The multiple facets that have to be managed in large-scale industrial SC network activities make the effective use of information technology systems very necessary as enhancing enablers. This will effectively and efficiently enables industries requiring a complex mix of planning, evaluation and decision-making to communicate effectively and strategically to enhance their SC network industrial competitive advantage. An effective and efficient communication network is thus, seen as the vehicle by which this coordination could be achieved. Moreover, communication itself is influenced by many different factors that are connected (Maier, et al., 2008)

Figure 1 illustrates some key attributes of effective communication in industrial organizations, which contributes directly to the flexibility of industrial organizations for a sustainable competitive advantage in their SC network operations with the help of the requisite IT systems enablers. The following section below *(Methodology)* will utilize an industrial use case example in terms of the data collection and analysis

Figure 1. Effective Communication in Industrial Organizations

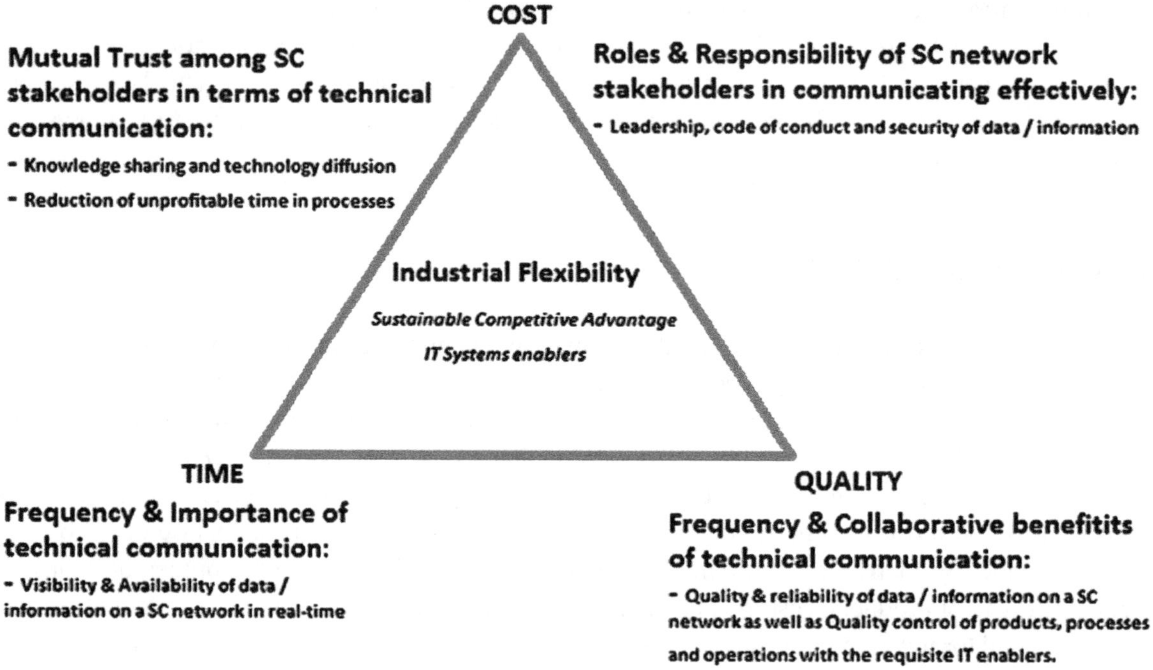

approach adopted to simulate the technical communication an industrial case study example using social network theory (SNT) analysis and tool (UCINet 6.0). The aim of this research is to (1) scientifically analyse complex / new-product development (C/NPD) engineering-design and delivery technical communication network by using Social Network Theory (SNT) analysis, (2) identify the bottle-necks in C/NPD engineering-design and delivery technical communication network. (3) Outline effective and efficient strategies for high-level managerial decision making that will enhance industrial flexibility in SCM for sustainable competitive advantage. According to Wang (2014) competitive advantage is achieved when an industrial organisation develops or acquires a set of attributes or executes actions (such as effective and efficient communication network) that allows it to outperform its competitors.

The quest by industrial organisations for competitive advantage in recent times is arguably the central theme of the academic field of strategic management (Furrer et al., 2008; Hoskisson et al. 1999; Porter 1996). Hence, this quest has led to the dynamic capability of industrial organizations renewing their advantages over time by innovation in various aspects of their operations including innovation in their communication networks (Peteraf 1993; Powell 2001). Research have identified that, there were two main theories of competitive advantage, thus: the *Market-Based View (MBV)* and the *Resource-Based View (RBV)*. Furrer et al. (2008), states that, strategic management was initially a body of knowledge that would underpin practical advice to managers, but developed into the endeavour to identify a theory with explanatory and predictive power (i.e., effective and efficient communication network). Grant (1991) argued that (i.e. Capability-based theory) capabilities are the source of competitive advantage while resources are the source of capabilities. Haas and Hansen (2005), as well as Long and Vickers-Koch

(1995), supported the importance of capabilities and suggest that a firm can gain competitive advantage from its ability to apply its capabilities to perform important activities within the firm. Even more compelling for this research, is Lee et al. (2001)'s discussion that, the influence of internal capabilities and external networks on firm performance. Dyer and Singh (1998) in their research have discussed a relational-based view of competitive advantage that focusing on the network routines and processes as an important unit of analysis for understanding competitive advantage.

The relational-based view of strategy has become increasingly popular (Ahuja 2000; Dyer & Singh 1998; Gulati 1998; Gulati et al. 2000; Seidmann & Sundararajan 1997; Wang 2004). A number of researches have discussed inter-firm collaboration (Easton & Araujo 1997; Ebers 1999), business interactions (Wang 2004), relationships (Walter et al. 2001) and networks (Ahuja 2000; Gulati et al. 2000). An inter-organisational network involves relationships between two or more firms both in the micro-level and macro-level contexts (Ebers 1997). The micro-level context involves resources flows, information flows and flows of mutual expectations between firms. According to Wang (2004), the relational view of strategy is also inter or intra-organisational, and the unit of analysis is, if anything, even more coarse-grained for the purposes of interaction-level analysis. Therefore, this research is underpinned more by the micro-level aspect as it is centred on the network of the SC stakeholders and also with a keen view on the strategic management of the industrial communication network aspect. The following section elaborates more on the methodology employed in the analysis of this research.

METHODOLOGY

Closed Questionnaires

Questionnaires are said to be of two types: open-ended (i.e., "open questions" – where the researcher does not provide the respondent with a set of answers from which to choose) and closed-ended (i.e., "closed questions" – where the researcher provides a suitable list of responses (e.g., Yes, No, or Maybe) (Robson, 2011; Mellenbergh, & Adèr, 2008). Questionnaires are the most commonly used tool in survey research. However, the findings and results of a particular survey are worthless if the questionnaire is written inadequately without taking the respondents into consideration. Questionnaires should produce valid and reliable demographic variable measures and should produce valid and reliable individual differences that self-report scales generate (Shaughnessy, Zechmeister, & Jeanne, 2011; Robson, 2011).

Questionnaires have advantages over some other types of surveys in that they are cheap, do not require such a strenuous effort as verbal or telephone surveys, and often have standardized answers that make it simple to compile data. However, questionnaires are also somewhat limited, in some sense, by the fact that respondents must be able to read the questions and respond to them effectively. A real-life industrial-based case study was employed to collect empirical data for an inbound and outbound industrial manufacturing SC project in the form of a closed-end questionnaire. These were disseminated by using online electronic forms (e-forms) designed in five parts. Thus; Part 1: Company's Design Team Rep. Background Information; Part 2: Frequent Concurrent Design Teams' Communication Interface. Part 3: Importance of Concurrent Communication within Enterprise Industrial Manufacturers' Design

Teams'; Part 4: Effective and Efficient Design Teams Relational Concurrent Communication; and finally, Part 5: Teams' Concurrent Communication Interfaces Review & Redesign. The questionnaires were then sent to the various partners (design teams) on the SC globally (Finland, Norway, Germany, Italy, India, Turkey and The Netherlands). There were eight (8) design teams with at least five (5) members in each team involved this research's technical communication analysis, hence making a total sample size of forty(40) respondents. The research recorded a 100% response rate with the help of the executive management team of our industrial partners. Finally, the questionnaires were then statistically tested for reliability using the Cronbach Alpha Analysis test, and the results were (0.8 and above) indicating the standard acceptable range. Table 1 outlines the statistical Cronbach Alpha Analysis test results of the questionnaires used. According to Tavakol and Dennick, (2011), there are different reports about the acceptable values of alpha, ranging from 0.70 to 0.95.

UCINet 6: Social Network Theory (SNT) Analysis

UCINET 6 is a Windows software package for the analysis of social network data. It was developed by (Borgatti, Everett, & Freeman, 2002) and has evolved over time, in terms of upgrades and revised versions. It comes with the NetDraw network visualization tool. Social network could be defined as a set of people, groups or teams of people with some form of relationship or interactions between them (Scott, 2000). Although only 18% of managers believe that effective social network is important to their businesses today, more than 63% assert that, Social networks will be important for businesses by 2015 - *within the next three years* (Kiron, Palmer, Phillips, & Kruschwitz, 2012).

Chui, et al., (2012) State that, there have been estimates that the economic impact of social network on business may exceed $1 trillion, most of which is gained from more efficient communication and collaboration within and across industrial organizations' SC network activities. Therefore, an effective and efficient SC communication network in industrial organization operations is very important and seems inseparable from a robust business environment (Puvanasvaran, Megat, Hong, & Razali, 2009) as well as organization operations' research and development (R&D) activities. Therefore, this paper adopted the social network theory (SNT) analysis approach to analyse and propose an effective strategic communication network flow within and across an industrial SC network, in terms of frequency, importance and level of collaboration among the SC network systems-design teams as well as the scale / level of mutual trust, and of their roles and responsibilities. According to Otte, and Roussaeu, (2002), Social network analysis (SNA) is not a formal theory in sociology but rather a strategy for investigating social structures such as in industrial organizations, etc., as it is an idea that can be applied in many fields. Social network theory and social network analysis are used interchangeably in this research paper.

Table 1. Questionnaire Cronbach Test Analysis Test Result

Mean	Variance	Std. Deviation
8.89	38.361	6.194

Social Network Theory (SNT) refers to the mechanisms and/or processes that interact with network structures to yield certain outcomes for individuals and groups (Borgatti & Halgin, 2011). In addition, according to Brass (2002), social network theory is about the consequences of network variables, such as having many ties or being centrally located. In contrast, the theory of networks refers to the processes that determine why networks have the structures they do and the antecedents of their structural properties. According to Kadushin (2004), social network theory (SNT) analysis was one of the few, perhaps the only theory in social science that is not reductionist. The theory applies to a variety of stages of analysis from individuals or small groups to entire global systems. Hence, in order to be sure, there are emergent properties at different system levels, but these are extensions of what can be done at a lower level and not entirely different forms of organization (Kadushin, 2004). It is important to note that, Social network theory and social network analysis are used interchangeably in this research article. It may also be referred in other parts of this paper as social network theory analysis, which also means the same as the former.

Therefore, getting a clear understanding of the technical communication network processes in complex engineering design and delivery and new/complex product development. It has become imperative to study information exchanges and communication networks among geographically distributed new/complex product development systems-design teams because of the complex nature of industrial engineering- design and delivery SCM network in organizations (Addo-Tenkorang, 2014). Furthermore, employing electronically-based communication information systems technologies have changed how new/complex product development and introduction design teams communicate on an enterprise SC network.

UCINET 6: ANALYSIS AND RESULTS

Newman, (2003) refers to a network is a set of items, which we will call vertices or sometimes nodes, with connections between them, called edges. Social network is a set of people or groups of people with some pattern of contacts or interactions between them (Scott, 2000; Wasserman, & Faust, 1994). Hence, the UCINet 6 simulation tool is employed in this paper to assess an industrial case' complex product-development systems-design teams' frequency of technical communication. In order to enhance a robust industrial sustainable competitive advantage by identifying the damning bottle-necks that exists within industrial SC networks and recommends strategic practical and managerial action plans, including the use of the requisite IT systems enablers.

Effective and efficient study of any aspect of organization or industrial SC network operations could not be robust enough without critically looking at the flow and exchange of data / information (communication network) within the industrial SC network partners, teams, or stakeholders involved. The product-development (PD) systems-design teams studied in the industrial case example conducted and adopted in this paper include eight (8) systems-design teams as follows:

1. Automation Systems / PLC.
2. Auxiliary Systems.
3. Combustion Systems – (Internal / Air).
4. Electrical & Instrumentation Systems / PLC.

5. Mechanical Systems – Coupling & Mounting.
6. Mechatronic Systems.
7. Noise and Vibration Systems.
8. Power Transmission Systems.

In collecting the required data from the product development system-design teams listed above, a closed-end questionnaire was formulated and sent through to the persons involved. The questionnaire was disseminated by way of e-forms, which was assessed to be the most optimum method to use, looking at the persons involved and their very busy time schedules. In this research paper, only two analyses were conducted to ascertain the "Isolation analysis" and "Harmonic-closeness analysis" in terms of technical communication among the teams, partners or stakeholder on an industrial SC network. The Figures 2 and 3 illustrate the UCINet 6 SNT analysis graphical representation of the above systems-design teams technical communication on an industrial SC network for a complex product-development.

Figure 2 illustrates a UCINet 6 isolates analysis. This isolates analysis shows which of the nodes is isolated from the other nodes and why? Thus, from Figure 2, Mechatronic Systems node and Combustion Systems nodes stand out clearly isolated from the other nodes. Therefore, this implies that in a real-life enterprise manufacturing environment, the Mechatronic Systems-design teams and Combustion Systems design teams are most likely or mainly independent in the concurrent or collaborative system design

Figure 2. UCINet 6 SNT analysis

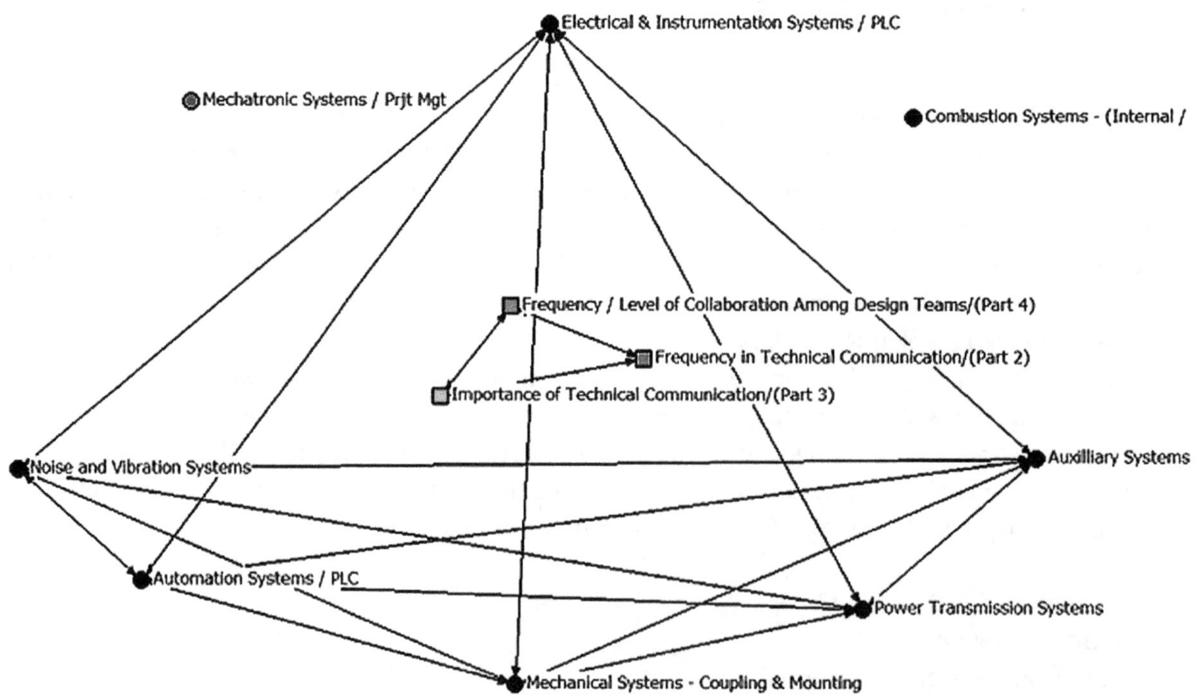

Figure 3. UCINet 6 harmonic closeness analysis

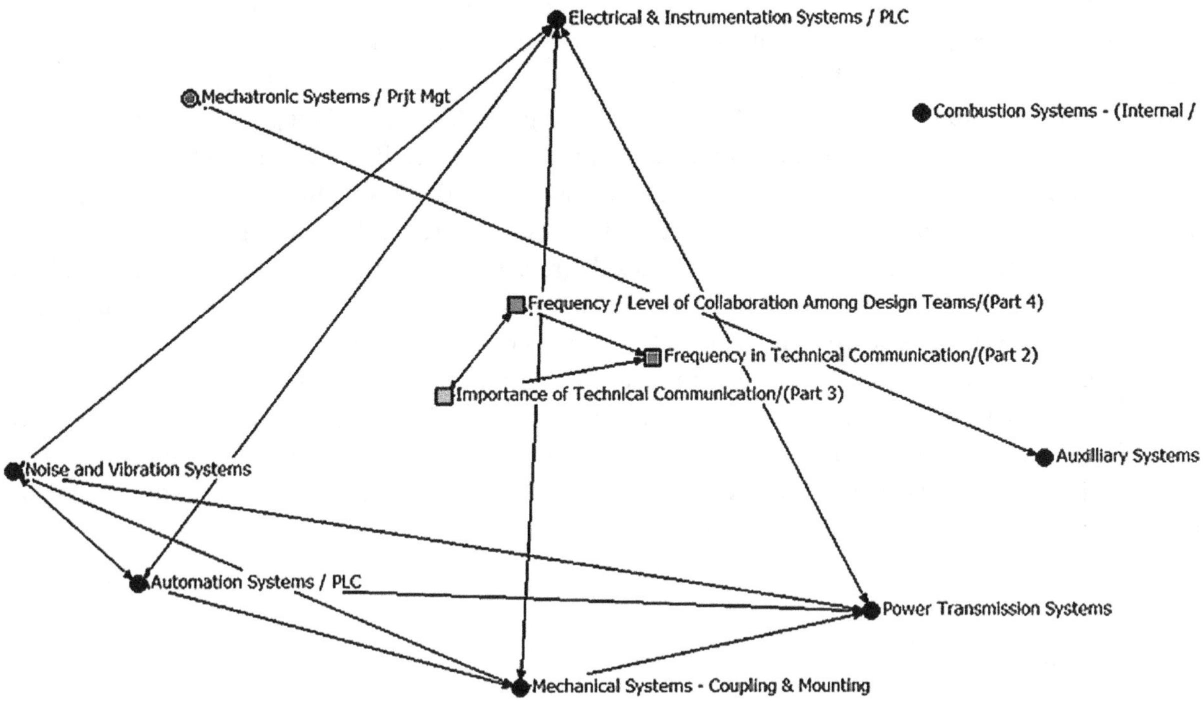

activities in complex engineering-design and delivery product-development processes, because their output products interface easily to other systems without any issues or with very trivial issues.

Figure 3 illustrates an UCINet 6 harmonic closeness analysis. This harmonic closeness analysis show which of the nodes is harmonically close to each other beyond all odds, and why they are so. Thus, from Figure 3, the Mechatronic Systems node and auxiliary Systems nodes are harmonically closer than the others, while the Noise and Vibration System, Automation Systems, Mechanical Systems – Coupling & Mounting System and Power transmission Systems is another cluster of linked / tied nodes, which is shown to be harmonically close as illustrated in Figure 3. This implies in a real-life industrial manufacturing environment that system-design teams are most likely or mainly harmonically close in the concurrent or collaborative system design activities in complex engineering-design and delivery PD processes. However, one very significant observation is still the isolative nature of the Combustion Systems node even in the harmonic closeness analysis. This indicates that the Combustion Systems node is the most independent node among all. This, analysis could be validated in a real-life industrial complex PD manufacturing environment. In a real-life industrial manufacturing environment, as in the case adopted in this paper; a ship's power engine could be designed and assembled or manufactured and then has the Combustion System (i.e. the exhaust) chosen and easily interfaced with the main product based on the calculation of the engine capacity. Alternatively, an exhaust could be designed separately for the engine, based on the calculated engine capacity.

The various systems-design teams have been colour coded and are also represented by different shapes to facilitate easy identification of the findings and results from the simulation analysis from the UCINet 6 simulation tool. The technical communication Frequency, Importance, Level of Collaboration, Level of Mutual Trust, and the Level of Roles and Responsibility within the industrial SC network are represented on a scale between (1) and (3): with (1) being "Low" and (3) being "High." Figures 2 – 3 illustrate the simulation results on the individual factors / cases in detail. According to Newman, (2003) the study of networks is by no means a complete science yet, and many of the possibilities have yet to be explored in depth. Hence, this paper's attempt within an industrial SC network perspective marks an important beginning of research in this broad and still green area in industrial SC network and information technology systems research for complex PD. The results in the graphical representation in Figures (2 and 3 indicate the that effective and efficient technical communication as well as data / information flow within an industrial SC network is critical in achieving real-time information and just-in-time (JIT) product development as well as enhanced industrial SC visibility (Musa, et al., 2014; Plex White Paper, 2013). However, the results from this analysis have also revealed certain weakness within the industrial SC network that could seriously undermine SC network activities for a sustainable competitive advantage in their operation if not checked and addressed accordingly.

Figure 4. Mechanism Network Model

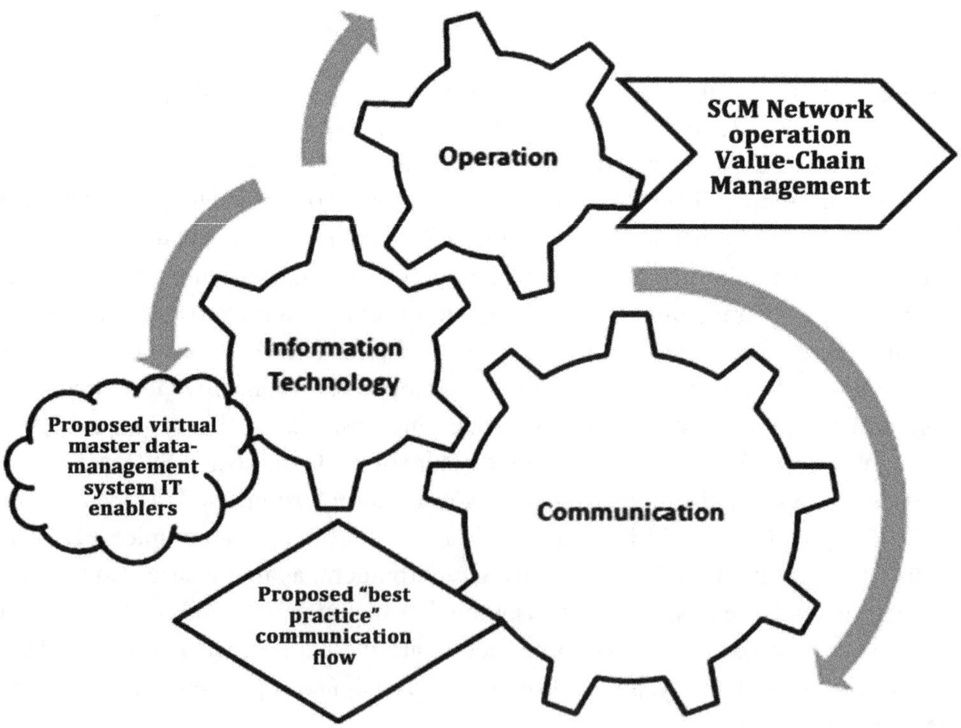

CONCLUSION AND RECOMMENDATION

Technical communication among design teams, partners or stakeholders on the adopted case's industrial SC network was investigated to assess whether their technical communication frequency, importance, level of collaboration, level of mutual trust, roles and responsibilities of the SC network technological complexity affects the SC network design-teams communication. In this respect, Roberts, Cheney, and Sweeney, (2002) came to the conclusion that in moderately complex product development projects, data / information sharing is greater than in highly complex projects, although there is a greater need in the latter for data / information communication and exchange. However, Chui, et al., (2012) also concludes that the type and structure of teams' organization impact the communication network. Tiernan, Flood, and Murphy, (2002) detected that changes to organizational structure also affect the collaborative design of product-development. While the importance of communication is generally acknowledged, there is a little consensus on how it can be directed or, at best, systematically improved.

Figure 4 illustrates a mechanism network model of a basic operational mechanism of the competitiveness of effective and efficient technical communication network among enterprise industrial stakeholders (system-design teams) (Addo-Tenkorang, 2014). This mechanism model illustrates that a sustainable technical communication network process would contribute positively to value-chain management within an industrial enterprise SCM networks in concurrence with the requisite IT enablers. In order for industrial SC networks to be competitive and sustain competitiveness, key manufacturing industrial SC network activities (i.e. information exchange, inbound & outbound deliveries, data storing and management, IT systems-integration, effective communication, etc.) are positively associated with industrial sustainable competitive advantage within the context of manufacturing industries globally. Thus, the industrial managerial focus of this research paper would be, to serve as a working R&D and/or tested in any industrial SC network. Also, this research's investigations revealed some interesting bottle-neck observations, which could practically be managed to enhance the effects competitively; these include the following in Table 2.

Table 2. Practical and/or managerial findings and observations

#	Bottle-Neck
1	Not all partners collaborate effectively from their end across the SC network, and the observation is quite significant. This observation is identified in the simulation illustration of the "Scale / Level of Mutual Trust"; "Scale / Level of Roles and Responsibilities." Therefore, ensuring partners / systems design teams on an enterprise SC networks are motivated and encouraged to communicate effectively will greatly enhance the level of collaboration on the industrial SC network.
2	This research paper's observation also identified that, some of the SC network partners / complex product-development systems-design teams work together more closely while others do not see the need or benefits of belonging to an enterprise SC network. Therefore, they operate in isolation within SC network, which implies that vital technical communication may not be coming from them to the other partners on the SC. Hence, they may also not be attentive to accept any such vital technical communication for the SC systems-design partners.

REFERENCES

Addo-Tenkorang, R. (2011). Concurrent Engineering (CE): A Review Literature Report. *Proceedings of the World Congress on Engineering and Computer Science Vol IIWCECS 2011*, San Francisco, USA.

Addo-Tenkorang, R. (2014). *Conceptual Framework for Large-Scale Complex Engineering-Design & Delivery Processes: A Case of Enterprise SCM Network Activities and Analysis* [Doctoral Dissertation].

Addo-Tenkorang, R., & Eyob, E. (2012). Engineer-to-Order: A Maturity Con-current Engineering Best Practice in Improving Supply Chains. In E. Eyob, & E.G. Tetteh (Eds.), Customer-Oriented Global Supply Chains: Concepts for Effective Management (Ch 7, pp. 112-122). doi:10.4018/978-1-4666-0246-5.ch007

Ahuja, G. (2000). The duality of collaboration: Inducements and opportunities in the formation of interfirm linkages. *Strategic Management Journal, 21*(3), 317–343. doi:10.1002/(SICI)1097-0266(200003)21:3<317::AID-SMJ90>3.0.CO;2-B

Borgatti, S. P., Everett, M. G., & Freeman, L. C. (2002). UCINET for Windows: Software for Social Network Analysis. Harvard: Analytic Technologies. Cambridge, MA.

Borgatti, S. P., & Halgin, D. S. (2011). *On Network Theory. OrganizationScience, INFORMS.*

Brass, D. J. (2002). *Social networks in organizations: Antecedents and consequences* (Unpublished). University of Kentucky, Lexington.

Brentani, U., Kleinschmidt, E. J., & Salomo, S. (2010). Success in Global-New Product Development: Impact of Strategy and the Behavioral Environment of the Firm. *Journal of Product Innovation Management, 27*(2), 143–160. doi:10.1111/j.1540-5885.2010.00707.x

Chandra, C., & Grabis, J. (2007). *Supply chain configuration*. New York: Springer.

Chui, M., Manyika, J., Bughin, J., Dobbs, R., Roxburgh, C., Sarrazin, H., & Westergren, M. et al. (2012). *The Social Economy: Unlocking Value and Productivity through Social Technologies*. McKinsey Global Institute.

Dyer, J. H., & Singh, H. (1998). The relational view: Cooperative strategy and sources of Inter-organisational competitive advantage. *Academy of Management Review, 23*(4), 660–679.

Easton, G., & Araujo, L. (1997). Interfirm responses to heterogeneity of demand over time. In M. Ebers (Ed.), *The Formation of Inter-organisational Networks* (pp. 66–94). Oxford: Oxford University Press.

Ebers, M. (1999). The dynamics of inter-organizational relationships. In *Research in the Sociology of Organisations* (Vol. 16, pp. 31–56). Stamford: JAI Press.

Furrer, O., Thomas, H., & Goussevskaia, A. (2008). The structure and evolution of the strategic management field: A content analysis of 26 years of strategic management research. *International Journal of Management Reviews, 10*(1), 1–23. doi:10.1111/j.1468-2370.2007.00217.x

Geunes, J., Pardalos, P., & Romeijn, H. E. (2002). *Supply chain management: models, applications, and research directions*. Dordrecht: Kluwer academic publishers.

Grant, R. (1991). The resource-based theory of competitive advantage: Implications for strategy formulation. *California Management Review*, *33*(3), 114–135. doi:10.2307/41166664

Gulati, R. (1998). Alliances and networks. *Strategic Management Journal*, *19*(4), 293–317. doi:10.1002/(SICI)1097-0266(199804)19:4<293::AID-SMJ982>3.0.CO;2-M

Gulati, R., Nohria, N., & Zaheer, A. (2000). Strategic networks. *Strategic Management Journal*, *21*(3), 203–215. doi:10.1002/(SICI)1097-0266(200003)21:3<203::AID-SMJ102>3.0.CO;2-K

Haas, M. R., & Hansen, M. T. (2005). When using knowledge can hurt performance: The value of organizational capabilities in a management consulting company. *Strategic Management Journal*, *26*(1), 1–24. doi:10.1002/smj.429

Holweg, M. (2005). The three dimensions of responsiveness. *International Journal of Operations & Production Management*, *25*(7), 603–622. doi:10.1108/01443570510605063

Hoskisson, R. E., Hitt, M. A., Wan, W. P., & Yiu, D. (1999). Theory and research in strategic management: Swings of a pendulum. *Journal of Management*, *25*(3), 417–456. doi:10.1177/014920639902500307

Hsu, C., & Wallace, W. A. (2007). An industrial network flow information integration model for supply chain management and intelligent transportation. *Enterprise Information Systems*, *1*(3), 327–351. doi:10.1080/17517570701504633

Kadushin, C. (2004). Introduction to Social Network Theory. In *Some Basic Network Concepts and Propositions* (Ch. 2). http://melander335.wdfiles.com/local--files/reading-history/kadushin.pdf

Kiron, D., Palmer, D., Phillips, A. N., & Kruschwitz, N. (2012). *Social Business: What Are Companies Really Doing?" MIT Sloan Management Review*. Social Business Global Executive Study and Research Project.

Lee, C., Lee, K., & Pennings, J. M. (2001). Internal capabilities, external networks, and performance: A study on technology-based ventures. *Strategic Management Journal*, *22*(6/7), 615–640. doi:10.1002/smj.181

Li, S. H., Rao, S. S., Nathan, R. T., & Nathan, B. R. (2006). The impact of supply chain management practices on competitive advantage and organizational performance. *Omega*, *34*(2), 107–124. doi:10.1016/j.omega.2004.08.002

Long, C., & Vickers-Koch, M. (1995). Using core capabilities to create competitive advantages. *Organizational Dynamics*, *24*(1), 7–22. doi:10.1016/0090-2616(95)90032-2

Maier, A. M., Kreimeyer, M., Hepperle, C., Eckert, C. M., Lindemann, U., & Clarkson, P. J. (2008). Exploration of Correlations between Factors Influencing Communication in Complex Product Development. *Concurrent Engineering: Research and Applications*, *16*(1), 37–59. doi:10.1177/1063293X07084638

Mellenbergh, G. J., & Adèr, H. J. (2008). Tests and Questionnaires: Construction and administration. In H. J. Adèr & G. J. Mellenbergh (Eds.), *Advising on Research Methods: A consultant's companion* (pp. 211–236). Huizen, The Netherlands: Johannes van Kessel Publishing.

Molina, A., Aca, J., & Wright, P. (2005). Global collaborative engineering environment for integrated product development. *International Journal of Computer Integrated Manufacturing*, *18*(8), 635–651. doi:10.1080/09511920500324472

Murman, E., Rebentisch, E., & Walton, M. (2000). Challenges in the better, faster, cheaper era of aeronautical design, engineering and manufacturing. Lean Aerospace Initiative. USA: MIT. Retrieved from web.mit.edu/lean

Musa, A., Gunasekaran, A., & Yusuf, Y. (2014). Supply Chain Product Visibility: Methods, Systems and Impacts. *Expert Systems with Applications – An International Journal,* 41(1), 176–194.

Newman, M. E. J. (2003). The structure and function of complex networks. *SIAM Review*, *45*(2), 167–256. Retrieved from[REMOVED HYPERLINK FIELD] http://arxiv.org/abs/cond-mat/0303516v1 doi:10.1137/S003614450342480

Otte, E., & Roussaeu, R. (2002). Social network analysis: A powerful strategy, also for the information sciences. *Journal of Information Science*, *28*(6), 441–453. doi:10.1177/016555150202800601

Peteraf, M. A. (1993). The cornerstones of competitive advantage: A resource-based view. *Strategic Management Journal*, *14*(3), 179–192. doi:10.1002/smj.4250140303

Improved Data Visibility Drives Manufacturing Performance. (2013). *Plex White Paper*. Retrieved from http://www.plex.com/wordpress/wp-content/uploads/2013/06/Manufacturing_Visibility_WP.pdf

Porter, M. E. (1996). What is Strategy? *Harvard Business Review*, *74*(6), 61–78. PMID:10158474

Powell, W. W. (2001). Competitive advantage: Logical and philosophical considerations. *Strategic Management Journal*, *22*(9), 875–888. doi:10.1002/smj.173

Puvanasvaran, P., Megat, H., Hong, T. S., & Razali, M. M. (2009). The roles of communication process for an effective lean manufacturing implementation. *Journal of Industrial Engineering and Management*, *2*(1), 128–152. doi:10.3926/jiem.2009.v2n1.p128-152

Roberts, T. L., Cheney, P. H., & Sweeney, P. D. (2002). Project Characteristics and Group Communication: An Investigation. *IEEE Transactions on Professional Communication*, *45*(2), 84–98. doi:10.1109/TPC.2002.1003690

Robson, C. (2011). *Real world research: a resource for users of social research methods in applied settings*. Chichester: Wiley.

Scott, J. (2000). *Social Network Analysis: A Handbook* (2nd ed.). London: Sage Publications.

Seidmann, A., & Sundararajan (1997). Building and sustaining inter-organizational information sharing relationships: the competitive impact of interfacing supply chain operations with marketing strategy. *Proceedings of the 18th International Conference on Information Systems (ICIS-97)*, Atlanta (pp. 205-222).

Shaughnessy, J., Zechmeister, E., & Jeanne, Z. (2011). *Research methods in psychology* (9th ed.). New York, NY: McGraw Hill.

Swafford, P., Gosh, S., & Murthy, N. (2006). A framework for assessing value chain agility. *International Journal of Operations & Production Management, 26*(2), 118–140. doi:10.1108/01443570610641639

Tavakol, M., & Dennick, R. (2011). Making sense of cronbach's alpha. *International Journal of Medical Education, 2*, 53–55. doi:10.5116/ijme.4dfb.8dfd

Tiernan, S. D., Flood, P. C., Murphy, E. P., & Carroll, S. J. (2002). Employee Reactions to Flattening Organizational Structures. *European Journal of Work and Organizational Psychology, 11*(1), 47–67. doi:10.1080/13594320143000843

Walter, A., Ritter, T., & Gemunden, H. G. (2001). Value creation in buyer-seller relationships: Theoretical considerations and empirical results from a supplier's perspective. *Industrial Marketing Management, 30*(4), 365–377. doi:10.1016/S0019-8501(01)00156-0

Wang, H. L. (2004). A framework to support and understand strategic decision-making in business-to-business electronic commerce. *Proceedings of the International Workshop on Business and Information (BAI2004)*, Taipei.

Wang, H.-L. (2014). Theories for competitive advantage. In H. Hasan (Ed.), Being Practical with Theory: A Window into Business Research (pp. 33–43). Wollongong, Australia: THEORI. Retrieved from http://eurekaconnection.files.wordpress.com/2014/02/p-33-43-theoriesof-competitive-advantage-theori-ebook_finaljan2014-v3.pdf

Wasserman, S., & Faust, K. (1994). *Social Network Analysis*. New York: Cambridge University Press. doi:10.1017/CBO9780511815478

Winner, R. J., Pennell, J. P., Bertend, H. E., & Slusarczuk, M. M. G. (1988). The Role of Concurrent Engineering in Weapon System Acquisition. *IDA Report R-338*, Institute for Defense Systems Analysis, Alexandra, VA, USA.

Zhou, H., & Benton, W. C. Jr. (2007). Supply chain practice and information sharing. *Journal of Operations Management, 25*(6), 1348–1365. doi:10.1016/j.jom.2007.01.009

KEY TERMS AND DEFINITIONS

Complex / New Product-Development (C/NPD): This is the process or concept of designing, developing and introducing a new or complex product to the market "Just-in-time" to fulfill the quest for return-on-investment. Hence, this process consists of the design of the product right from the conceptual drawing phase of the product through to the detail design phase and then the finished final product ready for the market.

Concurrent Engineering: Concurrent engineering in this sense is moderate collaborative engineering approach that is much more focused on the engineering product design process of concurrent cross-functional partner organizations together in order to create products that are much better, much affordable and with a much reduced lead-time.

Engineer-to-Order: ETO is a product development process, which starts with a product specification and finishes with an engineering design as its deliverable.

Enterprise Supply-Chain Management (eSCM): Enterprise Supply-Chain Management in this paper refers to an information technology enabled/embedded or an integrated supply-chain management organizational activity.

ETO – CE SCM Integration Process: This is where some ERP / IT systems are integrated in the ETO – CE supply chains to enhance effective and efficient improvement in the entire ETO supply chain management.

Industrial Flexibility: Industrial flexibility in this paper implies the agility of an enterprise industry to be capable of providing a common highly integrated platform for multiple industrial enterprise applications. This approach is or model of enterprise industries are very imminent in today's global era to enhance substantial reduction in new/complex product-development time as well as promote operational "best practice" processes.

Social Network Theory (SNT) Analysis: Refers to the mechanisms and/or processes that interact with network structures to yield certain outcomes for individuals and groups *(industries / organization / teams, etc.).*

Supply Chain: Supply Chain is a system of technology, activities, information, people as well as organizations and resources involved in moving a product or service from supplier to customer at the right place and at the right time.

Supply Chain Management: Supply Chain Management is the management of the entire interconnected chain of an organization including the activities, technology, information, the people as well as the all the other resources involved.

Sustainable Competitive Advantage: This is an industrial or organizational capability to have some kind of operational or performance advantage which is also sustainable over its competitors. This advantage allows the organization or industry to be more successful and able to sustain its success over its competitors for a long period of time.

APPENDIX: LIST OF ABBREVIATIONS

CE: Concurrent Engineering.
C/NDP: Complex / New Product Development.
ERP: Enterprise Resource Planning.
ESA: Enterprise Service Architecture.
eSCM: Enterprise Supply-Chain Management.
eSC: Enterprise Supply-Chain.
ETO: Engineer To Order.
IDA: Institute for Defense Analysis.
IPD: Integrated Product Development.
IT: Information Technology.
JIT: Just In Time.
LM: Lean Manufacture.
NPD/I: New Product Development/ Introduction.
PD: Product-Development.
PLC: Product Life Cycle.
R&D: Research & Development.
SC: Supply Chain.
SCM: Supply Chain Management.

Section 2

Models and Modelling in Supply-Chain Management and Engineering-to-Order Processes

Chapter 4
Improving Construction Supply Chains through Collaborative Modelling, a case of South Africa:
Construction Supply Chain and Collaboration Modeling

Clinton Aigbavboa
University of Johannesburg, South Africa

Neo Malose Masemeni
University of Johannesburg, South Africa

Wellington Thwala
University of Johannesburg, South Africa

ABSTRACT

This chapter report on some of the challenges faced and successes achieved during the construction of the recently completed Department of Environmental Affairs (DEA) headquarters in South Africa. Qualitative interviews where conducts with five (5) professionals which were part of the construction supply chain team that completed the DEA project in South Africa. Findings summary were that the South African construction industry exists in a very dynamic and volatile environment with common challenges as evident in other countries. This challenge, necessitates the use of collaborative means of managing the various phases of the supply chain. It was also found that the use of collaborative models such as Public Private Partnership (PPP) can prove rewarding when stakeholders involved in a project have an understanding of the model and its dynamics. This chapter contributes to the use of PPP collaborative models in construction project delivery in South Africa.

DOI: 10.4018/978-1-5225-0021-6.ch004

INTRODUCTION

Supply Chain Management (SCM) is a research area that has attracted the attention of many researchers for more than 20 years. According to Kopczak and Johnson (2003) and Kulp et al. (2004), developments of supply chain management (SCM) in the 1990s have been broadly studied in a range of research areas such as: information technology (IT), product variety, supply-chain costs, real-time information sharing, coordination, coordinated production planning, replenishment scheduling, and decision-making among trading partners. Likewise, Subramani, (2004) and Wang et al. (2006) further informed that SCM has extended from the focal firm's benefits to the entire supply chain members' benefits, with assistance and support from major manufacturers or retailers to reduce operational costs, improve customer service, and result in a comparative advantage for supply chain members.

However, SCM in the construction industry, is an inter-linked set of relationships connecting the construction industry client, suppliers, perhaps through a number of intermediate stages such as manufacturing, warehousing and distribution. The reason or considering the process as a single chain is so that the flow of materials, money and information can be effectively managed to meet the construction requirements. Also, Narasimha, Kamath and Roy (2007) informed that SCM is primarily concerned with cost effective way of managing materials, information and financial flows from the point of origin to the point of consumption to satisfy the clients' requirements. Likewise, La Londe (1998) posited that SCM is the delivery of enhanced customer and economic value through synchronized management of the flow of physical goods and associated information from sourcing through consumption. Whilst, Johnston (1995) further conceptualised SCM as the process of strategically managing the movement and storage of materials, parts and finshed inventory from suppliers, through the firm and to customers. The various definitions that have been proposed by researchers signify that SCM stipulates organisational restructuring, extended to the achievement of a company-wide collaborative culture. The key objective of supply chain management is to offer better underlying value to a client than the competition. This is done through a combination of: defining client value; establishing supplier relationships; integrating activities; managing costs collaboratively; developing continuous improvement; mobilising and developing people

Dubois and Gadde (2000) and Bankvall et al. (2010) posted that there are at least three types of supply chains associated with construction projects, which includes: temporary supply chains, framework specific supply chains and company strategic supply chains (permanent supply chains). However, Vrijhoef and Koskela (2000) contended that construction specific SCM (cSCM) is about the supervision of inter and intra-firm interactions in a construction project and hence, suggested a cSCM framework that is concerned with the interface between the supply chain and construction site; reducing costs related to logistics, lead-time and inventory on specific project supply chains; transferring activities from the site to earlier stages of the supply chain; and, integrated management of the supply chain with emphasis on improvement of supply chain and the site production (Pala, et al, 2014). However, Pala et al (2014) informs that an idea way to oversee construction supply chain is through a relationship-centric perspective where the aim is to maintain an effective operational and strategic engagement with suppliers. In this study, the improvement of construction supply chains through collaborative modelling is considered. This chapter report on the challenges and successes which were achieved in the construction of the recently completed Department of Environmental Affairs (DEA) headquarters in South Africa.

Collaboration is one of the most talked about topics in the model construction industry. In this regard, Anthony (2000) indicated that collaboration is defined as two or more companies sharing the responsibility of exchanging common planning, management, execution, and performance measurement information. Collaboration is about entities working together and emerges when firms come to the reliasation that working alone is not sufficient to resolve common problem and to being about sustainable growth for their firms in every aspect. When supply chain stakeholders integrate and act as a holistic entity, there is an improvement in the flow of goods and services, better financial returns, good leverage for information sharing, and improved performance (Simatupang & Sridharan, 2004). The key dimensions of collaborative relationships are information sharing, incentive alignment and decision harmonization. The performance of the construction industry lies on information sharing. Information sharing involves obtaining and disseminating timely and appropriate information to supply chain stakeholders in order to enable informed decision making. While decision synchronisation is focused on shared decision in supply chain activities. The remainder of this chapter is arranged in the following manner. Section two discusses collaborative in the construction with emphasis on partnering and alliancing. Section three is a discussion of the methodology adopted for the study. Section four discuss the case study and the findings, while conclusion and recommendations are given in Section five.

COLLABORATION IN THE CONSTRUCTION INDUSTRY

In recent years, collaborative models have been consistently used to deliver large and complex infrastructure project in the construction industry (Love et al., 2004; Manley and Chen, 2015). Morwood et al. (2008) informs that thee increased use of these models reflects their advantages in managing and delivery high levels of complexity in construction projects. Construction infrastructure projects are typically complex compared to other project-based activities, such as housing construction. The construction sector of any country is an important contributor to the gross domestic product (GDP) and is the subject of significant government and private sector investment. It is the complexity and uncertainty surrounding construction infrastructure project delivery that makes collaborative arranges amongst its stakeholders and the supply chain process more important than in other more straightforward types of projects.

Typical examples of collaborative models in the construction industry include: partnering, early contractor involvement (ECI) and project alliance (Love et al., 2011). Partnering in the construction sense can be seen as a project management technique that seeks to effectively create a nurtured environment where two or more organisations harmoniously co-exist. The emergence of partnering came after a need was seen to better manage particular processes in areas such as those of the construction industry. The construction industries that took the lead in the adaptation of this form of collaborative model were Australian, Japan and USA (Rowlinson & Cheung, 2002; Naoum, 2003). In the USA for instance, the approach was first introduced to by the U.S. Army Corps in the mid- to late 1980s. Literature on the subject suggests that the approach seems most popular in China as evidence from literature (Bayliss et al., 2004; Cheng & Li, 2002). Partnering was introduced in Japan in the mid-90s in the form of a health care facility where the partnering arrangement was adapted through a design and build contract. Popularity in Australia grew in the 1980 while in South Africa very little literature has been recorded on the subject as it has not been holistically adopted by the construction industry; howbeit, fragmented used.

According to Chueng et al. (2003) the goal of partnering is to improve relations between stakeholders on both a short term basis (single project) and a long term basis (multi projects). It is regarded as an effective way for management to improving quality, programming and reducing clashes. Individuals are required to provide support to their fellow associates creating an environment of commitment and mutual support, necessitating the abandonment of traditional procurement habits (Briscoe and Dainty, 2005). Exercising this will leave very little room for beyond traditional client and contractor relations allowing for total organisational involvement. The success of the relationships within a partnering arrangement is determined by the manner in which the organisations treat each other. In order to enforce a successful partnering relationship, objectives and goals of the project must be mutually agreed upon, problem solving mechanisms must be determined, inter-firm trust established and maintained, and benchmarking exercises and continuous and measurable improvement initiatives incorporated into the team culture (Rowlinson & Cheung, 2002; Naoum, 2003). Partnering is about looking out for your own interests and equally for those of other stakeholders making up the team (Bygballe et al., 2010). It is very difficult to work and think collaboratively when individuals in the team look to gain at the expense of others. The notion is that when everyone works together to help share pain-gain the team is placed at a competitive advantage to achieve the set project targets. Rowlinson and Cheung (2002) define partnering as "a structured management approach to facilitate team working across contractual boundaries". The creation of a "virtual" organisation comes about by the coming together of a multitude of individual organisations forming an inter-organisation with unlimited collaborative lines. The complexities that come with the concept of partnering are often misunderstood and as a result what partnering is and what it is intended to achieve can be somewhat confused. Partnering may be seen from a philosophical outlook, with a particular set of beliefs (Naoum, 2003). From a SCM point-of-view partnering is taken as relationship model (Chueng et al., 2003). Although it is conventionally looked at from a functional point-of-view as informed by Chan et al. (2004), who explained partnering to be the formation of good relationships between parties to a construction contract, with the goal of minimizing job costs and schedule overruns. The most often cited definition of partnering is that of the Construction Industry Institute (CII). The CII's (1991:5) defined partnering as: "a long-term commitment by two or more organisations for the purpose of achieving specific business objectives by maximising the effectiveness of each participant's resources. This requires changing traditional relationships to a shared culture without regard to organisation boundaries. The relationship is based upon trust, dedication to common goals, and an understanding of each other's individual expectations and values. Expected benefits include improved efficiency and cost-effectiveness, increased opportunity for innovation, and the continuous improvement of quality products and services."

Partnering and alliancing are regarded as two commonly used collaborative models of doing business collaboratively in the construction industry. Confusion between the two is common as a result the terms are often misused in literature and practice. The main similarity between partnering and alliancing is that they both seek to do business in a collaborative environment implemented through integrated supply chains. The major difference however is the manner in which the contracts are executed. The latter is inherently complex with the contract being executed formally, therefore arrangements are expressed in the form of binding contractual obligations (Xue et al., 2010; Davis & Love, 2011). Unlike partnering where the principal contract between stakeholders gives the client the advantage of reverting back to the contract should the partnering agreement fail; alliancing is seen as a formal approach to collaboration in business

(Anvuur & Kumaraswamy, 2007). Partnering is therefore not a contract as such but rather a commitment and open communication endeavour / collaboration to establish and sustain non-adversarial relations (Cheung et al., 2003). Walker and Hampson (2003) state that the two concepts can be distinguished by certain components that make-up the relationships which exist in each, under the following categories:

- The amount of trust and commitment.
- The extent to which the relationship is looked after and not forced upon the stakeholders by the conditions of contract.
- The manner in which the relationship is developed, taken care of and maintained as a good part of the overall contract.
- The extent to which openness/ transparency is maintained throughout the duration of the project.
- The manner in which "gain and pain" shared.

In a report by the Auditor General of the Australian National Audit Office (AGANO) (2000), alliancing is suggested as a form of contracting process that delivers, within a guarded time frame, a cost-effective way of sharing project risks and rewards. The report further informs that the process uses an integrated team employing the services of key stakeholders (i.e. the architect, the project manager, building and services contractors, and other parties) most likely to have an effect the outcome of the project. Much like the application of SCM principles the first recorded business alliances were in the manufacturing and automotive industries (Walker & Hampson, 2003). By forming alliances, what organisations are essentially doing is creating a project team extending beyond traditional organisational boundaries – sharing all losses and profits, risk and rewards – in a "supreme" collaborative manner. In so doing, individually the very same organisations establish a competitive advantage of furthering their business with clients and customers. Alliancing is especially beneficial in projects that appear to be extreme in terms of complexity, in situations where a mix of traditional construction and non-construction related professions are required to achieve common ground. Overall, alliancing provides a platform for high levels of performance across a wide range of expertise in various construction and engineering related fields (Walker & Hampson, 2003).

RESEARCH METHODOLOGY

The study adopts the use of a single case study because at the centre of the case study as a research method lie context-dependent knowledge and experience, which are central to expert activities (Flyvbjerg, 2011). The objective of the study and any exploratory studies is not to test or to confirm a specific hypothesis. Rather it is done to familiarise the researcher with the subject matter so as to make it more explicit. According to most qualitative approach authors, the case study must be used when the objective is to know more about a specific topic in which control of the events are limited. Hence, data for the study were collected through on-site face-to-face interviews in the month of May in 2015. The data were tape recorded before transcription. The data were analysed in a thematic manner based on the interview questions. The interviews were designed to collect data on the challenges and successes which were achieved in the construction of the recently completed Department of Environmental Affairs (DEA) headquarters in South Africa. Five construction professionals which were part of the construction supply chain team that completed the DEA headquarters in South Africa were engaged for the study. The next sections provide an overview of the background of the case study and the interviewee as well as the practice based responses that were recorded in the fieldwork.

FINDINGS - THE CASE STUDY: DEPARTMENT OF ENVIRONMENTAL AFFAIRS (DEA) HEADQUARTERS

The new Department of Environmental Affairs (DEA) head office in Pretoria, South Africa, is the first government building in South Africa to achieve a 6 Star Green Star SA rating, as well as being the first 6 Star rated green building in the City of Tshwane, Pretoria, South Africa. A further stellar accomplishment was that it achieved the highest score for a large commercial office space of this magnitude awarded by the GBCSA to date (Green Building Council South Africa, 2015). The DEA has taken the lead, and is demonstrating commitment to market transformation in the built environment in South Africa. This is an exceptional illustration of the public and the private sector working together to deliver an outstanding example of green building.

The DEA Headquarters was completed in 2014 and designed and constructed by the Aveng Grinaker-LTA and Keren Kula joint venture. The building, which is the result of a public-private partnership between the Imvelo Concession and the DEA was designed to allow a significant amount of natural light, external views and open space whilst concurrently providing repetitive forms for lower wastage and easier maintenance. Adding to the "greenness" of the building is the country's largest rooftop photovoltaic installation as well as solar thermal systems for hot water. This shortlisting is not the first accolade for this building, which was also awarded the highest score for a large commercial office space of this magnitude by the Green Building Council of South Africa. The new building will cover 34 143 square meters of office space in Gross Lettable Area with an overall size of 3.1309 hectares, and accommodate 1 305 employees. The parking requirement for the project is 598 parking bays: 4 for VIPs, 438 for staff and 156 for visitors.

SUCCESS ACHIEVED IN THE CONSTRUCTION OF THE DEPARTMENT OF ENVIRONMENTAL AFFAIRS (DEA) HEADQUARTERS IN SOUTH AFRICA

Selection of the DEA Headquarters Consortium

Collaborative tools relating to Africa have rarely been studied in academia. Although it is still not clear as to what extent these type of models have been used in construction supply chain in African. But, very little is known when compared to other developing economics. Africa is a continent with an abundance of developing economies with great potentials for PPP for infrastructure development. Economies such as South Africa, Ghana and Nigeria are amongst the most thriving and attractive developing economies.

Economies such as these provide a desirable breeding-ground for partnering projects between public and private sectors. Such projects are popularly referred to as Public-Private Partnerships (PPP) and are most popular in developing countries; with a stable and sizable macro-economy (Sharma, 2011).

South Africa's isolation from international scene saw the country trailing behind on the use of innovative construction collaborative models (PPPs in particular) which were being practised around the world. It was only after the apartheid period that PPPs were developed to assist government and in it growing municipalities in particular in the delivery of infrastructure services (Hlahla, 1999). An interdepartmental team was appoint in April, 1997 to facilitate and develop institutional reforms, policy and legislation on the use of PPPs in South Africa, marking the beginning of what has been (to date) a fruitful series of projects (PPP Unit, 2014). This supports Sharma's (2011) observation that PPPs have helped developing countries improve the quantity and quality of public infrastructure. The first PPP projects in South Africa began in 1997, which was used for the improvements to the N3 and N4 toll roads as well as the construction of two maximum security prisons. Since then PPP models has been use to roll out infrastructure projects in railways, airport, health, energy, education, accommodation, water and sanitation. Landmark projects where PPP arrangement was used include:

- Gautrain: 80 Km mass rapid railway system linking Pretoria and Johannesburg, largest PPP in South Africa to date.
- Department of Environmental Affairs (DEA) Headquarters: First building in to achieve a six (6) green star rating in South Africa, marking the beginning of government's commitment to green building and sustainable development.

Selection of an appropriate consortium for the DEA construction took a period of 12 months and a further period of 18 months of negotiation before construction commenced on May 2012. Bid documents for construction of the DEA Headquarters were first submitted in 2009 with submissions coming from four of the following consortiums: Murray & Roberts Construction Company, Group Five, Basil Read and Imvelo Concession. Selection process for the project had multiple stages (initiation, submission, selection and negotiation). This process of selection supports Kumaraswamy et al., (2000) findings that multiple stage selection is important for selection in these types of models. In 2010 Imvelo Concession was announced the preferred bidder succeeding which negotiations of terms of contract took place. According to Imvelo Concession the bid document took three months to prepare and submission done once. The Imvelo concession consisted of the following companies:

- Aveng Grinaker-LTA; part of The Aveng Group of companies, one of the leaders' of infrastructure developments on the African continent. Although this entity's reputation is built on its ability to build distinct infrastructures, for this particular project they had to take responsibility for the overall design as well (landscape, structural, architectural, and electrical, amongst others). Aveng Concessions are specialist PPP contractors and were responsible for putting together the winning bid. Aveng Grinaker-LTA was for the sake of the PPP structure identified as the D&C contractor.
- Kagiso Tiso Group; an all-woman, black-owned finance-house. Because of the dimensions of the PPP agreement this group played an instrumental role in achieving the set targets for community development.

- Old Mutual is one of the biggest investment companies in South Africa with portfolios in, inter alia, long-term and short-term insurance, property and facilities management. However for this PPP agreement Old Mutual provided the function of facilities management.
- Wiphold: Women's Investment Portfolio Holdings Limited (WIPHOLD) is an investment and operating company owned and managed by black women, who played the same role as Kagios Tiso Group on this project.
- ABSA is a major finance-house in the country; ABSA was used as the most suitable salient financer contributing 80% towards of the project.
- National Treasury was indirectly part of the Imvelo Concession, in that they actually fell under the clients' side. Their involvement however is that they invested R300 million into Imvelo concession to assist with the commencement of the project. However, the contact was worth about $70-million.

The Joint Venture agreement between the companies that made-up Imvelo was required to meet strict contractual requirements that include completing the project within a fixed price and fixed duration and to deliver a building of the highest quality. This were all meant based on the bi-lateral agreement and team spirit displayed by all parties to the collaborative model.

DEA Public Private Partnership (PPP) Agreement

PPP is a type of partnering where the parties to the agreement come from two sectors of the economy (namely private and public). In the case of the DEA headquarters a partnering agreement was made between the Department of Environmental Affairs (a state public entity) and Imvelo concession (a privately own and run entity). This arrangement, agrees with Walker and Hampson (2003) definition of the PPP agreement; which described PPP as a collaborative model with features emanating from Alliancing and Partnering arrangement in order to create a viable project delivering entity. When construction projects are undertaken this way risks are mostly carried by the concession entity. However, like in all collaborative models, when executed properly, PPPs generate prolific rewards for all parties concerned. A significant benefit of this PPP structure is that contractual obligations for all participants ensured that all cost, timeline, and green specifications were clearly outlined, and successfully met during design and construction. It will also ensure that the building is operated optimally over the next 25 years to ensure that it stays green throughout its lifespan, and achieves the environmental and resource savings envisaged at the outset.

In 2010 Imvelo concession went into an agreement to; finance, design and construct (D&C) the aforementioned project for the DEA. In the same contract it was agreed that upon completion, Imvelo would take on the rights to operate the commodity (the buildings to be constructed) on behalf of DEA. The duration of the operational period was 25 years from occupation of the premises. This type of PPP agreement is referred to as a design-build-operate (DBO) (Masterman, 2002).When Imvelo concession was awarded the contract it was agreed by all parties concerned that DEA would take on occupancy of the premises on 01 August 2014. This date was cast-in-stone, failure of which would result in penalties for the Imvelo Concession. Although this type of (penalty oriented) contracting is against the spirit of win-win scenario, according to Imvelo Concession (2014) working towards a set milestone is essential in any construction project. However, this milestone was achieved in 2014. Once construction was completed Imvelo Facilities Management, which is owned by the Imvelo concession group, took over the role of running the premises from day-to-day and this will culminate in 31 July 2039 (a total period of 25 years). For this service Imvelo FM receives an agreed premium.

The contracting parties were bound together by both informal and formal ties. Informal in that communication channels played a key role in getting the job done and took preference over formal lines of communications. The latter on the other hand translates that parties were being regulated by a contract drawn up by legal personnel specializing in this form of construction model. The bi-lateral agreement comprised of legal representation, with each party employing services of different legal firms. The Imvelo concession used the services of Levison legal team while the DEA employed their own government-sourced legal team. This is not unusual with collaborative models, because of their unaccustomed nature and complexity, will always require the use of external service providers, particularly with regards to legal circumstances. The manner in which dispute resolution is undertaken is an import part of the collaborative agreement (Ling et al., 2006). For the Imvelo concession team, a PPP fast-track dispute resolution agreement was drawn up with all consulting firms contracted to the consortium (architectural, structural, mechanical, and electrical engineering firms). The function of this resolution agreement was that disputes do not exist for too long such that the project performance is obscured in any way.

DEA PPP Incentive Structure

In PPP forms of constructions the use of incentive structures can be abstruse in that the price and dates negotiated at tender are cast in stone. As such the parties are not motivated to perform for any additional perks either than those agreed on at negotiation stage. This was the case in the DEA PPP agreement. Incentive structures are important for motivating parties to maintain their commitment to complete the "exchange" (Ling et al., 2006) and to reinforce the spirit of collaborative. Unfortunately in PPP incentives are not easy to incorporate because of the high-risk nature of the agreement. Fortunately for the Imvelo Concession this contradictory factor did not affect the spirit of collaboration as there was an ambient flow of information across all parties and ultimately no cost overruns and no delays were experienced.

Challenges Faced and Lessons Learnt in the Construction of the Department of Environmental Affairs (DEA) Headquarters in South Africa

Challenges are status quo in construction projects; and as such the PPP arrangement for the DEA construction was no different. Some of the challenges faced served as informed by the construction professionals interviewed serves as lessons which in future can be appropriately dealt with in projects of similar nature. The interviewees informed that some of the changes were out of the control on any individual, however as the appointed consortium for the project, Imvelo concession had to mitigate any problems, ameliorate the challenge and steer the status quo to the benefit of the project in order to meet the project objectives and contractual agreements. Findings form the interview were analysed in a thematic manner based on the interview questions and are presented in the subsequent sessions:

Limited PPP Knowledge of the Concession Team and Other Contract Parties

Because other joint partners on the Imvelo Concession have only been exposed to traditional contracting strategies to deliver a construction project and not PPP arrangement; as a consequence were unfamiliar with the PPP practise. To remedy to this challenge which was obvious from the very beginning, all key personnel from Imvelo concession underwent workshops (over two separate days) to assist in establishing an understanding of the legal framework of PPP projects. Despite this intervention, parties to the joint venture agreement still experienced difficulties with the uncharted, unconversant nature of PPP environment. Therefore more mentoring and education is necessary.

Likewise, the PPP community engagement tasked Imvelo to employ the services of local businesses and individuals. This was major challenge in that many of the businesses and individuals in the region lacked the required level of competence to achieve the standards set by the concession as they lack the fundamental understanding of a typical PPP collaboration. Hence, training was provided to local businesses and individuals and at the conclusion of the project the community engagement requirement set by government was exceeded. This success as informed by one of the interview respondent, "is vital to the community and the South Africa construction industry as it has enable and capacitated a number of local business on the contractual nature of PPP which will benefit the economy in time to come."

Industrial Action

South Africa has in recent years been plague by labour disruptions in all sectors of the economy due to on-going socioeconomic negotiations taking place in the country since the end of the apartheid government. Although there were many instance where problems were encountered during the construction of the DEA building due to industrial strike action. However, in August of 2013 when the construction industry was hit by industrial action for a duration of three (3) weeks, construction works were affected in the DEA project. But, this delay encountered was remedied by the Imvelo Concessions as they had to increase resources in an attempted to make up for lost time at their expense (as per PPP agreement) without any recourse to the government.

Stakeholder Involvement and Management

PPPs are about changing the way a public service is delivered. This creates uncertainly for end users and uncertainty for the workers providing the public service. PPPs disrupt the (often comfortable) status quo. People need to be respected, consulted, and heard, especially in times of uncertainty and change. Getting and managing stakeholders (i.e. consulting, involving, respecting, and communicating with affected people) is critical to the success of all PPP projects, particularly high profile, large, complex projects such as the DEA project. PPP projects often fail because of poor stakeholder management. Successful stakeholder management starts with the early identification of stakeholders and the deliberate implementation of tailored strategies for each group. For the DEA project, stakeholders were identified early and procedure for management were properly defined. However, because the DEA, as government entity is run by politically driven individuals, there were bottle-necks in some aspects. During the construction process, the Imvelo had to constantly engage and negotiate intermittently with community members, community leaders and politicians alike, to avoid major disruption. Analysis should be done early in the project cycle – after project identification, but well before the formal transaction development process begins.

Design and Subcontract Management

Imvelo concession experience problems which are normal in traditional contracts with quality maintenance of designs. During the construction process, problems were encountered with electrical and mechanical consultants that were responsible for designs, as such it was impossible to complete the dependent tasks as at when needed. Also, some sub-contractors also proved a challenge, by performing poorly on their scheduled task.

Vis Major

A challenge beyond the control of Imvelo concession as informed by the interview respondents was that of seasonal rain. Although this was allowed for in the initial contract, no-one was able to foresee the large amount of rain experienced in early-to-mid 2014, which was the largest amount of rain experienced in the region in years. Although it is impossible to mitigate time lost due to rain (vis-major), it was the responsibility of Imvelo concession to accelerate with the intention of completing on the agreed date of 01 August 2014 (without adjusting the contract, as would be the case in traditional contracting).

CONCLUSION AND RECOMMENDATIONS

The reviewed literature and the qualitative interview have shown that there were challenges encounter during the use of the PPP collaborative model in the construction of the DEA building. The findings reinforced the perception that there certainly challenges that needs to the addressed when in a collaborative model supply chain undertaking used in the development of projects, either infrastructure or residential projects. The findings revealed that efforts was made by the concession company to ensure that the PPP agreement was honoured as stipulated. The inference of the research findings is that there exist long-term values and benefits in the use of collaborative models to deliver state infrastructure. PPP collaboration can be deemed responsible for and/or contribute to success of state infrastructure project as the complexities associated with construction projects and proliferation are managed through the bilateral agreement and responsibility of each party to the project. This thus eliminate usual construction snags and there side effects which has labelled the construction chain as irresponsible to the need of the 21st performance driven construction industry.

In the context of the research findings, presented in this discourse, it can be argued that the management of infrastructure projects using collaborative models is dependent on the 'holistic understanding' of all aspects of the partnership contract by all key project stakeholders in project; in this instance South

Africa. It can therefore be advised that for performance improvement purposes of a PPP collaborative model, stakeholders to a PPP collaborative agreement should be well trained and exposed to all aspects of the model in order to benefit from the advantages inherent in the model. This will ensure that there is a maximum promotion of the assignment of collaboration and improvement the supply chain process and responsibilities to appropriately meet the need for infrastructure development. In addition, both the public and private sectors should avoid the use of collaborative models when the fundamentals to fully engage in such process may be considered immature or not practicable by the stakeholders. Furthermore, PPP stakeholders should place emphasis on the development and giving back to the community by way of training for the community members to be involved in the collaborative supply chain so the industry is sure of the sustainability of the use of collaborative models in the construction industry.

When these steps are followed, the beneficial rewards of collaborative models will be fully harness in the South Africa construction industry and the entire economy. The relative increase in total project cost that may arise from the adoption of these recommendations could be cancelled from the increased buffer in total infrastructural project success delivery that will be associated with increased number of completed delay-free, cost-free and defect-free, projects that require minimal maintenance cost.

REFERENCES

Anvuur, A. M., & Kumaraswamy, M. M. (2007). Conceptual Model of Partnering and Alliancing. *Journal of Construction Engineering and Management*, *133*(3), 225–234. doi:10.1061/(ASCE)0733-9364(2007)133:3(225)

Construction of the National Museum of Australia and the Australian Institute of Aboriginal and Torres Strait Islander Studies. Audit Report (No.34). (2000Auditor-General of the Australian National Audit Office (AGANAO). Canberra, Australia: Australian National Audit Office.

Bygballe, L. E., Jahre, M., & Swärd, A. (2010). Partnering relationships in construction: A literature review. *Journal of Purchasing and Supply Management*, *16*(4), 239–253. doi:10.1016/j.pursup.2010.08.002

Bayliss, R., Cheung, S., Suen, H. C. H., & Wong, S. (2004). Effective partnering tools in construction: A case study on MTRC TKE contract 604 in Hong Kong. *International Journal of Project Management*, *22*(3), 253–263. doi:10.1016/S0263-7863(03)00069-3

Briscoe, G., & Dainty, A. (2005). Construction supply chain integration: An elusive goal? *Supply Chain Management: An International Journal*, *10*(4), 319–326. doi:10.1108/13598540510612794

Cheng, E. W. L., & Li, H. (2002). Construction Partnering Process and Associated Critical Success Factors: Quantitative Investigation. *Journal of Management Engineering*, *18*(4), 194–202. doi:10.1061/(ASCE)0742-597X(2002)18:4(194)

Cheung, S., Ng, T. S. T., Wong, S., & Suen, H. C. H. (2003). Behavioral aspects in construction partnering. *International Journal of Project Management*, *21*(5), 333–343. doi:10.1016/S0263-7863(02)00052-2

Construction Industry Institute (CII). (1991). In search of partnering excellence. Special Publication No. 17-1, Rep., Partnering Task Force of CII, Austin, Tex.

Davis, P. & Love, P. (2011). Alliance contracting: Adding value through relationship development. *Engineering, Construction, and Architectural Management*, *15*(11), 444–461.

Flyvbjerg, B., Skamris Holm, M. K., & Buhl, S. L. (2003). How common and how large are cost overruns in transport infrastructure projects. *Transport Reviews*, *23*(1), 71–88. doi:10.1080/01441640309904

Gadde, L.-E., Huemer, L., & Hakansson, H. (2003). Strategizing in industrial networks. *Industrial Marketing Management*, *32*(5), 357–364. doi:10.1016/S0019-8501(03)00009-9

Hlahla, M. (1999). The Municipal Infrastructure Investment Unit: The government's PPP-enabling strategy. *Development Southern Africa*, *16*(4), 565–583. doi:10.1080/03768359908440102

Johnston, P. (1995). Supply chain management: The past, the present and the future. *Manufacturing Engineer*, *74*(5), 213–217. doi:10.1049/me:19950510

Kopczak, L. R., & Johnson, M. E. (2003). The supply-chain management effect. *MIT Sloan Management Review*, *44*(3), 27–34.

Kulp, S. C., Lee, H. L., & Ofek, E. (2004). Manufacturer benefits from information integration with retail customers. *Management Science, 50*(4), 431–444. doi:10.1287/mnsc.1030.0182

La Londe, B. (1998). *Supply Chain Management: An Opportunity for Competitive Advantage*. Cincinnati, OH: Department of Transport and Logistics, The Ohio State University.

Bankvall, L., Bygballe, L. E., Dubois, A., & Jahre, M. (2010). Interdependence in supply chains and projects in construction. *Supply Chain Management: An International Journal, 15*(5), 385–393. doi:10.1108/13598541011068314

Ling, F. Y. Y., Rahman, M. M., & Ng, T. L. (2006). Incorporating Contractual Incentive to Facilitate Relational Contracting. *Journal of Professional Issues in Engineering Education and Practice, 132*(1), 57–66. doi:10.1061/(ASCE)1052-3928(2006)132:1(57)

Love, P. E. D., Davis, P. R., Chevis, R., & Edwards, D. J. (2011). Risk/reward compensation model for civil engineering infrastructure alliance projects. *Journal of Construction Engineering and 616 Management*, 137(2), 127-136.

Love, P. E. D., Irani, Z., & Edwards, D. J. (2004). A seamless supply chain management model for construction. *Supply Chain Management: An International Journal, 9*(1), 43–56. doi:10.1108/13598540410517575

Manley, Karen & Chen, Le. (2015). Collaborative learning model of infrastructure construction: A capability perspective. *Construction Innovation: Information, Process, Management, 15*(3), 355–377.

Masterman, J. W. E. (2002). *Introduction to Building Procurement Systems*. London: Spon Press.

Pala, M., Edum-Fotwe, F., Ruikar, K., Doughty, N., & Peters, C. (2014). Contractor practices for managing extended supply chain tiers. *Supply Chain Management: An International Journal, 19*(1), 31–45.

Morwood, R., Scott, D., & Pitcher, I. (2008). *Alliancing A Participant's Guide: Real Life Experiences for Constructors, Designers, Facilitators and Clients*. AECOM.

Naoum, S. (2003). An overview into the concept of partnering. *International Journal of Project Management, 21*(1), 71–76. doi:10.1016/S0263-7863(01)00059-X

Kamath, , N., B., & Roy, R. (2007). Capacity augmentation of a supply chain for a short lifecycle product: A system dynamics framework. *European Journal of Operational Research, 179*(2), 334–351. doi:10.1016/j.ejor.2006.03.045

About the PPP Unit. (n. d.). *Public Private Partnership*. Retrieved from http://www.ppp.gov.za/Pages/About.aspx

Rowlinson, S., & Cheung, F. (2002). *Review of the concepts and definitions of the various forms of relational contracting (2002-022-A)*. Brisbane, Australia: CRC for Construction Innovation.

Sharma, C. (2011). Determinants of PPP in infrastructure in developing economies. *Transforming Government: People, Process and Policy, 6*(2), 149–166.

Simatupang, T. M., & Sridharan, R. (2004). Benchmarking supply chain collaboration: An empirical study. *Benchmarking, International Journal (Toronto, Ont.)*, *11*, 484–503.

Subramani, M. (2004). How do suppliers benefit from information technology use in supply chain relationships? *Management Information Systems Quarterly*, *28*(1), 45–73.

Vrijhoef, R., & Koskela, L. (2000). The four roles of supply chain management in construction. *European Journal of Purchasing & Supply Management*, *6*(3-4), 169–178. doi:10.1016/S0969-7012(00)00013-7

Walker, D., & Hampson, K. (Eds.). (2003). *Procurement Strategies: A Relationship-based Approach.* Oxford: Blackwell Science.

Wang, E. T. G., Tai, J. C. F., & Wei, H. L. (2006). A Virtual Integration Theory of Improved Supply-Chain Performance. *Journal of Management Information Systems*, *23*(2), 41–64. doi:10.2753/MIS0742-1222230203

Xue, X., Shen, Q., & Ren, Z. (2010). Critical Review of Collaborative Working in Construction Projects: Business Environment and Human Behaviors. *Journal of Management Engineering*, *26*(4), 196–208. doi:10.1061/(ASCE)ME.1943-5479.0000025

Chapter 5
Innovation Diffusion in the European Ceramic Tile Industry Supply Chain

Daniel Gabaldón-Estevan
University of Valencia, Romania

ABSTRACT

Following Nelson's primary typology of enterprises based on a characterization of their technical change process (bulk commodities; complex systems producers; and chemical products producers) this chapter presents a case study on the innovation diffusion in the European tile industry based on the analysis of its manufacturing process and supply chain. The producers of ceramic tiles, the machinery producers and the frits and glazes producers fit each of Nelson's categories. Further, the European ceramic tile industry is clustered in two southern European regions: Emilia Romagna (Sassuolo – Italy) and Valencia Region (Castelló de la Plana – Spain) with a different configuration of the type of predominant companies within the cluster but also regarding surrounding clusters. This relative specialization, the interactions between surrounding clusters, and the strong interlink supplier-producer between both ceramic tile clusters make the study of the innovation diffusion in the European ceramic tile industry supply chain an interesting case.

INTRODUCTION

Ceramic tile production despite having a highly integrated production process it certainly matches Nelson's (1993) primary typology of enterprises. Nelson distinguishes three types of industries (bulk commodities; complex systems producers; and chemical products producers) based on a characterization of their technical change process. The producers of ceramic tiles, the machinery producers and the frits and glazes producers fit each of those categories. Further, another characteristic of the European ceramic tile industry is that it is strongly clustered in two southern European regions: Emilia Romagna

DOI: 10.4018/978-1-5225-0021-6.ch005

(Sassuolo) in Italy and the Valencia Region (Castelló de la Plana) in Spain with, however, a different configuration not only of the type of predominant companies within the cluster but also regarding surrounding clusters. This relative specialization, the interactions between surrounding clusters, and the strong interlink supplier-producer between both ceramic tile clusters make the study of the innovation diffusion in the European ceramic tile industry supply chain a very interesting case. The objective of this chapter is to present the case study on the innovation diffusion in the European tile industry based on the analysis of its manufacturing process and supply chain and identify the mechanisms through which innovation is produced and disseminated in the industry.

BACKGROUND

As Fagerberg (2003) points out, innovation has a clear impact on economic and social development. In the first place innovation introduces novelty (variety) in economic activity, which is necessary for economic growth in the long run. Also innovation tends to group in particular industries or sectors, which, as a consequence, develop faster, provoke structural change to production and demand and, exceptionally, also promote institutional and organizational changes. As many authors point out (Lundvall, 1993; Carlsson & Stankiewicz, 1995; Edquist, 1997), companies do not innovate in isolation. Innovating is a process in which companies interact with each other, as well as with organizations such as universities, research centres, public administration, and financial institutions and so on. Therefore, many authors explain the outcome of the innovation process by studying the innovation system in which the innovation takes place. Innovation, therefore, is to be understood as a collective and interactive process among a wide variety of actors.

A general definition of Systems of Innovation has been given by Edquist (2004, p. 182) as "all important economic, social, political, organizational, institutional and other factors that influence the development, diffusion and use of innovations". This theoretical approach to innovation considers relationships and networks as key elements of the innovation and production processes (Edquist, 1997).

Following Edquist (2004) there are 6 main strengths on the Systems of Innovation's approach. In the first place, it places innovation and learning at the core of the analysis, acknowledging that the production of knowledge, or the combination of already existing knowledge in new ways, is central to innovation. Innovation is therefore considered to be an endogenous process. Secondly, the Systems of Innovation's approach is both holistic and interdisciplinary, i.e. it tries to cover all the important determinants of innovation and incorporate perspectives from many different fields within the social sciences. Thirdly, under this richer, but also more complex, approach the notion of optimality becomes irrelevant as it adopts an historical and evolutionary perspective where innovation is affected by many factors and feedback processes. A concrete innovation system can be compared with other real systems as well as with target systems but not between real and optimal systems as those cannot be specified. Forth, this approach accentuates the interdependence of innovating firms to other organizations in complex relational patterns far from the simplistic linear model. Fifthly, this approach allows a more comprehensive innovation concept that encompasses not only technological process innovations and goods products innovations, but a wider spectrum of innovations to include also non-technological and intangible innovations such as service products and organizational process innovations. And last but not least, this approach reveals the weight of institutions' influence on innovation processes.

Nevertheless the Systems of Innovation's approach presents also some weaknesses as conceptual diffuseness regarding core terms like institution, i.e. for some authors it refers to institutional actors while for others it is associated with being 'the rules of the game' (Edquist, 2004) but also an absence of consensus regarding the boundaries of the analysis. The lack of well-established empirical regularities has also been signalled; consequently Edquist (2004) indicates that Systems of Innovation should be labelled an approach or a conceptual framework rather than a theory. "If all factors that influence innovation process are not included in a definition, one has to argue which potential factors should be excluded –and why. [...] This is quite difficult, since, at the present state of the art, we do not know the determinants of innovation systematically and in detail. It seems dangerous to exclude some potential determinants, since these might prove to be very important, once the state of the art has advanced" (Edquist, 2004, p. 183).

As we have already mentioned, one of the main goals of this perspective is that it places innovation and learning at the core of the analysis. Three kinds of learning can be distinguished (Edquist, 2004). Firstly innovation, it happens mostly in firms and creates 'structural capital', which is a knowledge-related asset controlled by firms (as opposed to 'human capital'); it's a matter of organizational learning. Secondly 'research and development' is carried out in universities and public research organizations, in addition to firms, leading to publicly available knowledge along with knowledge owned by firms and other organizations, as well as by individuals. Thirdly 'competence building' is training and education, which occurs in schools and universities as well as in firms, and leads to the creation of 'human capital'. Since human capital is controlled by individuals, it is a matter of individual learning.

In this perspective, interaction includes (OECD 2002a:15, in Edquist, 2004) (i) Competition, which is an interactive process wherein the actors are rivals, and which creates or affects the incentives for innovation. (ii) Transaction, which is a process by which goods and services, including technology-embodied and tacit knowledge, are traded between economic actors. And (iii) networking, which is a process by which knowledge is transferred through collaboration, cooperation and long term network arrangement.

National System of Innovation

Systems of Innovation first appeared in the form of 'national system of innovation' by Lundvall in 1992, and has been widely accepted, and used since, although "the idea actually goes back at least to Friedrich List's conception of 'The National System of Political Economy' (1841), which might just as well have been called 'The National System of Innovation'" (Freeman, 1995:5). He defined it as "the network of institutions in the public and private sectors whose activities and interactions initiate, import and diffuse new technologies" (Freeman & Soete, 1987). National systems of innovation's focus, however, is on national boundaries and non-firms organizations and institutions (Freeman & Soete, 1987; Nelson, 1993) but is the common starting point for the following developments of the Systems of Innovation we focus in this section: such as Sectoral System of Innovation, Regional System of Innovation, Technological Systems, the Functions of Innovation Systems Approach, and District Innovation System.

Sectoral System of Innovation

Sectoral Systems or the Sectoral approach by Breschi and Malerba (1997) focuses on a group of firms that develops and manufactures the products of a specific sector and that generates and utilizes the technologies of that sector:

A sectoral system has a specific knowledge base, technologies, inputs and demand. Agents are individuals and organizations at various levels of aggregation. They interact through processes of communication, exchange, co-operation, competition and command, and these interactions are shaped by institutions. A sectoral system undergoes change and transformation through the co-evolution of its various elements. (Malerba, 2002, p. 247)

According to Malerba (2002) there have been two main traditions dealing with sectors: The industrial economics literature and a more heterogeneous and case-studied-based one. Studies within the former have examined the structure of sectors in terms of concentration, vertical integration, diversification, and have considered sectoral boundaries static and delimited in terms of similarity in techniques or similarity in demand. However, no much emphasis has been paid to the role of non-firms organizations, to knowledge and learning processes by firms, to the wide range of relations among the agents, or to the transformation of sectors in their boundaries, actors, products and structure. Studies within the latter are seen as a richer body of research but also more heterogeneous, eclectic and dispersed where mostly operated through case studies focused on a single dimension, asking different research questions, done with different methodologies and different level of aggregation in terms of unit of analysis.

A sectoral system of innovation and production is defined as:

A set of new and established products for specific uses and the set of agents carrying out market and non-market interactions for the creation, production and sale of those products. Sectoral systems have a knowledge base, technologies, inputs and demand. The agents are individuals and organizations at various levels of aggregation, with specific learning processes, competencies, organizational structure, beliefs, objectives and behaviours. They interact through processes of communication, exchange, co-operation, competition and command, and their interactions are shaped by institutions. A sectoral system undergoes processes of change and transformation thought the co-evolution of its various elements. (Malerba, 2002, p. 261)

Malerba (2004) states that the innovation in a particular sector is affected by three main factors: (1) knowledge and technology; (2) actors and networks; and (3) institutions. Following the evolutionist literature, from a sectoral perspective, knowledge is considered to be at the heart of technological change. Since knowledge and technology at firm level are highly idiosyncratic, they do not diffuse freely or automatically among companies, and can only be reabsorbed through the different abilities that companies accumulate over time. In sectors where innovations are frequent, the knowledge and technology frontiers are rapidly changing.

From a sectoral perspective, knowledge as well as learning processes and technologies are of great relevance. As Malerba (2004) points out, it is common that in a particular industrial sector more than one technology may be relevant and the same is true for a particular technology (i.e. it can be relevant for more than one sector). From this it follows that in a given sector it may be that some companies need to master more than one technology. But it also implies that within a given sector, and especially in relation to large companies, there will be few if any differences in their technology diversification profiles. Alongside evolutionary theory, the sectoral approach states that there will be differences between the knowledge bases and learning processes of sectors and technologies related to innovation.

Other dimensions, highlighted in Nelson and Winter (1982) and their notion of technological regimes, are useful to understand innovative activity in sectoral systems. A first dimension refers to appropriability or the opportunity to obtain knowledge external to the company. It implies that the more accessible is this knowledge the lower will be the industry concentration. Also, accessible knowledge can be internal or external to the sector. Easy accessibility to internal knowledge leaves little room for appropriability (it is easy for competitors to imitate products and processes), while easy accessibility to external knowledge is related to the levels and sources of scientific opportunities that enable external factors to affect companies through their human capital or the scientific knowledge developed in non-firm organizations (Malerba, 2004). Furthermore, as both Freeman (1982) and Rosenberg (1982) pointed out, the sources of technological opportunity differ greatly among sectors. Technological opportunities in some sectors are related to scientific progress at universities, while in others they depend on scientific progress made in research and development (R&D) departments in firms. In other sectors, the opportunities for innovation come from providers, through capital goods or instrumentation, or clients. Therefore, the ease with which external knowledge is assimilated and applied in an industry will depend on its availability and on the chances of it being transferred. If these are high, innovation entry rate will be equally high; if, on the contrary, specific advanced capacities are needed to assimilate it, the industry will more likely be configured around big well established companies.

The actors involved can be individuals or organizations, and companies, universities, financial organizations, unions, etc. or sub-organizations such as R&D departments, or groups such as producers' associations. Agents, together with companies and non-firm organizations, include providers and clients. Depending on the type of industry their roles can be more or less important. All those heterogeneous agents, who are characterized by particular learning processes, abilities, beliefs, objectives, structures and behaviours, relate to each other through market and non-market interactions. The types of networks and relationships vary between sectors according to knowledge base characteristics, learning processes, basic technologies in use, demand characteristics and dynamic complementarities.

Agents' knowledge is built upon the institutions, which include norms, routines, habits, practices, rules, laws, standards, etc. some of which are imposed and some of which are optional, some of which are formal and some of which are informal, some of which are national and some of which are sectoral.

Finally, demand in a sectoral system is made up of individual consumers together with companies and institutions, each characterized by particular knowledge, learning processes, competencies and objectives, and subject to social factors and institutions (Malerba, 2004). Therefore, demand cannot be considered an aggregate of homogeneous buyers but as comprising heterogeneous agents which interactions with producers conformed by institutions.

Regional System of Innovation

Close to the previous theoretical proposal is the Regional Innovation Systems by Cooke (Cooke et al., 1997) in which the focus or boundary is not the nation or the sector but the region and has specially fitted regions of big countries which are in fact equivalent to some "smaller" nations.

Therefore the three perspectives –national, sectoral and regional- might be clustered as variants of a single generic "systems of innovation" approach as stated by Edquist (2004)

...the different variants of systems of innovation coexist and complement each other. Whether the most appropriate conception of the system of innovation, in a certain context, should be national, sectoral or regional, depends to a large extent on the questions one wants to ask. (Edquist, 2004, p. 184)

And,

The concept of national [or regional, sectoral, technological] innovation systems rests on the premise that understanding the linkages among the actors involved in innovation is key to improving technology performance. Innovation and technical progress are the result of a complex set of relationships among actors producing, distributing and applying various kinds of knowledge. The innovative performance of a country [or region, sector, technology] depends to a large extent on how these actors relate to each other as elements of a collective system of knowledge creation and use as well as the technologies they use. These actors are primarily private enterprises, universities and public research institutes and the people within them. The linkages can take the form of joint research, personnel exchanges, cross-patenting, purchase of equipment and a variety of other channels. There is no single accepted definition of a national [or regional, sectoral, technological] system of innovation. What is important is the web of interaction or the system [...] (OECD, 1997, pp. 9)

Technological System (of Innovation)

Within the other specifications of systems of innovation Carlsson and colleagues focus on 'technological systems' arguing that these are unique to technology fields (Edquist, 2004; 183). Consequently the technological systems perspective (Carlsson & Stankiewicz, 1995; Huges, 1984; Callon, 1992), in which the focus is mainly on networks of agents for the generation, diffusion and utilization of technologies, ads to the map of the Innovation Systems approaches. In Carlsson's words:

This is a study of technological innovation as the primary source of economic growth in an advanced economy. The purpose is to show how technological innovation which may occur at many levels –in individual firms or in a various groupings of firms- gives rise to economic growth at the aggregate (macro) level. This process involves interaction; both planned and unplanned, via markets as well as via other mechanisms, among a variety of economic agents [...] There is no shortage of studies of economic growth. What is unique and novel about the work presented here is the approach, viewing the economy as a set of dynamic and complex systems, partially overlapping and interacting with each other. Innovations –new combinations of products, processes, organizations, buyers and suppliers- are at the core of the analysis. They propel the economy forward. Sometimes they improve and expand existing systems, and sometimes they generate new systems. The technological possibilities are virtually limitless, but the ability of individuals, firms, and other entities to identify and exploit the resulting business opportunities is strictly limited (Carlsson, 1997, p. 1)

And:

The central units of our analysis are technological systems. We have previously defined technological systems as knowledge and competence networks supporting development, diffusion and utilization of technology in established or emerging fields of economic activity. (Carlsson & Stankiewicz 1991, p. 2)

Consequently they consist of networks of firms, research and development (R&D) infrastructures, educational institutions, and policy making bodies.

Technological systems can be seen as mediating structures between the knowledge base of a sector and the firms active in it. Their quality influences the degree to which the firms can exploit the emerging technological opportunities. Hence, the importance of technological systems increases with the complexity and heterogeneity of the knowledge base of innovation. (Carlsson, 1997:2)

The Functions of Innovation Systems Approach

A recent development in the innovation systems literature is to focus on the functioning of innovation systems (Negro and Hekkert, 2006, Hekkert et al., 2007, Bergek et al., 2008). The basic idea is that a well-functioning innovation system requires that a number of key activities take place: "Functions analysis is based on idea that a well-functioning innovation system requires that a number of key activities takes place. If this occurs, output of innovations is higher" (Gabaldón-Estevan & Hekkert, 2013, p. 151).

Many studies adopt a focus on how innovation systems function (Suurs & Hekkert, 2007; Suurs, 2009, Negro, 2007; Negro & Hekkert, 2008; Negro et al., 2007, 2008a, 2008b; Alkemade et al., 2007). Most of these case studies are related to specific technological trajectories and very often the focus is on emerging technologies and, therefore on the formative stage of innovation systems, i.e. in the first stages of Roger's S-curve of innovation adoption (Rogers, 2010).

District Innovation System

Originally the concept of "District Innovation System" has been brought about to emphasize the relevance of territory when an industry adopts the industrial district form but is also dependent on other elements in the innovation system (Gabaldón-Estevan et al., 2012).

The Industrial Districts perspective is derived from the new economics of Marshall developed since the seminal work of Becattini in 1979 (Becattini, 1979, 1990, 2002; Bellandi, 1996, 2002, 2006; Brusco, 1990; Brusco & Pezzini, 1990; Pyke & Sengenberger, 1992). This new perspective changed the focus of the industrial development analysis from the industrial sector to the industrial district and, by doing so they moved as well the stress from the sector to the region. The benefits from adopting this perspective have improved our understanding of regional development, especially when analysing relations between enterprises.

The main characteristics of an industrial district, as enumerated by Porter (1998), are flexible specialization, cooperation between enterprises, geographic proximity, and the social relations of the enterprises. That perspective implies that the aggregation of a variety of small and medium enterprises that produce in the same goods chain and in a concrete area can cause innovation by means of the special synergies driven by a cooperative versus competitive tension. These synergies elicited by close relations among elements within a district affect innovation production and innovation transfer by several means. These means include incrementing qualitative and quantitative transfer of information, increasing confidence between the different elements in the district, lowering transaction costs, or improving access to experienced and high-skilled human capital. In addition, industrial districts also promote innovation in a variety of forms, because of the privileged perspective of those inside the district, the social capital available inside the district, the pressure of competitiveness, and the availability of resources inside the district. Following Becattini, the characteristic element of an industrial district is the consciousness of subjects of pertaining to a specific industry, 'the sense of belonging'.

District Innovation System is defined as a system of relationships within an industrial district where externalities facilitate firms' innovation processes (Gabaldón-Estevan et al., 2012). An industrial district is understood as a social entity that, while linked to a territory, usually does not conform to the limits of a specific administration. District Innovation System emphasizes the relevance of territory, that is, when an industry adopts the district form, but is also highly dependent on other elements in the innovation system. Consequently the Distritual Innovation System or District Innovation System is formed by a set of institutions, firms and promotion mechanisms that offer continuous support to district firms. The DIS involves interconnections and cooperation among elements within the same environment for the purposes of innovation.

The advantage of the District Innovation System approach is that it overcomes the potential limitations of the district concept to capture and explain the innovative processes occurring within it but, at the same time, does not ignore the unique specificities that characterize and differentiate an industrial district from other levels of analysis.

Summing up, the theoretical and analytical approach from which this paper builds up comes from the relevance of the concept of innovation systems, that has been growing during the last decades as has it has become clear that technological change cannot longer be considered as exogenous factor when analysing economic growth. And innovation systems perspective is in debt with the evolutionary theory from where central concepts such as learning, knowledge, competencies and a major focus on dynamics comes from. The main contributions to this debate can be summarized in Figure 1.

Figure 1. Main Contributions

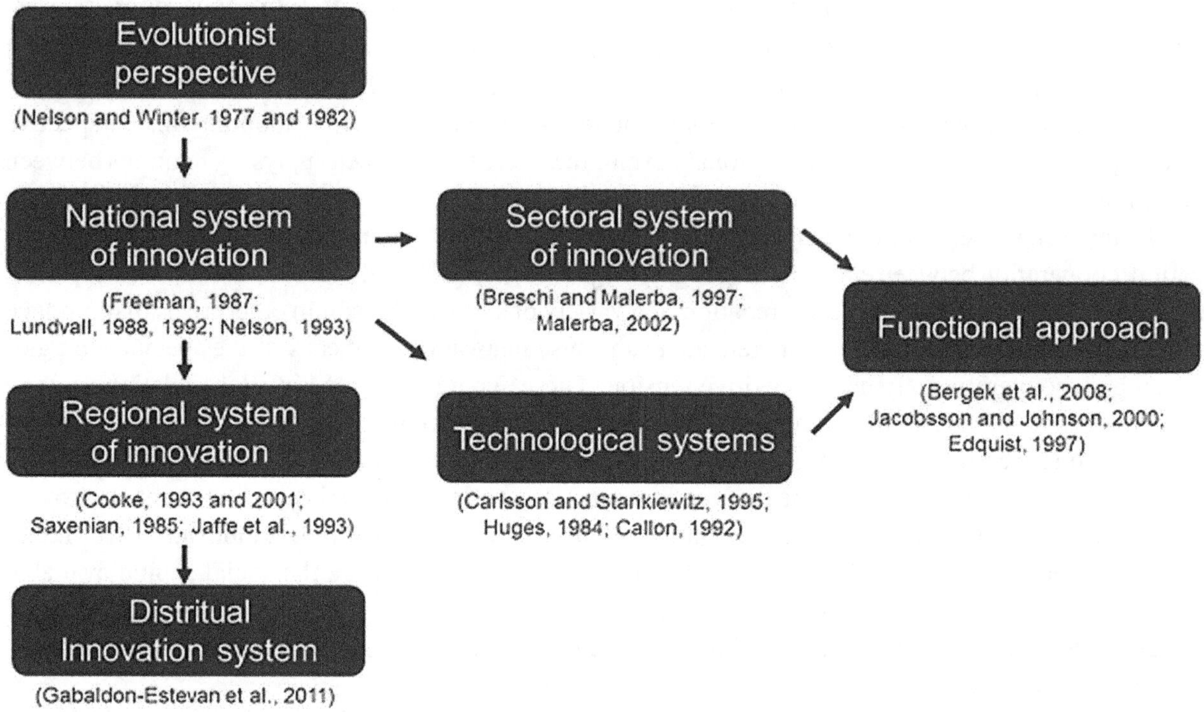

RESEARCH METHODOLOGY AND RESEARCH QUESTIONS

This chapter builds and revisits the evidence from previous research accumulated during the last ten years, on the European ceramic tile industry (Fernández et al. 2005; Fernández de Lucio and Gabaldón-Estevan, 2007; Gabaldon-Estevan, 2011; Gabaldón-Estevan and Hekkert, 2013; Gabaldon-Estevan et al. 2007, 2008, 2009, 2012, 2014a, 2014b; Tortajada-Esparza et al. 2008a, 2008b, 2008c, 2009; Monfort et al. 2014).

More specifically, some of this previous research consisted in a case study analysis of the ceramic tile industry located in the north of the autonomous region of Valencia, and its counterpart in the north of Italy. The emphasis at Gabaldon-Estevan et al. (2007) was on understanding the links between innovation and cooperation among the elements of a given industrial district and the presence of enterprises and providers of technology and horizontal advanced services serving several districts This research was built up by revising previous research on the European ceramic tile industrial districts, by doing compilation and analysis of secondary statistical data on the sector and, finally, by carrying out more than two dozen semi-structured interviews with representatives of the ceramic industrial districts of Sassuolo in Italy and Castellon in Spain. Some interviewees were managers from either ceramic, electro-mechanical or glaze companies; others were representatives of employers and workers associations; some were representatives of public institutions specialized in technology or trade; still others were in charge of research institutions directly responsible for I+D for the industry, or academics whose work had intensively been focused on the issue. In the interviews information was obtained on the tile process, on the relevant elements and the role of each element within each tile district, on how innovations were produced and disseminated through agents in the market, on how the different agents did participate in the innovation processes, and on how innovation was stimulated in the sector, apart from more general matters about the sector evolution such as global trends in production, new competitors or trade affairs.

At Gabaldón-Estevan et al. (2009) we applied a systemic approach to the analysis of the Spanish ceramic tile industrial district. We were particularly interested on the analysis of the interaction between the productive-technological environment and the scientific environment by an examination thru the analysis of research contracts and patents. Our analysis revealed that R&D activities were mainly conducted by providers, whereas final producers' focussed their innovation efforts towards no technological aspects and as their differentiation strategy.

For a subsequent research the methodology proposed by Bergek et al. (2008) was partially adopted for analysing the dynamics of the functional pattern, which included defining the innovation system in focus, identifying its structural components, and mapping the functional pattern of the innovation system. Data availability influenced the methodologies used for this research and qualitative tools substitute for some of the quantitative sources proposed in Bergek et al.'s model. To complete these research 15 semi-structured interviews were carried out with representatives of the Spanish tile innovation system. The interviewees were chosen on the basis of their expert knowledge on the Spanish tile industry, and included managers of ceramic, electro-mechanical, glaze companies or atomisation plants; representatives of employers' or workers' associations; directors of research institutions responsible for R&D for the industry; academics and consultants. The interviews provided information on different aspects of industry activity and the production process. The interest was on how the innovation system performed. The interviews were structured in six sections according to six functions that we consider to be most relevant for our analysis of the innovation system (Knowledge development and diffusion, Influence on the direction of search, Entrepreneurial experimentation, Market evolution and competition, Legitimation / counteracting resistance to change, and Resource mobilisation) which were taken from Johnson (2001). A full summary of this research is found at Gabaldon-Estevan and Hekkert (2011).

Latest research has been oriented towards sustainability issues regarding the ceramic tile industry, particularly on how environmental issues affect their development (Gabaldon-Estevan et al, 2014a) where the main interest was to establish the interactions between the innovation system and the new environmental requirements. A specific analysis on energy consumption and energy efficiency was shown in Monfort et al. (2014) were we analysed how the specific characteristic of the ceramic tile manufacturing process and supply chain affects its innovation capacity and is also threaten by environmental and energy challenges. Finally, at Gabaldon-Estevan et al. (2014b) an analysis on EU energy and environmental policy and its effect on the European ceramic tile sector were presented.

All in all, all this pieces of research serves both as the starting point for this chapter and as the case study material where to select the evidence that will be use to respond to the research questions this chapter tries to respond to:

- Through which mechanisms innovation is produced and disseminated in the ceramic tile industry?
- How is the structure of its supply chain affecting its evolution?
- What are the challenges the industry is facing?

THE EVIDENCE

In this section we present the main evidence that will allow us to answer the above stated questions regarding the European ceramic tile industry. For this purpose this section is divided in three that will subsequently address the description of the industry main descriptors; the supply chain; and challenges the industry is actually facing.

The European Ceramic Tile Industry

Ceramic tiles have many functional characteristics for indoor and outdoor including hygienic and safety characteristics, resistance to light exposure and chemical attacks, and their ease of maintenance make them an attractive solution for construction purposes. According to the European Ceramic Industry Association wall and floor ceramic tiles are the biggest sector in European ceramic industries measured by turnover, with total sales in 2013 estimated at around €9 billion.

The European ceramic tile industry is the world leader for design and innovation, setting trends for new designs and functionalities of tiles. As a consequence around a third of ceramic tile production is exports resulting in a largely positive trade balance. However, after decades of leadership in the world tile market and partially due to the financial crisis, European tile manufacturing in world production decreased from 16.6% in 2008 to 10.5% in 2012.

The European ceramic tile industry, which includes production of ceramic floor and wall tiles, decorative pieces, frit and glaze, machinery and equipment as well as other activities related to the ceramic process, has adopted an industrial district agglomeration and is strongly clustered in two southern European regions: Emilia Romagna (Sassuolo) in Italy and the Comunitat Valenciana (Castelló de la Plana) in Spain with, however, a different configuration not only of the type of predominant companies within the cluster but also regarding surrounding clusters. Within the EU, the Italian leadership in production (until 2005) and on exports, has been due not only to its high quality but to innovations in the functionality which its tiles are designed to supply. In these respect, the role of the Sassuolo district is pre-eminent. The growth in tile production was for a long time relied on the expansion in the demand for tiles, in Europe and later worldwide. The roots of tile production in the Castellon de la Plana district can began be traced back to the XIII century, nevertheless it was during the 1970s, 1980s but specially 1990s when the take-off of the industry took place. Initially, the tile manufacturers that established their activities in the area were serving the internal demand for tiles, which was growing in parallel with the expansion of the construction industry. Subsequently, they consolidated their European presence in competition with Italian ceramic tile producers.

A transfer of knowledge from Sassuolo to the Castellon de la Plana district also contributed to the latter's growth. Such transfers have taken place through three main channels. First, Italian exports to Spain of machinery for tile production allowed district actors to benefit from the knowledge about production technology incorporated in the machines themselves although developments (like mono-pore or atomization) were developed by the Italian machinery producers in collaboration with the Spanish chemistry technicians (which is, indeed, an example of the rich interaction between both districts). Second, in the 1990s numerous Italian companies producing colours and glazes moved part of their production activities to Spain, and with them they transferred to the area their knowledge about the physical and chemical properties of colours and glazes this process was reinforced by the presence of the ITC (Ceramic Technology Institute) by chemistry graduates and by the delocalization of international companies. Third, frequent contacts with technicians working at Sassuolo helped the Spanish producers develop, in the course of several decades, the in-depth knowledge of tile production processes needed in order to be able to perform 'reverse engineering' on new types of tiles, that is, to reconstruct, from the new tiles shown in exhibitions, the production processes through which they were obtained.

Several ideas emerge from the comparative analysis of the tile industry districts in Castellon and Sassuolo (Gabaldón-Estevan et al. 2007) a strong role of competition in the Spanish district, which is not accompanied by similarly strong cooperation. Interviews show that the level of competition in the Spanish and Italian is similar, but there is less active cooperation in the Spanish district. Also that the scarcity of technology and advanced services providers in the Spanish district suggests that the important process and product innovations were made in Italy, with Spain adopting a follower role in most areas. And finally, that the existence of horizontal technology enterprises increases the competition for innovation due to technology diffusion across neighbouring districts specialized in other sectors, favoured by the mobility of qualified workers.

With the arrival of Chinese producers to ceramic tile world markets, the transformations precipitated by the ongoing process of European integration, the growth differential among Italy and Spain, the social and institutional differences among the two districts, open up new possible scenarios, in which the districts may have different development opportunities.

The Supply Chain

This section provides a descriptive analysis of these elements measuring their contribution to the value chain. Figure 2 depicts the six central elements in the tile production process.

The most basic activity in tile production is raw materials extraction (mainly clay, but also silica and other elements). The clay quality, like for instance its organic composition, determines its appropriateness for tile production. As ceramic tiles are a heavy product in relation to its value, its final cost is very much dependent on the distance from the mine to the plant. Simplifying, two types of clay are used for tile making, red and white clays. Since white clay is not available in either of the abovementioned regions, it has to be imported from Eastern Europe (mainly Germany and Ukraine). However, as it offers the possibility of colouring the tile without adding frits (technique porcelain) its demand grew during the years previous to the crisis. Porcelain tiles were an innovation developed by Italian machinery producers as a way to weaken the frits and glaze producers' leadership in the decoration of tiles. In Spain however, the close proximity of red clay mines and the preeminence of frits and glaze producers waken the introduction of porcelain tiles. Atomizers are companies that require enormous investments to get started but once they start producing, are able to satisfy the demands of several medium sized tile producers. Many were set up by groups of tile companies, with only the biggest tile producers having their own atomizer plants. For the atomizer companies the main efforts have been focused on increasing the capacity to process white clay since demand for this product by the tile companies has been increasing steadily. This requires modifications to the plant, such as new infrastructure, and imports of white clay from Eastern Europe involving negotiation with foreign mine and transport companies.

Figure 2. Central Elements in the Tile Production Process

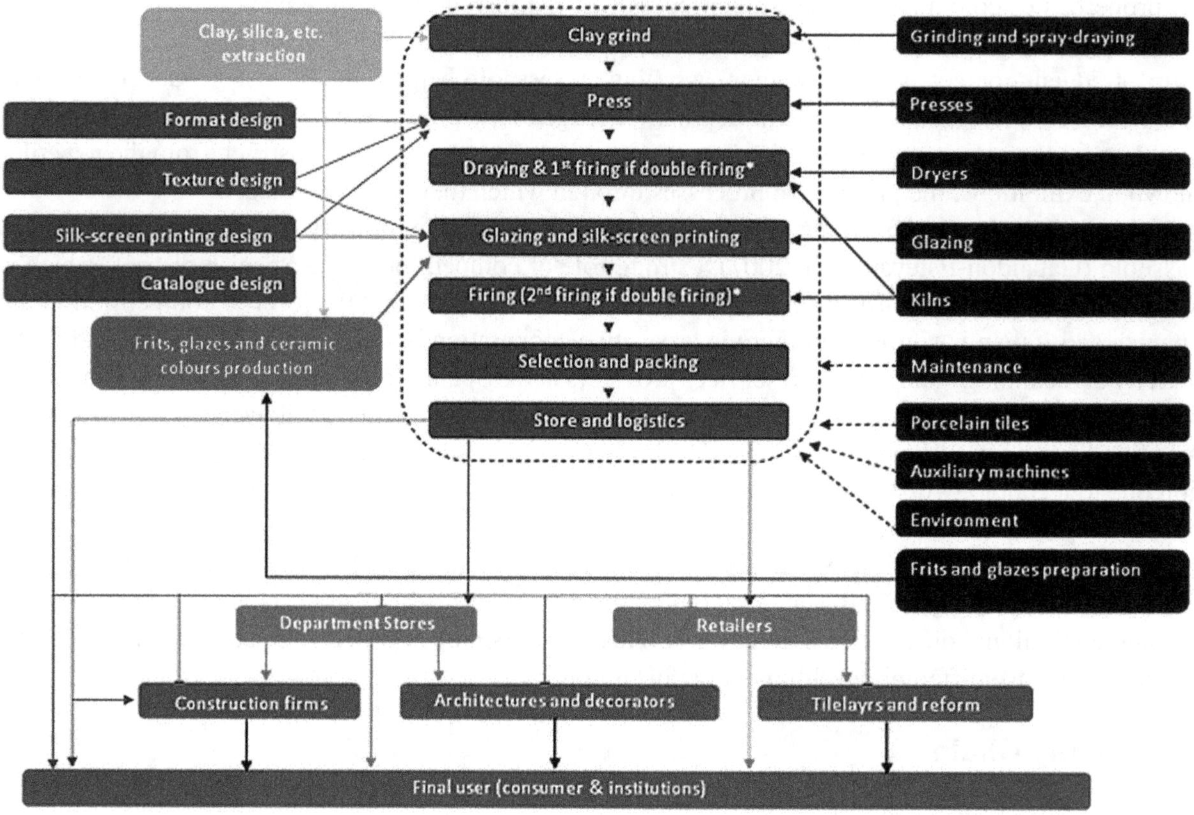

Machinery and equipment are needed for all the activities of tile production - from clay grinding to storing, including pressers and kilns, and machinery for frits and glaze production. The acquisition of new machinery is how most tile companies innovate (Fernández et al. 2005). Particularly important, as signalled by Russo (1996), are the generic innovations that emerge from neighbouring districts, such as the case of the atomizer machine, which was previously adapted from the pharmaceutical industry to the food industry for the production of powdered milk, or the furnace tunnel that was adapted to the ceramic tile industry from the food industry for the production of cookies. Therefore, the relationship between tile and machinery manufacturers is one of strong cooperation (Meyer-Stamer et al. 2004). The majority of machinery providers are Italian, although Spanish machinery providers are reasonably well specialized in machinery for glazing, machinery for frits and glaze preparation, and equipment maintenance.

Glaze production is of great importance in tile production. Aesthetics are becoming increasingly relevant for product differentiation as markets mature. The glaze is responsible for the final properties and appearance of the product, and allows the addition of a variety of decorations. Glaze companies produce innovations in design and applications (Meyer-Stamer et al. 2004). The most important glaze companies are in Spain and Italy (Tortajada-Esparza et al., 2008b,c).

Design is of major importance for almost every consumer good, and especially for decorative goods such as tiles. Tile design can be decomposed into three basic appearance properties i.e. size, texture and colour. Variations in these properties are developed by designers in collaboration with tile, machinery and glaze producers.

Production of floor and wall tiles is a highly integrated process because of the costs associated with the breaking down of the various processes involved. Consequently, tile enterprises, on average, are large sized. Tile production has evolved greatly since the 1960s due to increased mechanization, reductions in the time for processing, and simplification of the process (i.e. single firing or Laminam[1])

Finally, distribution is largely delegated to mediators and construction companies, especially in Spain. Only during the previous years to the economic crisis have some of the leading tile companies been recognizing the importance of direct relationships with end users.

From the specific analysis of the research contracts and patents within the Spanish ceramic tile innovation system (Gabaldón-Estevan et al. 2009) the main findings are as follows. Technological innovation within the district is mostly achieved by glaze and frit companies. The evidence shows an intense relationship between firms, particularly from the frit and glaze subsector, with university departments and research institutes. This type of cooperation is focused mostly on R&D projects. The ceramic tile producers focus their innovation on non-technological developments. Other activities do not play a role in the innovation process. Patenting to protect innovation is generally not used by companies in the ceramic tile industry where innovation diffuses to other firms. Many companies prefer alternative non-contractual means of protection. However, data on patenting for the period considered shows there is a lack of cooperation between glaze and frit producers and limited cooperation between firms and other elements in the DIS.

Our analysis shows that R&D in the Spanish tile ceramic tile innovation system is mainly carried out by providers of frits, glaze and colours, and that ceramic tile producers tend to focus on non-technological innovations. It shows weak use of patenting to protect innovations, low levels of cooperation between the frits, glaze and colours providers, and very limited cooperation with other elements of the DIS. We can conclude that technological innovations diffuse easily within the DIS which is the reason why ceramic tile producers focus on non-technological innovations as their differentiation strategy. The lack of homogeneity within the DIS is particularly relevant considering that external networks with enterprises and agents are a strategic asset.

The Challenge

The need for mature industries to innovate in order to survive is intensifying in an increasingly open world where trade and knowledge borders are becoming more porous. In mature industries, past success has often rested partially on comparative advantages such as pioneering activity, protected markets or, as in the case of the Spanish tile industry, because of favourable environmental regulation, raw materials transport and labour cost differentials (Fernández et al. 2005). With globalization and the entry of new producers some of those earlier comparative advantages are becoming comparative disadvantages, as newcomers are operating under less restrictive legislation and lower costs (Gabaldón-Estevan et al. 2014).

The challenge for mature districts depends increasingly in their ability to upgrade and innovate to avoid 'getting locked into a race to the bottom' (Schmitz, 2004). An example here is the Spanish tile industry where, during the early phases of its development, legislation was favourable compared to competitors from Italy in terms of employment and environmental issues. However, Spain's membership of what was the European Economic Community now the European Union, has progressively reduced these differences to the point that competition currently comes from new entrants from China, Brazil, Mexico, etc. and is similar to that faced by the Italian tile industry. Consequently there is a need for mature industries to be innovative in order to face these challenges. The question we analyse in this chapter is whether innovation is the primary strategy being used by mature industries to deal with these challenges.

Along the supply chain, the most recent advances are related to reductions in the costs of materials, treatment of mass (mixing humid and dry mass) and tile decoration (fast prototype and inject, laser and dry decoration). New and bigger formats are demanding innovations in presses (double press). These new developments are slow to diffuse through the industry because of the low added value they provide and the difficulty to make the market value these technical improvements, situation that has worsen significantly since the start of the economic crisis. Also logistics and management of stocks, storing and distribution absorb the resources of companies as orders become smaller. Besides, most innovations in the European ceramic tile industry are incremental[2].

A process optimization, in terms of both materials required and energy consumed (reutilization of oven gases and improved combustion), is one of the main concerns in the industry given the increasing environmental constraints (Gabaldón-Estevan et al. 2014; Mezquita et al. 2014; Minguillón et al. 2013).

There is a will to transfer new scientific developments to the tile industry, such as nanotechnology, and tenacious and advanced materials, however is at odds with the low added value of the tile production process. With new technologies and new applications for ceramics, the demand for new applied scientific knowledge is growing together with a deeper development of the chemistry. Some of the new knowledge demanded by the tile industry is related to the areas of energy, biology and medical science and health. And this demand is increasing the need for new specialists to join the innovation system. However, there is a parallel process of externalization, which implies that many enterprises do not develop most of this scientific work on their own but outsource it to providers and specialized service providers. As a consequence of this externalization and the mechanization of activities, the scientific knowledge required by personnel in ceramic plants is being redirected towards keeping plants producing and being able to interact with outside specialists.

Regarding the functional analysis of the Spanish tile innovation system (Gabaldón-Estevan and Hekkert, 2013) the investigation corroborated the spread feeling that a technological leap is not on the horizon and that the crisis is mainly related to the end of a long period of growth where construction played a major role. But also due to other factors related to globalisation such as the increasing competition in markets, the delocalisation, or the environmental and energy constrains. Technical innovations were generally incremental, and the technological challenges such as the reduction in the materials and energy consumption were not matched by alternative production processes. Technology improvements are likely to spread rapidly to other countries and the comparative advantage will not be enough to guarantee survival.

A shared strategic view would launch a renewed innovative strategy would have a positive effect on a TIS that is lacking a clear objective. That shared strategic view might be "ecological or green tiles", as suggested by one interviewee, and might not only group different agents within the industry but also group several different sub-strategies under one umbrella. Such a strategy would need to include a revision of all the elements in the value chain, from raw materials extraction to packaging and transportation, including firing, water and energy consumption, etc. but also market strategies. This will require new resources devoted to R&D but in the run will guarantee the TIS the privileged position of being at the forefront of a new paradigm in the tile industry. (Gabaldón-Estevan & Hekkert, 2013, p. 158)

CONCLUSION

This study aimed to understand the links between innovation and cooperation involving the various elements in the European Ceramic Tile Supply Chain, the presence of enterprises and providers of technology and horizontal advanced services, serving several different types of industrial districts. The main conclusion from our research is that the specific characteristics of the ceramic tile innovation system need to be considered for systemic analysis of the innovation process.

The internal regime of accessing, transmission and exploiting knowledge and innovation determines this particular system. On the other hand, from a global perspective the existence of other districts allows interactions and a certain international division of labour among districts, which may influence the development of a particular district.

We can also conclude that innovation depends positively on the relative level of technology and the advanced service providers in a district, as well as the strength of the cooperation between these firms and similar firms in neighbouring districts.

Our analysis suggests also that the districts in both countries would benefit from stronger links with universities which would enhance the possibilities of developing radical innovations. The entry of new countries to industrial tile production has increased the threat for firms in both countries although more dramatically in Spain.

REFERENCES

Alkemade, F., Kleinschmidt, C., & Hekkert, M. P. (2007). Analysing emerging innovation systems: A functions approach to foresight. *International Journal of Foresight and Innovation Policy*, *3*(2), 139–168. doi:10.1504/IJFIP.2007.011622

Becattini, G. (1979). Dal Settore Industriale al Distretto Industriale. Alcune considerazioni sull'unitá di indagine in economia industrial. *Revista di Economia e Politica Industriale*, *1*, 7–14.

Becattini, G. (1990). The Marshallian industrial district as a socio-economic notion. In F. Pyke, G. Becattini, & W. Sengenberger (Eds.), *Industrial districts & inter-firm co-operation in Italy* (pp. 37–51). Genève: International Institute for Labour Studies.

Becattini, G. (2002). Industrial sectors & industrial districts: Tools for industrial analysis. *European Planning Studies*, *10*(4), 483–493. doi:10.1080/09654310220130194

Bellandi, M. (1996). Innovation & change in the Marshallian Industrial District. *European Planning Studies*, *4*(3), 357–369. doi:10.1080/09654319608720351

Bellandi, M. (2002). Italian industrial districts: An industrial economics interpretation'. *European Planning Studies*, *10*(4), 425–437. doi:10.1080/09654310220130158

Bellandi, M. (2006). A perspective on clusters, localities, & specific public goods. In C. Pitelis, R. Sugden, & J. R. Wilson (Eds.), *Clusters & Globalisation: the Development of Urban & Regional Economies* (pp. 96–113). Cheltenham: Edward Elgar. doi:10.4337/9781847200136.00012

Bergek, A., Jacobsson, S., Carlsson, B., Lindmark, S., & Rickne, A. (2008). Analyzing the functional dynamics of technological innovation systems: A scheme of analysis. *Research Policy*, *37*(3), 407–429. doi:10.1016/j.respol.2007.12.003

Breschi, S., & Malerba, F. (1997). Sectoral systems of innovation: Technological regimes, Schumpeterian dynamics & spatial boundaries. In C. Edquist (Ed.), *Systems of Innovation* (pp. 130–156). London: Frances Pinter.

Brusco, S. (1990). The idea of the industrial district: its genesis. In *Pyke, F., Becattini, G. & Sengenberger, W. Industrial districts & inter-firm co-operation in Italy* (pp. 10–19). Genève: International Institute for Labour Studies.

Brusco, S., & Pezzini, M. (1990). *Small-scale enterprise in the ideology of the Italian left' n Pyke. In F., Becattini, G. & Sengenberger, W. (Eds.), Industrial districts & inter-firm co-operation in Italy* (pp. 142–159). Genève: International Institute for Labour Studies.

Callon, M. (1992). The dynamics of techno-economic networks. In R. Coombs & V. Walsh (Eds.), *Technical change & company strategies: Economic & sociological perspectives* (pp. 72–102). San Diego: Harcourt Brace Jovanovich Publishers.

Carlsson, B. (Ed.). (1997). *Technological systems & industrial dynamics (No. 10)*. Springer Science & Business Media. doi:10.1007/978-1-4615-6133-0

Carlsson, B., & Stankiewicz, R. (1991). On the nature, function & composition of technological systems. *Journal of Evolutionary Economics*, *1*(2), 93–118. doi:10.1007/BF01224915

Carlsson, B., & Stankiewicz, R. (1995). On the Nature, Function & Composition of Technological Systems. In B. Carlsson (Ed.), *Technological systems & economic performance: the case of factory automation*. Boston, Dordrecht, London: Kluwer Academic Publishers. doi:10.1007/978-94-011-0145-5_2

Cooke, P., Uranga, M. G., & Etxebarria, G. (1997). Regional innovation systems: Institutional & organisational dimensions. *Research Policy*, *26*(4), 475–491. doi:10.1016/S0048-7333(97)00025-5

de Souza, D. M., Lafontaine, M., Charron-Doucet, F., Bengoa, X., Chappert, B., Duarte, F., & Lima, L. (2015). Comparative Life Cycle Assessment of ceramic versus concrete roof tiles in the Brazilian context. *Journal of Cleaner Production*, *89*, 165–173.

Edquist, C. (Ed.), (1997). *Systems of innovation; Technologies, Institutions, & Organizations*. London: Pinter.

Edquist, C. (2004). Systems of Innovation: Perspectives & Challenges. In J. Fagerberg, D. Mowery, & R. R. Nelson (Eds.), *Oxford Handbook of Innovation* (pp. 181–208). Oxford: Oxford University Press.

Fagerberg, J. (2003). Innovation: A Guide to the Literature. *Working Papers on Innovation Studies 20031012, Centre for Technology, Innovation and Culture*, University of Oslo.

Fernández, I. Gabaldón, D. & Gómez, C. (2005). *La innovación en el sector de pavimentos y revestimientos cerámicos de la Comunidad Valenciana*. Valencia: Alto Consejo Consultivo en I+D de la Presidencia de la Generalitat Valenciana.

Fernández de Lucio, I., & Gabaldón-Estevan, D. (2007). Nous reptes per a la indústria ceràmica. At Observatori Estadístic de l'EURAM; Institut d'Economia i Empresa Ignasi Villalonga (Eds.) Recerca i innovació a l'EURAM. València: Tres i Quatre, pp. 37-50.

Freeman, C. (1982). *The Economics of Industrial Innovation*. London: Pinter.

Freeman, C. (1995). The 'National System of Innovation' in historical perspective. *Cambridge Journal of Economics*, *19*(1), 5–24.

Freeman, C., & Soete, L. (Eds.). (1987). *Technical Change & Full Employment*. Oxford: Basil Blackwell.

Gabaldón-Estevan, D. (2011). *El sistema distritual de innovación cerámico de Castellón*. Universitat de València. Servei de Publicacions.

Gabaldón-Estevan, D., Criado, E., & Monfort, E. (2014a). The Green Factor in European Manufacturing: A case study of the Spanish ceramic tile industry'. *Journal of Cleaner Production*, *70*, 242–250. doi:10.1016/j.jclepro.2014.02.018

Gabaldón-Estevan, D., Fernández-de-Lucio, I., & Molina-Morales, F. X. (2012). Sistemas distrituales de innovación. *Arbor*, *188*(753), 63–73. doi:10.3989/arbor.2012.753n1005

Gabaldón-Estevan, D., Fernández de Lucio, I., & Tortajada-Esparza, E. (2007, June 18-20) How Does Structure Affect Industrial Districts Innovation? *Danish Research Unit for Industrial Dynamics Summer Conference.* Copenhagen, (Denmark). Retrieved from http://digital.csic.es/handle/10261/10172

Gabaldón-Estevan, D., & Hekkert, M. P. (2013). How does the innovation system in the Spanish ceramic tile sector function?' *Boletín de la Sociedad Española de Cerámica y Vidrio*, *52*(3), 151–158. doi:10.3989/cyv.202013

Gabaldón-Estevan, D., Mezquita, A., Ferrer, S., & Monfort, E. (2014b) Is European Union Environmental Policy Efficient at Promoting a Post-carbon Industry? The Case of Energy in the European Ceramic Tile Sector. *Proceedings of the 11th International Conference on Innovation & Management* (pp. 104-113). Wuhan University of Technology Press.

Gabaldón-Estevan, D., Molina Morales, F. X., & Fernández de Lucio, I. (2009). Research contracts and patents in the Spanish ceramic tile district innovation system. *INGENIO Working Papers Series.* 2009-1

Gabaldón-Estevan, D., Tortajada-Esparza, E., & Fernández de Lucio, I. (2008). *El papel de las fritas, esmaltes y colores en la evolución del sector cerámico de la Comunidad Valenciana.* Generalitat Valenciana.

Hekkert, M. P., Suurs, R. A. A., Negro, S. O., Kuhlmann, S., & Smits, R. E. H. M. (2007). Functions of Innovation Systems: A new approach for analyzing technological change. *Technological Forecasting and Social Change*, *74*(4), 413–432. doi:10.1016/j.techfore.2006.03.002

Hughes, T. (1984). The evolution of large technological systems. In W. Bijker, T. Hughes, & A. Johnson (Eds.), (2001). *Functions in innovation systems approaches. In electronic paper at the proceeding of the Nelson and Winter conference*, Aalborg. Retrieved from http://www.druid.dk/conferences/nw/paper1/a_johnson.pdf

Lundvall, B. A. (1992). User-producer relationships, national systems of innovation & internationalisation. *National systems of innovation: Towards a theory of innovation & interactive learning* (pp. 45-67).

Malerba, F. (2002). Sectoral systems of innovation & production'. *Research Policy*, *31*(2), 247–264. doi:10.1016/S0048-7333(01)00139-1

Malerba, F. (2004). Sectoral systems of innovation: basic concepts. In F. Malerba (Ed.), *Sectoral systems of innovation: Concepts, Issues & Analyses of Six Major Sectors in Europe* (pp. 9–41). Cambridge: Cambridge University Press. doi:10.1017/CBO9780511493270.002

Meyer-Stamer, J., Maggi, C., & Seibel, S. (2004) Upgrading in the tile industry of Italy, Spain, and Brazil: insights from cluster and value chain analysis. In H. Schmitz (Ed.), Local enterprises in the global economy: issues of governance and upgrading (pp. 174-199). Cheltenham (Reino Unido): Edward Elgar.

Mezquita, A.; Monfort, E.;Vaquer, E.; Ferrer, S.; Pitarch, J.M.; Arnal, M.A.; Cobo, F. (2014) Reduction of CO2-emissions in ceramic tiles manufacture by combining energy-saving measures. *Cfi Ber. DKG 85*, 91(5), E37-E42

Minguillón, M. C., Monfort, E., Escrig, A., Celades, I., Guerra, L., Busani, G., & Querol, X. et al. (2013). Air quality comparison between two European ceramic tile clusters. *Atmospheric Environment*, *74*, 311–319. doi:10.1016/j.atmosenv.2013.04.010

Monfort, E., Mezquita, A., Vaquer, E., Mallol, G., & Gabaldón-Estevan, D. (2014). La evolución energética del sector español de baldosas cerámicas. *Boletín de la Sociedad Española de Cerámica y Vidrio*, *53*(3), 111–120. doi:10.3989/cyv.152014

Negro, S. O. (2007). *Dynamics of Technological Innovation Systems - The case of biomass energy*. Innovation Studies.

Negro, S. O., & Hekkert, M. P. (2006). Benchmarking the Functions of Innovation Systems - The Success Story of the German Biomass Digestion Innovation System. *SPRU 40th Anniversary*. Retrieved from http://www2.unil.ch/easst2006/Papers/N/Negro%20and%20Hekkert%20Paper%20EASST%202006.pdf

Negro, S. O., & Hekkert, M. P. (2008). Explaining the success of emerging technologies by innovation system functioning: The case of biomass digestion in Germany. *Technology Analysis and Strategic Management*, *20*(4), 465–482. doi:10.1080/09537320802141437

Negro, S. O., Hekkert, M. P., & Smits, R. E. (2007). Explaining the failure of the Dutch innovation system for biomass digestion--A functional analysis. *Energy Policy*, *35*(2), 925–938. doi:10.1016/j.enpol.2006.01.027

Negro, S. O., Hekkert, M. P., & Smits, R. E. H. M. (2008a). Stimulating renewable energy technologies by innovation policy. *Science & Public Policy*, *35*(6), 403–416. doi:10.3152/030234208X323334

Negro, S. O., Suurs, R. A., & Hekkert, M. P. (2008b). The bumpy road of biomass gasification in the Netherlands: Explaining the rise & fall of an emerging innovation system. *Technological Forecasting and Social Change*, *75*(1), 57–77. doi:10.1016/j.techfore.2006.08.006

Nelson, R. R. (Ed.). (1993). *National Innovation Systems*. New York: Oxford University Press.

Nelson, R. R., & Winter, S. G. (1982). *An evolutionary theory of economic change*. Cambridge: Belknap Press.

OECD. (1997). *National Innovation Systems*. Paris: OECD Publications.

Penteado, C. S. G., de Carvalho, E. V., & Lintz, R. C. C. (2015). Reusing ceramic tile polishing waste in paving block manufacturing. *Journal of Cleaner Production*. doi:10.1016/j.jclepro.2015.06.142

Porter, M. E. (1998). Clusters & the new economics of competition. *Harvard Business Review*, *76*(6), 77–90. PMID:10187248

Pyke, F., & Sengenberger, W. (Eds.), (1992). Industrial districts & local economic regeneration. International Labour Organisation.

Rogers, E. M. (2010). *Diffusion of innovations*. Simon and Schuster.

Rosenberg, N. (1982). *Inside the Black Box: Technology & Economics*. Cambridge: Cambridge University Press.

Russo, M. (1996). *Cambiamento tecnico e relazioni tra imprese. Il distretto cerámico di Sassuolo*. Turin: Rosenberg y Sellier.

Schmitz, H. (2004). Globalized localities: introduction. In H. Schmitz (Ed.), *Local enterprises in the global economy: issues of governance and upgrading* (pp. 1–19). Cheltenham, United Kingdom: Edward Elgar. doi:10.4337/9781843769743.00008

Suurs, R. A. A. (2009). *Motors of sustainable innovation - Towards a theory on the dynamics of technological innovation sytems*. Department of Innovation & Environmental Sciences.

Suurs, R. A. A., & Hekkert, M. P. (2007). *Cumulative Causation in the Formation of a Technological Innovation System: Biofuels Development in the Netherlands*. Working Paper /Innovation Studies Utrecht (ISU) /working paper series.

Tortajada Esparza, E., Fernández de Lucio, I., & Gabaldón Estevan, D. (2008a). Competitiveness and Profitability – New Challenges in the Ceramic Frit. *Glaze and Colour Industry, Part 1 Interceram, 57*(6), 436–439.

Tortajada-Esparza, E., Fernández de Lucio, I., & Gabaldón-Estevan, D. (2008c) Competitividad y rentabilidad. Nuevos retos de la industria de fritas, colores y esmaltes cerámicos. At Qualicer 2008, X Congreso Mundial de la Calidad del Azulejo y del Pavimento Cerámico. Cámara Oficial de Comercio, Industria y Navegación de Castellón

Tortajada Esparza, E., Fernández de Lucio, I., & Gabaldón Estevan, D. (2009). Competitiveness and Profitability – New Challenges in the Ceramic Frit. *Glaze and Colour Industry Part 2 Interceram, 58*(1), 36–39.

Tortajada Esparza, E., Gabaldón Estevan, D., & Fernández de Lucio, I. (2008b). La evolución tecnológica del distrito cerámico de Castellón: La contribución de la industria de fritas, colores y esmaltes. *Boletín de la Sociedad Española de Cerámica y Vidrio, 47*(2), 57–80. doi:10.3989/cyv.2008.v47.i2.198

Woolcock, M. (2001). The place of social capital in understanding social & economic outcomes'. *Isuma: Canadian Journal of Policy Research, 2*(1), 11–17.

Zhou, J., Shu, Z., Tiantian, L. I., Dongxue, Y. U., Sheng, Z., & Wang, Y. (2015). Novel fabrication route for non-fired ceramic tiles only using gypsum. *Ceramics International, 41*(7), 9193–9198.

KEY TERMS AND DEFINITIONS

Bulk Commodities: Companies based on minimal product and process innovation, which exploit equipment and input suppliers as the sources of their innovation.

Complex Systems Producers: Companies whose technical advances tend to be incremental improvements to components and systems designs.

Chemical Products Producers: Companies that innovate mainly through the introduction of new products.

District Innovation System: A system of relationships within an industrial district where externalities facilitate firms' innovation processes.

ENDNOTES

1 Laminam, developed by a leading Italian machinery producer named System, is a recent product that has revolutionised the tile industry. The production process for Laminam, named Sinterflex, is fully automated, from the pouring of the clay into the line to storage. The product requires one third of the material inputs of conventional tiles including energy and water. It is 3mm thick and weighs only one-third of the weight of a conventional tile. It can be produced in bigger size, 1x3 metres, which can be cut in different shapes. Despite being a considerable innovation its success has been small with only about 5 enterprises producing it, due to the reluctance among tile producers to adopt this new technology.

2 One of the areas where R&D effort is concentrated is improvement to injection technology for tile decoration where obstruction is a problem. Another technology is laser also used in decoration. Other developments, e.g. domotic, heating or conductor properties, antibacterial, non-slip are being studied though the low estimated demand and the high costs of production render them not very appealing. All these developments are mostly developed by suppliers often in association with technological partners and the role of public R&D institutions is marginal.

Chapter 6
Modeling and Synthesis of Supply Chain Networks Using High-Level Petri Nets

Gen'ichi Yasuda
Nagasaki Institute of Applied Science, Japan

ABSTRACT

This chapter presents a systematic methodology for modeling and control of supply chain networks, especially focusing on formal representation and control synthesis aspects. A supply chain is represented as a discrete event dynamic system. The use of high-level Petri nets is proposed to formulate event related rules commonly seen in supply chains and analyze cause-effect relationships between events. Petri nets have been successfully introduced as an effective tool for describing control specifications and realizing the control in automated supply chain processes. The extended net representation of the productive task flows can provide more synthetic specifications for consistent management and control of supply chain systems by a top-down refinement methodology. Software implementation is described to simulate and control real supply chain processes.

INTRODUCTION

This chapter presents a systematic methodology for modeling and synthesis of supply chain networks, especially focusing on formal representation and control aspects. In the last two decades, manufacturing industry has experienced some notable changes from flexible manufacturing through lean manufacturing towards agile manufacturing, to seek to produce a high level of throughput with a minimum of inventory, lowering waste and enhancing productivity on the factory floor. The aim of Just-In-Time control is that in-process inventories have to be as small as possible, which means that at the exact time when a request is received by a machine, the required material and part has to be available in its input buffer,

DOI: 10.4018/978-1-5225-0021-6.ch006

ready to be pushed. This kind of optimization can only result from a real-time synchronization between the operations in the manufacturing system (Suda, 1989). As production systems move toward more flexible for smaller batch sizes and shorter product cycles to cope with highly demanding customer orders, more advanced control systems are needed. To develop optimized architectures to control and supervise the production system by organizing the cooperation between facilities that are geographically distributed, the system architecture has changed from centralized processing to distributed processing in order to meet the requirements of lean and agile software development. In place of early process oriented approach, following the data-oriented approach, in 90s the object-oriented approach was highlighted. Supported by the feasibility of novel distributed control architectures (Duffee & Prabhu, 1996; Zamei et al., 1997), advanced large and complex production systems have a holonic or self-organizing and self-adaptive hierarchical structure like bionic systems. According to not only abilities of computers and networks but also needs from the business environment based on the Internet popularization, the service oriented architecture has been successfully applied to supply chain management, order entry system, etc. However, failures may occur due to an inadequate specification of services with particular real-time requirements in a supply chain. In this context, a formal approach is strongly required to treat the integration of distributed system components loosely coupled in a heterogeneous environment such as a supply chain network. (Komoda, 2006; Vollman et al., 2004; Yu et al., 2014; Zeng et al., 2014 ; Zhang et al., 2004).

Yasuda is the first that originated conceptual and synthetic researches towards realization of non-centralized systems (Yasuda & Mori, 1974). Non-centralized systems or center-less systems are featured as self-coordinated systems without any main agent or any centralized decision-making mechanism against unpredicted, emergency situations. In supply chains, because each facility has some decision making abilities with respect to input and output of productive items as an independent agent, non-deterministic decision or task selection for cooperation should be essentially taken into account. So, the supply chain management system is one of typical examples of non-centralized systems, which can be commonly found in biological organisms and societies, although, in real biological systems, non-centralized and centralized systems alternate with each other as the system evolves. In this context, web services protocol through Internet can afford to provide communication tools for non-centralized realization of supply chain systems.

This chapter discusses the specification problem for real-time monitoring and control of different productive processes involved with distributed supply chains, aiming at realizing non-centralized control. Petri nets are a modeling tool for describing information processing systems characterized as being concurrent, asynchronous and nondeterministic, based on discrete events and state change. Beside the mathematical formalism, it can simulate the dynamic behaviors of the system graphically through the flow of tokens. Thus, a Petri net approach can be applicable to obtain an easily understandable and systematic procedure. In the chapter, the specification task is based on the characterization of automated supply chain processes as a discrete event system, and then the use of high-level extended Petri nets based on composed condition-event nets is proposed to formulate event related cause-effect rules commonly seen in supply chains, aiming to synthesize strategic and tactical operations in the hierarchical distributed control of supply chains in order to react events in a real-time manner. Because Petri nets can provide a structural design methodology in production systems through a top-down and/or bottom-up relationship between abstract specification and practical implementation (Moreira & Basilio, 2014), (Yasuda, 2011), a unified Petri net based methodology is proposed to design and implement distributed control software corresponding to the hierarchical structure of the overall supply chain system. Software implementations are also described to simulate and control real supply chain processes.

BACKGROUND: MAIN IDEAS OF HIERARCHICAL APPROACH TO MODELING SUPPLY CHAIN NETWORKS

Manufacturing supply chain networks are formed out of complex interconnections amongst various manufacturing and service providers, such as raw material vendors, manufacturers, and logistics and warehouses operators, distributors, and retailers. Supply chain processes encompass the full range of intra- and inter-company activities, located in geographically different places, beginning with raw material procurement, through manufacturing and distribution, and concluding with delivery of the products. The supply chain process transits several facilities and each time a transition is made, logistics is involved, taking the form of inventory packaging, warehousing, transportation, etc. Supply chain management can be treated as coordination of all the activities involved in the supply chain processes. In this chapter, a supply chain network is viewed as the top-level in a hierarchical production system, where the performance depends on all the constituents of the supply chain. The network design problems concern the establishment of the supply chain network at the strategic and tactical levels, including the long term decisions choosing order and stock policies and real-time operational policies of various facilities (Aalst, 1998; He et al., 2011). A supply chain is composed of generic objects, suppliers, manufacturers, retailers, etc., having all structural and functional characteristics of the relevant production resources. Because comprehensive models for supply chain management are very large scale, it is often difficult to formulate and solve optimization programs. So, supply chains need to be modeled at various levels of abstraction for strategic and tactical levels decisions. To cope with complexity of overall production system, a hierarchical architecture groups the elements of the system into levels: supply chain, factory, shop, cell, workstation and equipment levels. Inspired by biological evolution, in the hierarchical structure, each object is a recursive structure able to develop models of production flows and corresponding activities at different levels of abstraction and hierarchy, according to the level of detail needed in the model, at the same time as functions generated by lower-level objects. Each of the facilities in a supply chain performs its scheduling under autonomous control in a decentralized fashion, where materials and information flows are managed by rule-based procedures, negotiations, or decision making through interfaces between facilities (Clempner, 2014; Dong & Chen, 2001; Serral et al., 2015; Valckenaers et al., 1997).

In this chapter, the task flow net is proposed as a conceptual representation of the supply chain control system. Based on the hierarchical approach, an interpreted form of the Petri net which includes the required control conditions is introduced to efficiently detail the functions of the supply chain system's elements identified by the task flow net. The main advantages of modeling and simulation using the conceptual object in hierarchical supply chain systems can be summarized in the following points: the reusability of the generic objects by the systematic refinement over the different levels of abstraction and the modular modeling property considering the use of recursive structures of the overall system (Zhou et al., 1998). In actual supply chain systems, productive processing controllers are structured in a hierarchy, so there is a close relationship between the conceptual simulation objects and the actual objects. The hierarchical approach offers a simple model building and provides possibilities to reduce the size of the specifications. In this context, a hierarchical formalism should be adopted, which allows some structural and behavioral properties of a model to be formally proved.

Hierarchical and distributed control (Cho & Lim, 1999; Yasuda, 2011) is considered to provide a methodology for the design and operation of flexible, reconfigurable control in the supply chain. Conventional supply chain is controlled and managed based on predetermined resource functions, production sequences, centralized task-resource assignments, and fixed predetermined planning and scheduling strategies. For a supply chain to be well suited to the requirements in mass customization environments, the operation strategy should maximize robustness against breakdown and responsiveness to order variation based on the design of resource control functions specified independent of any products to be manufactured. Inspired by main concepts of non-centralized self-coordinated control behind biological, hierarchical autonomy and cooperation, hierarchical and distributed control represent no fixed relationship between control functions such as machine control, shop floor control, scheduling and planning. Rather, these control functions emerge through the interaction or negotiation between distributed decision making agents which correspond to the physical entities involved in the supply chain environment such as resources (robots and machines), products (recipes) and orders, for example forming a temporary supply chain route and adjusting that route in case of a resource failure (Jules et al., 2015; Tao et al., 2013). Thus, in place of solution specific software for planning, scheduling, execution and control in a solution-oriented supply chain environment, software capabilities for supporting distributed decision making embedded in resource agents for a pool of resources with known relevant capabilities are required.

PETRI NET MODELING OF MANUFACTURING SUPPLY CHAINS

From an abstract viewpoint, a supply chain system is characterized by the flow of discrete items such as raw materials, products, and information concerning order, stock, etc. which pass in ordered form through subsystems, or facilities, and receive appropriate operations to achieve the purpose of supply chain management (Dotoli et al., 2005; Viswanadham & Raghavan, 2000). Concepts such as non-determinism, concurrence, parallelism and synchronism are the basis on which real models of supply chains are depicted. Petri nets are a modeling technique that captures the causal relationships between events and conditions, firstly proposed by C. A. Petri for communication protocol between automata in the early 1960s (Petri, 1962). They are commonly used to model concurrent processes such as the information and control in programmable systems. A Petri net comprises two types of nodes, places representing conditions or states and transitions representing events, which are interconnected by directed arcs. Tokens, which reside at the places, are used to indicate the instantiation of a state. The current state of a net is represented by the distribution of tokens in the net, which is called the marking which maps places to the number of tokens in each place. In a Petri net, the places, transitions and tokens are represented by the circles, bars and dots respectively. In basic Petri nets (Peterson, 1981), a transition is enabled if each of its input places contains the specified number of tokens according to input arc weights. When a transition is enabled, it can possibly fire. If it fires, tokens are removed from each of its input places and placed in each of its output places according to input and output arc weights, respectively. The marking of a Petri net changes when a transition, which is enabled, eventually fires.

Because Petri nets are a generalized and uninterpreted technique to represent discrete event systems, to model a specific system, an interpreted form of the Petri net is necessary to clearly model the features of the system, both for the net structure elements and the structure composed of these elements. Condition-event nets, where each place can contain not more than one token, are very useful in the analysis of the qualitative behavior of multi-sequence control systems, because this allows a one-to-one semantic correspondence between places and conditions. Thus, a Petri net representation of a system can be used as a design schema as well as a supervising framework (Costa et al., 2010; David & Alla, 2005; Lianfeng et al., 2009; Villani et al., 2006), a lowest-level real-time control algorithm (Yasuda et al., 1996; 1997), etc. For modelling event relationships with respect to timing, or delays, in a supply chain by Petri nets, a time-interval of delay is associated with a transition, which indicates the minimum and maximum time the transition must wait for before firing after enabled. A workflow is represented using tokens of different colors for each order or product and arcs labelled with colors. Based on the above formalism, Petri nets are helpful to analyse performance measures such as cycle time, work-in-process or on-time delivery performance for different scheduling and work release policies, in supply chain management (Balsamo et al., 2012; Zegordi & Davarzani, 2012). The drawback in applying this technique to the modeling and analysis of supply chain systems is its expressiveness, principally when a big number of conditions are involved. Unfortunately, modeling real production systems frequently implies the representation of a great number of conditions, and leads to complex net structures. So, some high-level representation technique is required.

Similar to conventional Petri net based controllers (Murata et al., 1986), a gate is introduced to control the firing of an activity in a supply chain as a relation between the activity and an internal or external condition, whether the condition belongs or not to the system being modeled respectively, which is represented as another class of objects. An external condition is a rule that is not part of the system in the sense that its satisfaction is not controlled by the target system. Internal and external gates can still be classified as permissive or inhibitive according to whether it can enable an activity when it is marked or when it is unmarked. A permissive external condition can be interpreted as a relation between an external precondition and an activity element in the system. On the other hand, an inhibitive external condition would stand for a relation between an activity element in the system and an external post-condition. A new state in a conventional condition-event net depends linearly on the previous one and on the events currently enabled. Since all the gates with persistent marking can be included in the incidence matrix, the high-level net state equation is similar to the conventional one by following the definitions above, consequently preserving most of the formalism and properties of Petri nets and improving their expressiveness. In a conventional synchronized Petri net (David & Alla, 2005), an event is associated with a transition and the firing of this transition occurs if the transition is enabled and when the associated event occurs. With defined input and output signals of the controller, a deterministic, live and safe Petri net can be designed as a control Petri net, such that the net receives information from the environment and sends output signals to it, where inputs are associated with transitions and outputs are associated with places. In situations such as conflicts, the permissive or inhibitive arc is used effectively to enforce priorities. Conflicting situations can be solved by a concurrent firing if they are in the concurrent list of each other, by a firing policy deduced from the application, or with the inclusion of an arbiter to solve conflicting situations. Information for simulation or execution of a net model includes definition of logical variables (places, input and output signals and their initial values, and description of transitions with their pre-conditions and post-conditions, which describes basic elements of a net and shows how tokens flow through places as a rule based logical model for model verification.

Synchronization of activities between facilities or stations in a supply chain network can be performed with gate signals. Generally, decomposition of a task net model permits splitting the net into separate concurrent modules, or state machines, which can be implemented independently in various devices or subsystems. Several known methods of the decomposition of Petri nets can be applied, such as invariants computation using linear integer programming and reachability set analysis (Aalst, 2013; Wisniewski et al., 2014). In supply chain networks, a strategy for decomposing a net may be adopted according to the geographical distribution of facilities and stations or the similarities of the subnets. Synchronization of two decomposed modules is performed based on asynchronous communication; the transformation is intuitionally obvious considering a non-zero time delay in execution (firing) between the master transition and the slave transition involved in the same communication channel in different platforms with different execution ticks (Yasuda, 2002; Uzam et al., 2005). Synchronizing signals are issued from the input places of the shared transition (slave) in one of the modules as 'ready to synchronize' status. The shared transition (master) in the other module is guarded by the synchronizing signals, then, the output place generates the synchronizing signal for the first module. Finally the output transition is guarded by the negation of the first synchronization signal. For synchronization of more than two decomposed modules, a coordinator net for synchronization can be realized to generate the synchronizing signal that enables shared transitions in all decomposed modules. The coordinator net is realized as part of the higher-level net model, or it could be temporarily created in one of the decomposed modules with some selection rule such as the highest responsiveness. In the latter case, for every shared transition a separate coordinator is created, called distributed coordinators. With the above, the decomposed modules are not modified, maintaining their primary structure.

REPRESENTATION OF BASIC FACILITY MODULES IN SUPPLY CHAINS USING HIGH-LEVEL PETRI NETS

A supply chain system is composed of a set of resources for dedicated productive purposes, that is, physical transformations such as machining, assembling, transportation, and buffering operations. These resources are classified into two categories: active resources such as production machines and transportation vehicles and passive resources such as buffers and workplaces. Various resources are interconnected in a supply chain and controlled so that some items are passed and others are circulated to achieve the supply chain purpose. The functions of active resources can be viewed abstractly as activities. An activity starts when all the specified items are arrived, and after the working time it ends and as a result produces particular items at the same time. A sub-flow of items may intervene into an activity in a main flow. Sub-flows can be seen as involved with an activity in a lower level. After an item exits from an activity and then it enters the next activity, passing through a passive resource such as a buffer. Thus, the functions of passive resource can be viewed abstractly as distributions. An item can not change its physical characteristics in a distribution. Thus an activity is distinguished from a distribution. From the viewpoint of discrete event process control, an overall supply chain process can be represented as a set of distinct activities or events and distributions or conditions mutually interrelated in a complex form. The abstract concepts of activities and distributions are combined with the pure objects of conditions and events in the condition-event net. Thus, high-level activity-based Petri nets are defined to hierarchically extend the condition-event formalism to the modeling and design phases of supply chains. A high-level net is represented as a bipartite graph, similarly to conventional condition-event nets.

Based on the object-oriented approach, the class ‘activity’ is defined as the abstract active resource. Each activity is represented by an object characterized by a set of attributes. Thus, a simple activity is an alternative definition of an event, except that a reference list of other system processes that should be executed concurrently can be specified. A compound activity is defined by two simple activities, called start and end activities respectively, and an activity place which stores estimated time of processing and a pointer to another net representing the model of the specified supply chain process. Once an estimated time argument is provided, the model of an activity could be simulated as a compound activity with input and output events separated by the activity place.

The abstract activity is classified into the following levels: action, operation, and task. At the action level, primitive actions executed by machines and devices are considered. At the operation level, operations executed by one or several subsystems are considered. At the task level, integrated activities as supply chain behaviors are considered, such as processing, assembly, disassembly, treatment, packaging, etc.

The class ‘distribution’ is defined as the abstract passive resource. A condition in a conventional condition-event net is a simple object instance of the class ‘distribution.’ A passive resource is generally defined as the system workplace where items or parts can be temporary held or stored in a stable position. In a high-level net, each passive resource may be single or composed. Composed resources manage the information attached to a set of single resources closely located and functionally equivalent. A resource contains information about the work zone if it can be used as a mutual exclusion resource. Thus specific ways to store items, such as FIFO, LIFO, etc., can be represented in the subnet of the resource. Assembling and disassembling processes which only affect the item flow are abstracted in the subclasses of the class ‘distribution’. Thus, assembling stations and storage mechanisms such as intermediary buffers are considered static compound resources as macro-places and can be reduced to conditions in the condition-event net.

There are some subclasses of the class ‘distribution’. A subclass ‘timer’ inherits all the attributes of ‘distribution’ and has an extra parameter with the estimated time to enable the firing. Similarly ‘capacity’ is the maximum number of marks it could accept. There are also other subclasses for capacity; assembling and disassembly capacity according to input-output relation values. The attribute of the subnet is a pointer to a subnet. By defining subclass with other input-output relation values, the relationship between conceptual distribution and extended ones can be established. One of the features of extended

Petri nets as well as high-level nets is the capability of representing numerable items or conditions. The capacity is the basic element capable to hold multiple tokens. Any other places can be represented by a subnet of the basic elements, starting and ending with a condition.

Activities in a high-level net are redefined using the extended conditions, or distributions, as shown in Figure 4. In simple activities, only one activity element acts. In concurrent activities, up to N elements act independently and concurrently by starting and ending their execution through the common initial and final events. In assembling activities, each of N elements starts through the same event and ending simultaneously through the common final event. In disassembling activities, N elements start simultaneously at the common event and end separately through the same event.

Activities and distributions are hierarchically structured, and each activity can be substituted by a subnet starting and ending with an event, while each distribution element can be substituted by a subnet starting and ending with conditions. Hierarchy of activity is defined by introduction of substitution transitions. A substitution transition is a transition refined by another subnet with the common input and output transitions, which represents the detailed content of the activity. All the input and output places of the transition at the higher level are mapped into associated places in the lower-level subnet. For simulation, these corresponding places are glued together, since physically different but logical same places have always the same marking as a fusion set. Figure 5 also shows a hierarchical description of an activity, where the activity at the higher level constitutes a parallel construct with an activity path composed of several sub-activities at the lower level, since the net of the activity must be safe. In the hierarchical net model, if the activity content is refined, a pointer to the subnet is added to the object representation and internal gates connect and synchronize the aggregated subnet at the lower level, with two nets connected by gates and having two events (start and event) in common. Consequently, the separation of a large net into several subnets provides a representation amenable to a distributed implementation of a hierarchical net model.

Tokens in high-level nets are classified into flow and control tokens. Flow tokens describe the task flow in a supply chain. They are useful in representing input and output conditions for subtasks. Control tokens are used in representing actual system resources. The presence or absence of these tokens denotes resource availability. Communication messages for order are also indicated using control tokens. For real-time control, because the net models should be live such that their initial states are reachable repeatedly, a direct algorithm to find all the enabled events and some knowledge based approach as an arbiter, to choose a configuration for the enabling vector can be used.

Active resources, such as manufacturers and transporters in supply chains, return their initial states when the assigned activities end. In this way, at the start of an activity, the required resources must be available. Each active resource is represented as a unique distribution, where the number of tokens corresponds to the number of resources which can be assigned. The arc from the distribution to the input transition of the activity represents the request for the use of the resource when the activity starts, while the arc from the output transition of the activity to the distribution represents the release of the resource when the activity ends. Thus, such a distribution element represents the function of assignment of a resource involved with an activity.

CONTROL SYSTEM DESIGN FOR SUPPLY CHAINS

A particular implementation of the proposed procedure to define a supply chain process is presented through a simple example. The objective of the system is to automate and coordinate different activities or services executed in a distributed supply chain system composed of the following facilities: two suppliers, two manufacturers, and two retailers. These facilities are interconnected through a communication network to produce the customers' requested product. Figure 1 depicts an example representation of a production task flow using the high-level net.

Each supplier executes the service that carrying parts from a storage system in order in a pallet, and putting it into the output buffer. Then, the transporter moves the pallet to a manufacturer. When a pallet reaches the manufacturer, the manufacturing activity is started to produce the final product. The manufacturing service is carried out in three stages. Initially, the part is located in an appropriate base. Then, according with parts physical characteristics inspected using a camera, some identified parts are assembled. The assembled product is put on a free pallet to leave the system. Finally, products are carried out by the transporter. The type and quantity of products are defined and requested by a customer.

At the beginning, the supply chain system should be specified with the use of the high-level Petri net. The description permits further decomposition and formal verification of the whole net model. The whole design process can be fully automated. For the modeling and analysis, the supply chain process is hierarchically decomposed into constituent discrete sub-processes. By refinement of activities 'Supply', 'Manufacturing', 'Retailing', and 'Transportation' between facilities, the functional net models of the facility controllers are shown in Figures 2, 3, and 4, respectively. The functional net models are employed as the system controller to globally coordinate the activities of all the facilities in the supply chain. The conceptual net model composed of the manufacturing stations performing the machining task and assembly task is shown in Figure 5. The detailed net models of the machining and assembly tasks in the corresponding machining and assembly stations are shown in Figures 6 and 7, which are employed to locally coordinate the activities of the stations based on hierarchical and distributed control of supply chain.

Figure 1. Net representation of basic workflow in a supply chain

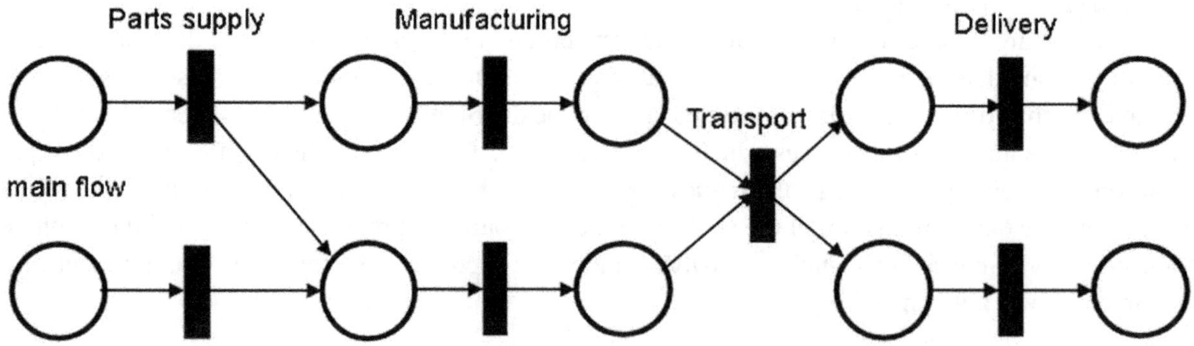

Figure 2. Net model of supply chain facilities: (a) 'Supplier', (b) 'Manufacturer'

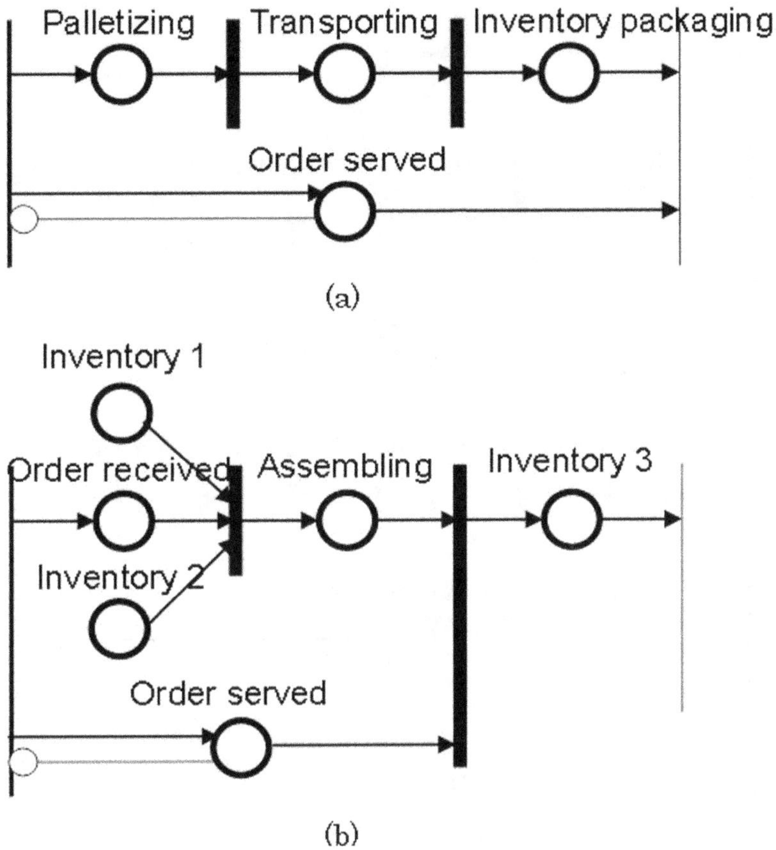

PETRI NET BASED IMPLEMENTATION OF SUPPLY CHAIN PROCESS MODULES

In supply chain simulation, on each supply chain step from supplier to retailer each product order agent should decide to which station or facility it should go. At the beginning of each step it contacts all relevant stations with a question which station could perform the next supply chain step. Then stations which can perform the step send the answer based on a situation of queue with following information: time to needed to perform the required operation, transport time to the station, time of waiting for the required operation. On the basis of the above information, the order agent decides which station to choose. Priority may be assigned to each order as importance of finishing supply chain processes. When the step is finished, the order agent is going to search for other station for the next supply chain step.

The proposed specification procedure is summarized as follows.

1. Specify task flows using activity elements and distribution elements.
2. Represent net models of the activity elements and distribution elements using extended high-level Petri nets.

Figure 3. Net model of supply chain facility 'Retailer'

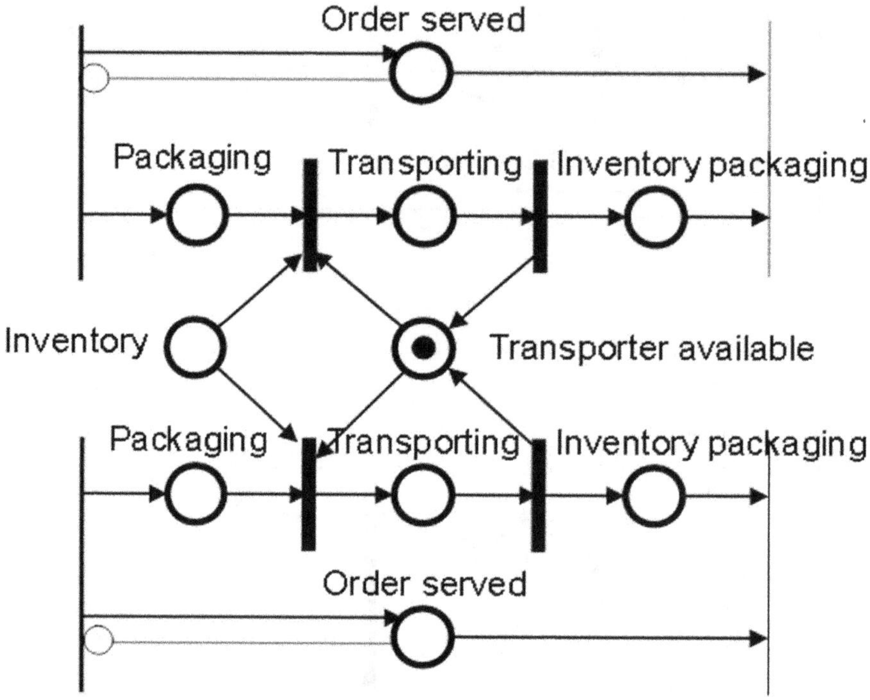

Figure 4. Net model of supply chain facility 'Distributor'

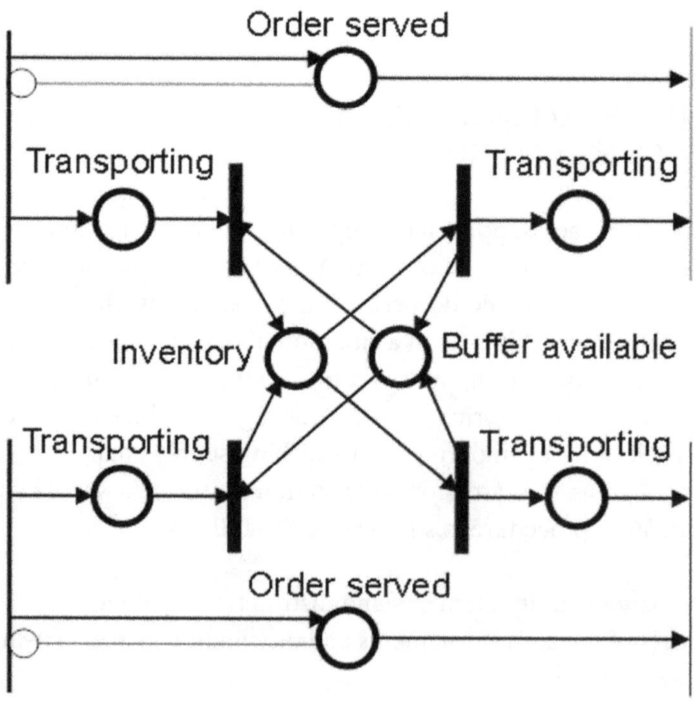

Figure 5. Conceptual net model of manufacturing stations: (a) machining task, (b) assembly task

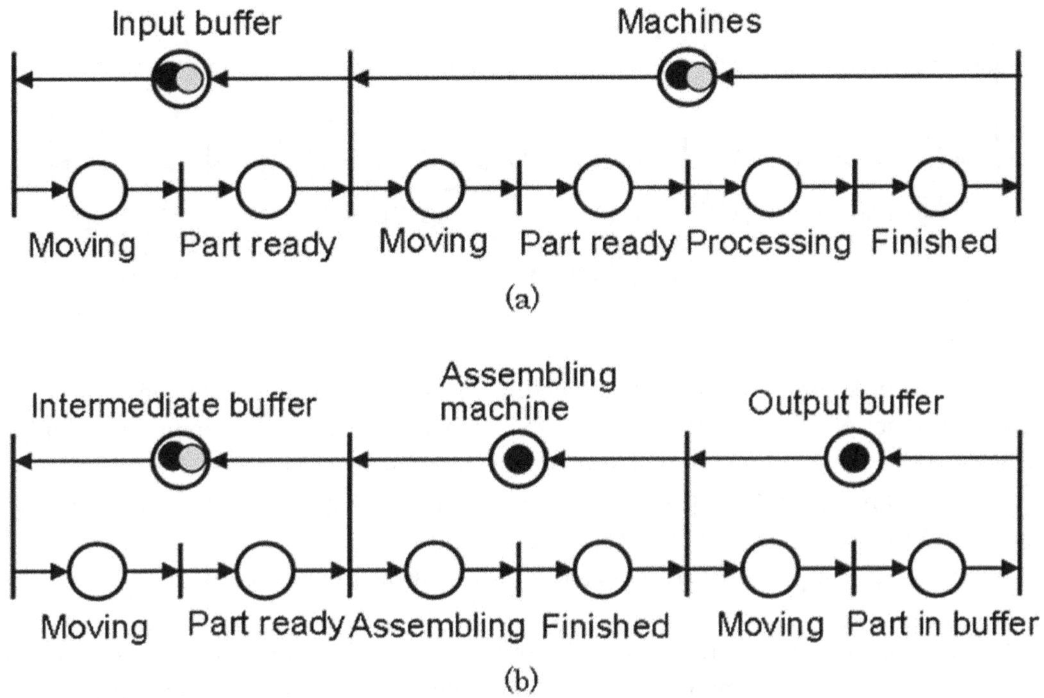

Figure 6. Net model of operation 'Moving part into buffer' in assembly task

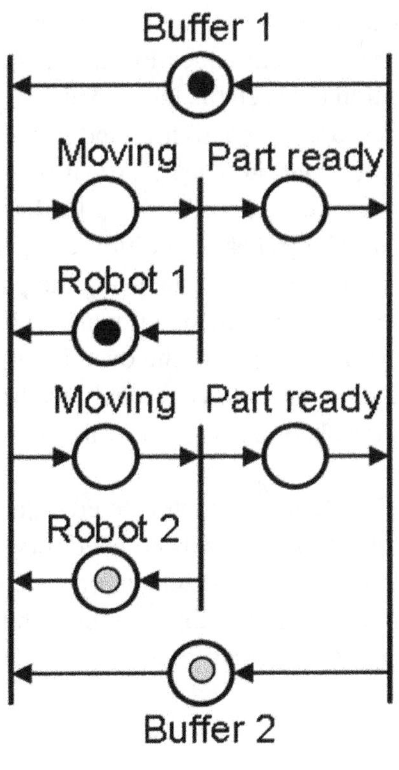

Figure 7. Net model of operation 'Inserting one part into another part' by one robot in assembly task

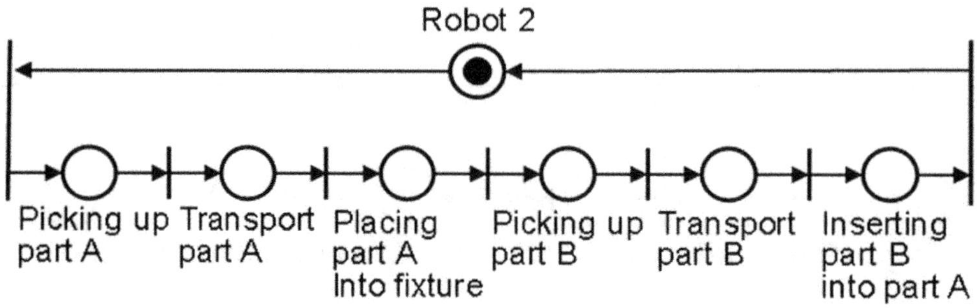

3. Transform the task flows into the detailed net models needed for simulation and control.
4. Add trigger functions for transition firing based on tactical operational conditions of supply chain modules.
5. Add the control places and net models that represent availability and exclusive assignment control (request and release) of the associated resources.
6. Add control places required for liveness and safeness of the overall net model.

The net based control architecture is composed of the system controller and distributed facility controllers. The system controller composed of the net executer and the net model, coordinates the activities of active resources in the production system based on transition firing rule and current marking. Facility controllers execute the commands from the system controller and send back the status response. In the hierarchical control structure, each net model of the two level controllers is implemented as an autonomously executable module with liveness, safeness and repeated reachability, and their concurrent activities are synchronized with inter-level communication based on the concept of command response.

Due to the complementary definition between the high-level nets and conventional Petri nets, different abstraction levels of composed resources, or activities and distributions, can be connected to single events and conditions in the same representation. The specifications of supply chain task represented by high-level nets using abstract resources are supposed to be filled later with proper detailed subnets in a top-down refinement methodology until it reaches a condition-event net form as the final representation. The verification of correctness of the net model is done using the plain condition-event net form at the final design stage. Aimed to validate and verify the functional requirements of the overall net model of the supply chain network, the functional modules in the overall net model are submitted to the model checking software for structural and dynamic behaviour analyses, based on Petri net properties such as deadlock reversibility, existence of unreachable states, where the formal analysis techniques are combined with simulation (Silva et al., 1996).

A prototyping distributed control system was implemented on a PC network over the Internet. The control software is integrated for a collection of loosely coupled service modules that communicate with each other using standard interfaces and message-exchange protocols based on the web service (Puttonen et al., 2013). Based on a set of standards to support interoperability among different platforms or operating systems, a web services engine enables different Service Oriented Architecture (SOA) imple-

Figure 8. Architecture of distributed controller for supply chain using the web service

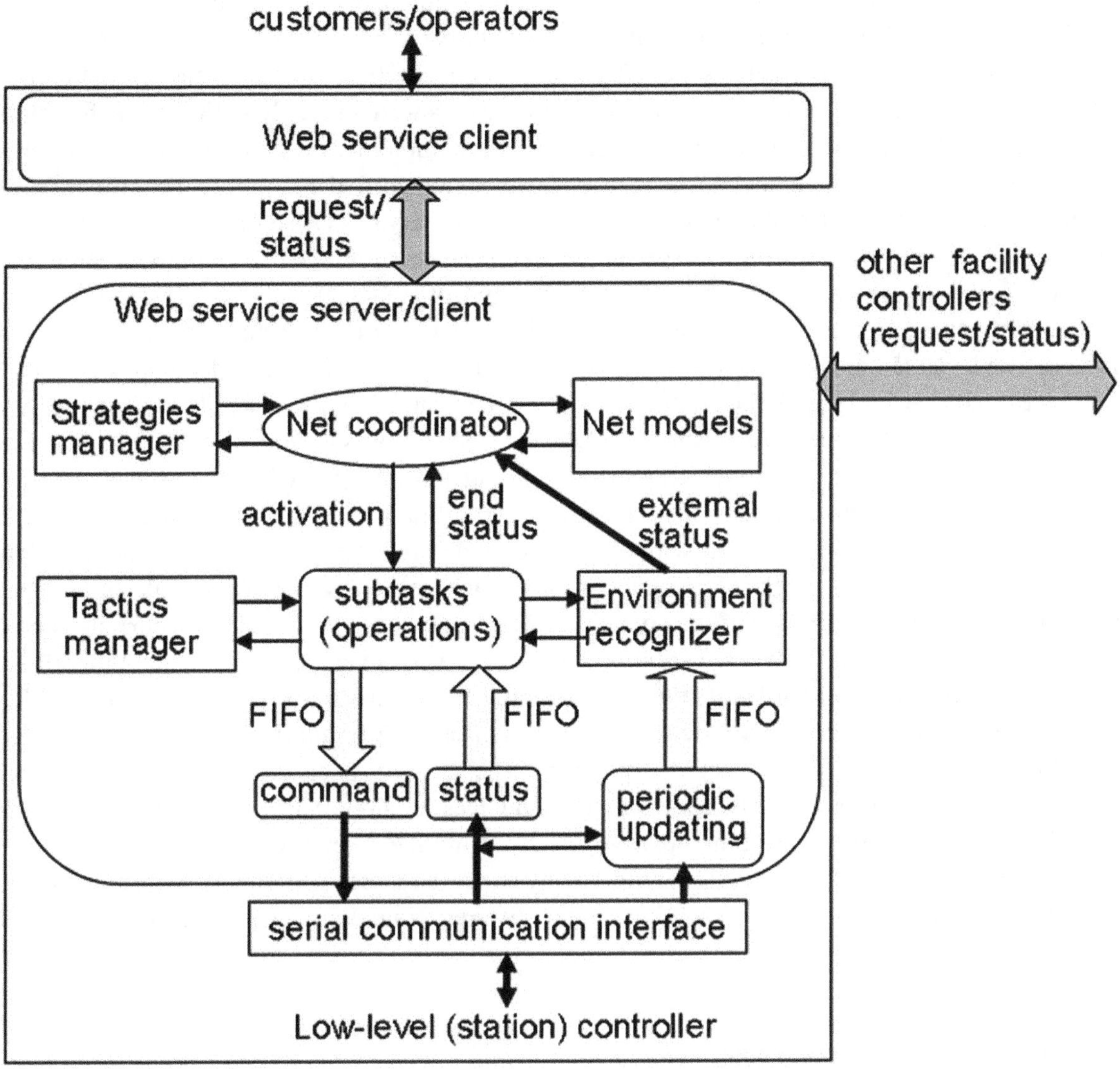

mentations using C on Linux. Figure 8 shows an architectural design of a distributed controller based on web service for non-centralized or autonomous distributed cooperative control of supply chains. One controller is connected to the web service client, which sends a set of commands on the service request specified by the customers or operators. The web service server program receives the requests and executes the specified service based on the detailed subnets of subtasks/operations with strategic/tactic control information. Cooperative task execution between two controllers is also established through the message exchange of execution request and notification response based on the web service, exchanging the roles of the web service client and server.

The supply chain tasks are presented as a set of functional operations, and each functional operation is mapped into the high-level net activity to build a task flow graph. Then, in accordance with the hierarchical approach, a refinement is automatically conducted to obtain its detailed functional model in an interpreted Petri net on a web service server. The net execution software with a conceptual net model which represents a supply chain task flow was implemented on a web server client PC, while the detailed functional net models for facilities/stations in the system are implemented and executed separately on different server PCs. Since command requests from the client are unified and different control programs can be implemented on local facility controllers, the implemented control system can be used to different kinds of supply chain systems, using the communication protocols for task execution between the web client and the web server using the command/response concept are shown.

The specification of the control processes in a supply chain is further decomposed into some smaller parts to be implemented as separate modules in a single facility controller. The decomposition allows dividing the controller into sequential automata; each of them can be implemented by a different device or hardware platform. Because the consistency between models on subsequent design stages is preserved with the use of model checking techniques, including analysis of reachability graphs, the finally implemented control software satisfies all behavioral user-defined requirements. Dedicated input and output signals used for communication with the environment are taken into account. The partitioned modules are synchronized although communication between more than two modules results in significant increase in the size and complexity of the controller. The proposed approach fits into current trends of model based development of embedded system, where early design verification is crucial for software productivity. A specification can be also written formally by requirements defined as temporal logic formulas, such as Linear Temporal Logic (LTL), Computation Tree Logic (CTL), or semi-formally by Unified Modeling Language (UML) activity diagrams. The verified specification net can be a basis for further steps, as synthesizable VHDL code for FPGA devices by direct hardware implementation of a net based controller with the translation into asynchronous circuit elements, or code implementation for microcontrollers. The approach can be used for supply chain processes which have to be implemented over the distributed facility controllers working in the different time domains with the proper synchronization of the whole system.

FUTURE WORK DIRECTIONS

Flexible and robust supply chains in mass customized environments require the capabilities to adapt quickly to changes and disturbances. The Petri net approach must possess the ability to reconfigure the net model of the existing supply chain and automatically recover from expected and unexpected errors or faults in condition-event relationships, such as timings of event occurrence. In place of pre-enumerating a large number of pre-specified error occurrences, intelligent agent techniques for unanticipated situations should be embedded in Petri net models. Future works include the modification of stochastic Petri nets to represent learning activities in transition firing combined with neuro-fuzzy networks for error and fault diagnosis, so that the subnet representing recovery steps to a normal state without deadlocks are generated and incorporated in the original net.

CONCLUSION

The Petri net approach was extended for hierarchical description, analysis and control of discrete supply chain processes to fit the high-level representation of task flows in supply chain systems; the concept of activity is extended to enclose the idea of events, while the concept of condition enclose several kinds of composite distribution elements. Approached in different levels of abstraction, supply chain task flows at a higher level are specified in a conceptual net model and functional models for facility control are developed in condition-event Petri nets through a top-down refinement technique. Based on the net modeling method, main control rules are identified and some practical solutions to problems, such as resource allocation and conflict arbitration, are introduced to assure that the resultant overall system is well-designed. The extended transition firing and state change rules are similar to the conventional synchronized Petri nets with transitions associated with events, and it preserves most of the formalism and properties of Petri nets and also improves their expressiveness. The current approach is more suitable to applications in supply chain systems where models are revised several times, and also improve reusability. It enables easy adaptation for launching of new products models, rapid adjustment of the productive system capability for variable quantities of products, rapid integration of new productive technologies into existing systems. Also, separating a large net in several subnets can produce a representation amenable to a distributed implementation of a high-level Petri net simulator.

REFERENCES

Balsamo, S., Harrison, P. G., & Marin, A. (2012). Methodological construction of product-form stochastic Petri nets for performance evaluation. *Journal of Systems and Software*, *85*(7), 1520–1539. doi:10.1016/j.jss.2011.11.1042

Cho, K., & Lim, J. (1999). Mixed centralized/decentralized supervisory control of discrete events dynamic systems. *Automatica*, *35*(1), 121–128. doi:10.1016/S0005-1098(98)00137-X

Clempner, J. (2014). Verifying soundness of business processes: A decision process Petri nets approach. *Expert Systems with Applications*, *41*(11), 5030–5040. doi:10.1016/j.eswa.2014.03.005

Costa, A., Barbosa, P., Gomes, L., Romalho, F., Figueiredo, J., & Junior, A. (2010). Properties preservation in distributed execution of Petri nets models. *Emerging Trends in Technological Innovation*, *314*, 241–250. doi:10.1007/978-3-642-11628-5_26

David, R., & Alla, H. (2005). *Discrete, Continuous and Hybrid Petri Nets*. Berlin: Springer.

Dong, M., & Chen, F. F. (2001). Process modeling and analysis of manufacturing supply chain networks using object-oriented Petri nets. *Robotics and Computer-integrated Manufacturing*, *17*(1-2), 121–129. doi:10.1016/S0736-5845(00)00045-4

Dotoli, M., Fanti, M. P., Meloni, C., & Zhou, M. C. (2005). A multi-level approach for network design of integrated supply chains. *International Journal of Production Research*, *43*(20), 4267–4287. doi:10.1080/00207540500142316

Duffee, N. A., & Prabhu, V. V. (1996). Heterarchical control of highly distributed manufacturing systems. *International Journal of Computer Integrated Manufacturing*, *9*(4), 270–281. doi:10.1080/095119296131562

Fischer, K. (1999). Agent based design of holonic manufacturing systems. *Robotics and Autonomous Systems*, *27*(1-2), 3–13. doi:10.1016/S0921-8890(98)00079-7

He, F., Dong, M., & Shao, X. (2011). Modeling and analysis of material flows in re-entrant supply chain networks using modified partial differential equations. *Journal of Applied Mathematics*.

Jules, G. D., Saadat, M., & Saeidlou, S. (2015). Holonic ontology and interaction protocol for manufacturing network organization. *IEEE Transactions on Systems, Man and Cybernetics. Systems*, *43*(5), 1240–1247.

Komoda, N. (2006). Service oriented architecture (SOA) in industrial systems. *Proceedings of the 2006 IEEE International Conference on Industrial Informatics* (pp. 1-5). doi:10.1109/INDIN.2006.275708

Lianfeng, Z., Xiao, Y., Jianxin, J., & Petri, H. (2009). Supply chain configuration with coordinated product, process and logistics: An approach based on Petri nets. *International Journal of Production Research*, *47*(23), 6681–6706. doi:10.1080/00207540802213427

Moreira, M., & Basilio, J. (2014). Bridging the gap between design and implementation of discrete-event controllers. *IEEE Transactions on Automation Science and Engineering*, *11*(1), 48–65. doi:10.1109/TASE.2013.2281733

Murata, T., Komoda, N., Matsumoto, K., & Haruna, K. (1986). A Petri net based controller for flexible and maintainable sequence control and its application in factory automation. *IEEE Transactions on Industrial Electronics*, *33*(1), 1–8. doi:10.1109/TIE.1986.351700

Peterson, J. (1981). *Petri Net Theory and the Modeling of Systems*. Englewood Cliffs, NJ: Prentice-Hall International.

Petri, C. A. (1962). *Kommunikation mit Automaten [Communication with automata]*. Schriften des Rheinisch-Westfalischen Institutes fur Instrumentelle Mathematik an der Universitat Bonn, Nr.2, (PhD dissertation, Darmstadt University of Technology, 1961.7).

Puttonen, J., Lobov, A., & Martinez Lastra, J. L. (2013). Semantics-based composition of factory automation processes encapsulated by web services. *IEEE Transactions on Industrial Informatics*, *9*(4), 2349–2359. doi:10.1109/TII.2012.2220554

Serral, E., Smedt, J. D., Snoeck, M., & Vanthienen, J. (2015). Context adaptive Petri nets: Supporting adaptation for the execution context. *Expert Systems with Applications*, *42*(23), 9307–9317. doi:10.1016/j.eswa.2015.08.004

Silva, M., Teruel, E., & Colom, J. M. (1996). *Linear algebraic and linear programming techniques for the analysis of net systems. Lectures on Petri Nets,* LNCS (Vol. 1491, pp. 309–373). Springer.

Suda, H. (1989). Future factory system formulated in Japan. *Techno Japan*, *22*(10), 15–25.

Tao, F., Laili, Y. J., Xu, L. D., & Zhang, L. (2013). FC-PACO-RM: A parallel method for service composition optimal-selection in cloud manufacturing system. *IEEE Transactions on Industrial Informatics*, *9*(4), 2023–2033. doi:10.1109/TII.2012.2232936

Uzam, V., Koc, B., Gelen, G., & Haken Aksebzeci, B. (2005). Asynchronous implementation of a Petri net based discrete event control system using a Xilinx FPGA. *Proceedings of the 35th International Conference on Computers and Industrial Engineering* (pp. 2025-2030).

Valckenaers, P., Van Brussel, H., Bongaerts, L., & Wyns, J. (1997). Holonic manufacturing systems. *Integrated Computer Aided Engineering*, *4*, 191–201.

van der Aalst, W. M. P. (1998). The application of Petri nets to workflow management. *Journal of Circuits, Systems, and Computers*, *8*(1), 21–66. doi:10.1142/S0218126698000043

van der Aalst, W. M. P. (2013). Decomposing Petri nets for process mining: A generic approach. *Distributed and Parallel Databases*, *31*(4), 471–507. doi:10.1007/s10619-013-7127-5

Villani, E., Miyagi, P. E., & Valette, R. (2006). *Modeling and Analysis of Hybrid Supervisory System: A Petri Net Approach*. London, UK: Springer Verlag.

Viswanadham, N., & Srinivasa Raghavan, N. R. (2000). Performance analysis and design of supply chains: A Petri Net approach. *The Journal of the Operational Research Society*, *51*(10), 1158–1169. doi:10.1057/palgrave.jors.2600063

Vollman, T. E., Berry, W. L., Whybark, D. C., & Jacob, F. R. (2004). *Manufacturing Planning and Control Systems for Supply Chain Management*. New York: Irwin/McGraw Hill.

Wisniewski, R., Stefanowicz, L., Bukowiec, A., & Lipinski, J. (2014). Theoretical aspects of Petri nets decomposition based on invariants and hypergraphs. Multimedia and Ubiquitous Engineering, LNEE (Vol. 308, pp. 371-376).

Yasuda, G. (2002). Hierarchical and distributed control of discrete event robotic manufacturing processes by extended Petri Nets. *Proceedings of the 4th Asia-Pacific Conference on Industrial Engineering and Management Systems* (pp. 709-712).

Yasuda, G. (2011). Petri net model based design and control of robotic manufacturing cells. In V. Modrak & R. Sudhakara Pandian (Eds.), *Operations Management Research and Cellular Manufacturing Systems: Innovative Methods and Approaches* (pp. 385–400). Hershey, PA, USA: IGI Global.

Yasuda, G., & Mori, M. (1974, February). Future concepts on grouping robots. *Presented at the Meeting on Neuro-Sciences,* Tokyo, Japan.

Yasuda, G., Takai, H., & Tachibana, K. (1996). Petri net based implementation and performance evaluation of parallel control architectures for distributed autonomous mechatronic systems. *Proceedings of the 3rd France-Japan Congress & 1st Europe-Asia Congress on Mechatronics* (pp. 99-104).

Yasuda, G., Takai, H., & Tachibana, K. (1997). Implementation and performance evaluation of parallel processing architectures for distributed autonomous motion control systems. *Proceedings of the 2nd Asian Control Conference* (Vol. 2, pp. 823-826).

Yu, W., Yan, C., Ding, Z., Jiang, C., & Zhou, M. (2014). Modeling and validating E-commerce business process based on Petri nets. *IEEE Transactions on Systems, Man, and Cybernetics*, *44*(3), 327–341. doi:10.1109/TSMC.2013.2248358

Zamei, E., Subias, A., Combacau, M., & De Bonneval, A. (1997). A hierarchical structure for control of discrete event systems and monitoring of process failures. *Studies in Information and Control*, 6(1).

Zegordi, S. H., & Davarzani, H. (2012). Developing a supply chain disruption analysis model: Application of colored Petri nets. *Expert Systems with Applications*, *39*(2), 2102–2111. doi:10.1016/j.eswa.2011.07.137

Zeng, Q., Lu, F., Liu, C., Duan, H., & Zhou, C. (2014). Modeling and verification for cross-department collaborative business processes using extended Petri nets. *IEEE Transactions on Systems, Man, and Cybernetics*, *45*(2), 349–362. doi:10.1109/TSMC.2014.2334276

Zhang, J., Chung, J. Y., Chang, C. K., & Kim, S. (2004). WS-Net: a Petri net based specification model for web services. *Proceedings of the IEEE International Conference on Web Services* (pp. 420-427). doi:10.1109/ICWS.2004.1314766

Zhou, M. C., & Venkatesh, K. (1998). *Modeling, Simulation and Control of Flexible Manufacturing Systems. A Petri Net Approach*. Singapore: World Scientific.

KEY TERMS AND DEFINITIONS

Agent: An independent program that manages the activities of related to a subtask in the system, including the knowledge of the subtask, the associated environment, the message handling with other agents, etc.

Business Process: Execution flow of services managed by a service coordination platform, including calling, routing, recovery, logging, directory management, etc.

Hierarchical Decomposition: Decomposition of a system by substitution of a node in the net at a higher level with the corresponding detailed subnet at the lower level.

Mass Customization: Control of flexible production systems with abilities to efficiently produce a wide range of products in relatively low volumes in a cost and quality effective manner.

Rule-Based Control: System control using a collection of rules, where each rule defines the actions with pre-conditions and post-conditions.

Service Oriented Architecture: A design framework for construction of information systems by combination of services, which are independently executed standardized program units on different hardware, OS, programming language.

Web Service: An autonomous software application or component with semantically well-defined functionality, uniquely defined by a Uniform Resource Locator (URL).

Chapter 7
The Production Capacity Planning and Scheduling Models in Term of Supply Chain Management

Dušan Malindžák
Technical University of Košice, Slovakia

Peter Kacmary
Technical University of Kosice, Slovakia

ABSTRACT

Each company is original from the point of production processes. Application of standard enterprise information system (SAP, proAlpha, Baan, etc.) need difficult adaptation, especially for conditions of small or medium enterprises (SMEs) and the price of that system is relatively high. Therefore, the proposed capacity planning model is much more suitable for the conditions, requires and demands of SMEs. Specifics of production processes of RS Ltd. are: - combinations of discrete and continuous processes, - this small company is connected to a supply chain (KANBAN) with its mother company, which defines the level of some chosen products in the expedition warehouse by the end of a month. For the above mentioned reason it had to be designed the new original production logistic system.

DOI: 10.4018/978-1-5225-0021-6.ch007

INTRODUCTION

The capacity production planning (CPP) is one level of operative planning (Malindžák, 1997; Rayward-Smith, Osman, Reeves & Smith, 1996). The main targets of CPP are:

- The records of customer orders and production tasks for the next planning period (some weeks) and production department.
- The order assignment to time unit(week) of the capacity plan.
- Capacity smoothing for machine and time unit.
- Creation of the internal orders – cumulated orders from the real customer orders.
- Creation of the production batches (Malindžák, 1983).

By these activity sequences it is created capacity plan of production department, which is input to the next step of operative planning to production scheduling (Malindžák, 1997).Each company is original from the point of production processes. Application of standard enterprise information system (SAP, proAlpha, Baan, etc.) need difficult adaptation, especially for conditions of small or medium enterprises (SMEs) and the price of that system is relatively high (Khouri, Kačmáry, 2009). Therefore, the proposed capacity planning model is much more suitable for the conditions, requires and demands of SMEs (Malindžák et al., 2011; Malindžák & Takala, 2005; Malindžák, Kačmáry & Moussttfa, 2012). Specifics of production processes of RS Ltd. are:

- Combinations of discrete and continuous processes (Malindžák, 1983).
- This small company is connected to a supply chain (KANBAN) with its mother company, which defines the level of some chosen products in the expedition warehouse by the end of a month (Johnson & Wood1996).

For the above mentioned reason it had to be designed the new original production logistic system. This system starts with order evidence, it includes the model of capacity planning, which respects KANBAN and model for production scheduling and operation evidence of the semi and final products (Malindžák, 1997).

Previous reasons forced authors to develop qualitatively new methods of problem solutions by computers, through modelling of various sides of creative human activities (Zimmermann, 2002). This approach is called heuristic approach. Heuristic approach, in contrary to accurate approach (analytic methods), uses modelling of processes of information processing realized by people on different stages of activity and during various tasks solution and development of heuristic models (HM) on this principle (Malindžák, Kačmáry et al., 2013; Michlowicz, 2002).

Design of the logistic system of RS Ltd. can be also applied in any SME, in the area of machinery industry connected to supply chain system with another company (Takala, Malindžák, Straka et al. 2007; Christopher 1997).

In this case, the RS limited is a daughter company of the big foreign company – Mother Company. Inside one company between two subjects (mother, daughter) the SCM can be understand as a KANBAN (Malindžák et al., 2011).

Model for CPP uses the information from SCM – KANBAN. KANBAN store may be used as a regulatory element in CPP model (Malindžák et al., 2011).

METHODOLOGY OF CPP MODEL CREATION

Both models are created on the heuristic approach i.e. there were analysed rules and limitations, which were applied before by logistic worker for preparation of the CPP and it was applied into algorithms of the models (Lenort, Staš & Samolejová, 2012). By this approach applied in the model were included human properties (all sequence of activities, knowhow, experience, knowledge of experts and people, who works in the company for a long time) (Malindžák, Kačmáry & Moussttfa 2012; Schroder, 1989).

When the model of CPP is created, it is necessary to solve the problem as multicriteria optimization (Levin, Rubin, Stinson & Gardner 1989). However, if it is considered larger number of criterion in operative plan, e.g. delivery terms, internal and external production, production and technological priority, directionally engaged priority, production continuity, etc., then these criterion will be arranged in order and they will be considered during process of initial variant in HM algorithm development and initial variant will be realized again in accordance with one variable criterion or to apply it in initial variant development process (Johns, Harding 1989; Malindžák1993).

Approaches to CPP

In the theory and practice were formed two basic approaches:

1. **Customer Oriented Model of CPP:** This approach is derived from an idea to satisfy all customers by their order (due date, order amount, etc.) without consideration of the capacity possibilities machine and workplace. Each order – product has technological prescription for its manufacturing, where it is defined sequence of operations as S_I and operational time of this product I on machine J $t_{I,J}$. It is created order sequence by define priority (due date, cost, profit, etc.) from the set of the orders. The products are assignment on the machine by the technological prescription applying the pull system. Result of this assignment is initial variant of CPP, which go to next step to capacity smoothing process (Figures 1 and 2).

$CN(J) = \Sigma OA_{I,J} \cdot t_{I,J}$

$OA_{I,J}$- orderamount of product I at machine J

$t_{I,J}$ – operation time of product I at machine J

Figure 1. Principles of customer oriented model CPP

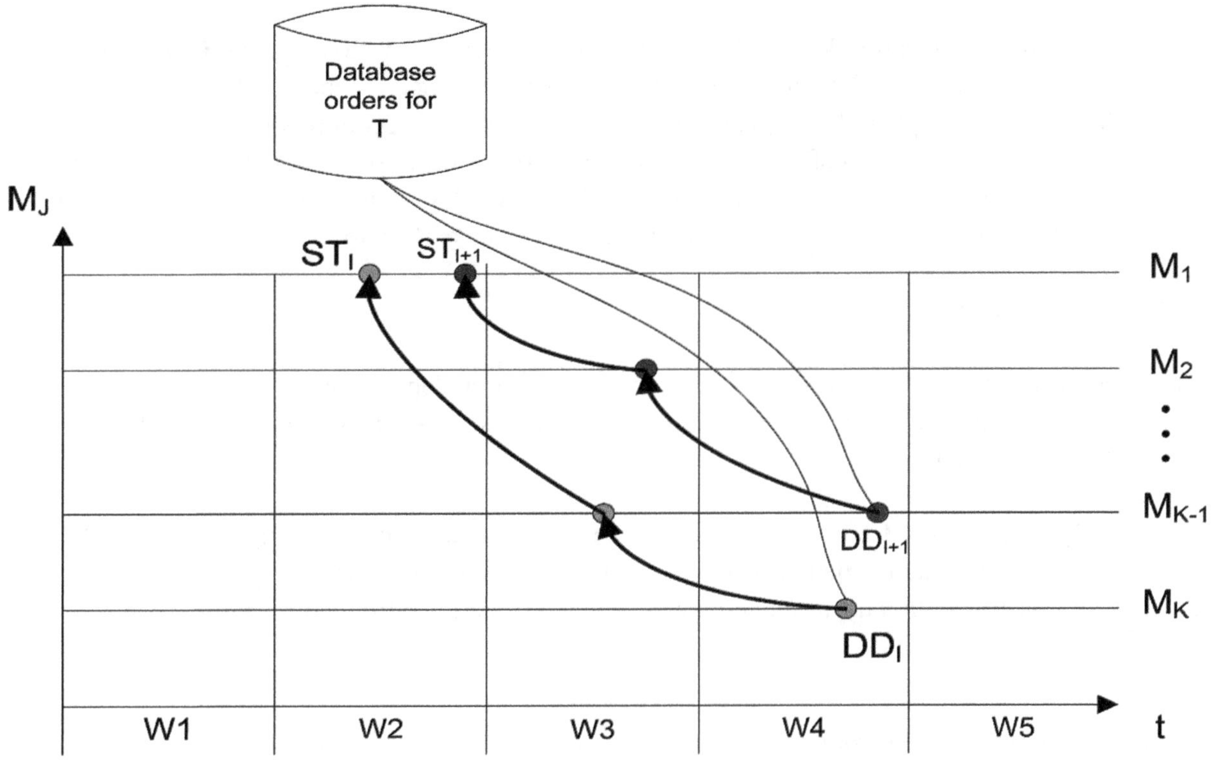

Figure 2. Capacity needs for machine J before smoothing

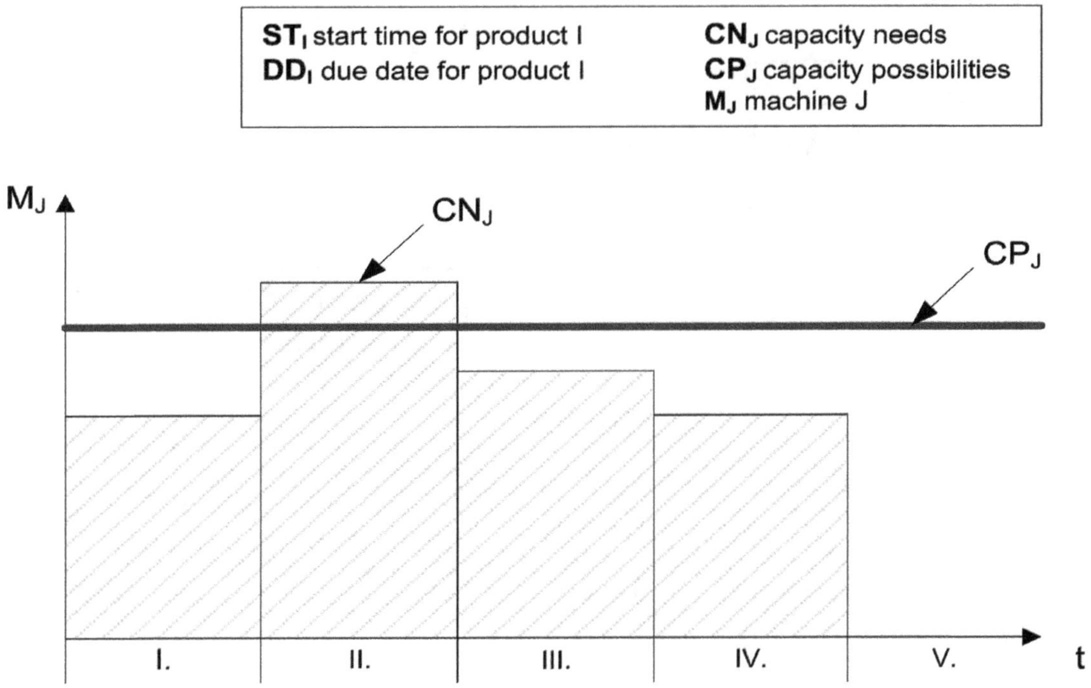

Follows the capacity smoothing, i.e. harmonisation of CN(J) ≠ CP(J).

2. **Manufacturing Oriented Model of CPP:** But also different approach is possible. Ordered products will be not allocated in planning periods, but products sequence will be developed in accordance with delivery date. From this sequence it should be taken such part of products, which will maximize capacity exploitation of production processes bottle neck. So it is possible to avoid capacity smoothing. It is necessary to realise that production processes bottle neck space can be changed in accordance with assortment and amount of ordered products (Figure 3).

Strategies of CPP

Very often manufacturing process is realized in some production departments. By the optimization criteria and technological, economical, environmental constrains, two strategies can be select (Schroder, 1989; Schmenner, 1990).

1. **The Open System of CPP:** Open system means that product in one planning period (week) pass through more production departments. Capacity planning in this case has to coordinate operation amount this production department to keep DDI. The creation CPP model on this strategy is very complicated.

Figure 3. Principles of production oriented model CPP

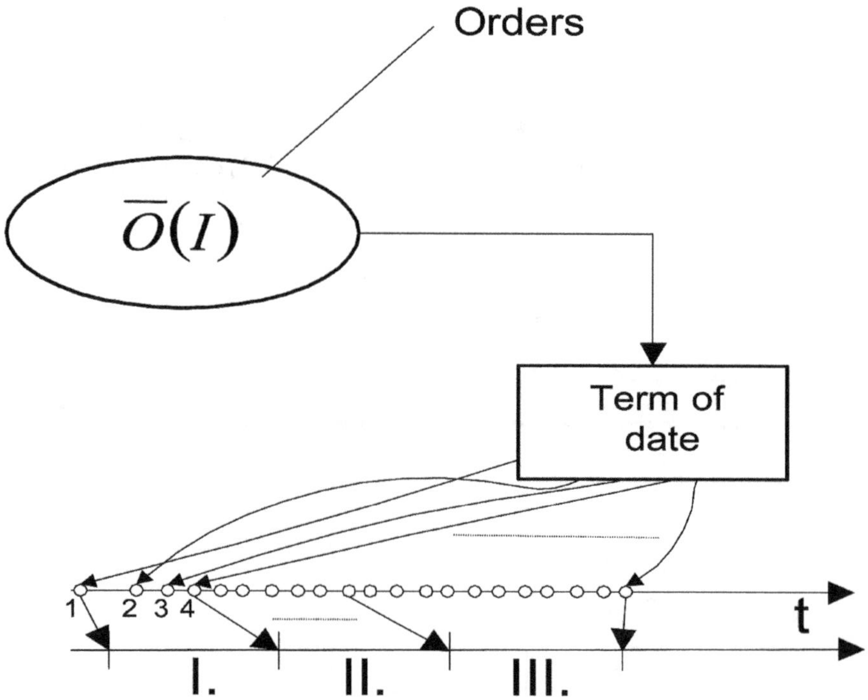

2. **The Closed System of CPP:** This strategy define that operations on some product is realized in one planning period (week) only in one production department.

The selection of the strategy is influenced from the production cycle, delivery cycle and from the type of dominant optimization criterion. When it is dominant technological optimization criterion, in this case, it is better apply close strategy. And vice versa, when it is dominant economic and trade optimization criterions, in this case, there is recommended open system (Bindzár & Kačmáry, 2009).

CPP Model in SCM Condition

Can be defined by free basic SCM system:

1. The SCM established of exchange of information about forecast or manufacturing plan among the member of the SCM agreement e.g. on the base of electronic data interchange (EDI). The forecast is the only information for supplier is not binding, it has only information character.
2. Enterprise N1 receives or gains the manufacturing plan forecast of enterprise N2 where N2 enterprise will specify the levels of ready-made products in output warehouse of S1, should be e.g. in (weeks)W_1, W_2 ......W_N period in the output warehouse of N1 enterprise. The same will apply for N_3 enterprise, which will specify the level of inventory stock for the weeks$W_{1,}$ W_2 ......W_N as in S2 of N2 enterprise etc. (see the Figure 4). This will enable to manage these enterprises in the long run and to increase the uniformity of production, to create optimum production batches, and to generate internal orders by accumulation of a larger number of orders for a longer period of time. This all, however, is just a forecast. As a part of this forecast business is carried out on the basis of particular orders. The supply chain is mostly initiated by the companies with the strongest position in the chain or based on the agreement of all companies becoming a part of this chain (Johnson & Wood, 1996; Malindžák, Lenort, &Mervart, 2011).

If subject of SCM N1 and N2 are member of one company (N1 is daughter, N2 is mother company) in this case SCM is similar as KANBAN system.

3. Demand chain. This philosophy has recently led mainly to bigger pressure from chain dominant enterprises that specify, for their sub-suppliers, the level of products and the period (W_1, W_2, ...,

Figure 4. The principle of SCM-KANBAN

I1, I2, I3 – theinputwarehouse of N1, N2, N3 enterprises
S1, S2, S3 – theoutputwarehouse of N1, N2, N3 enterprises

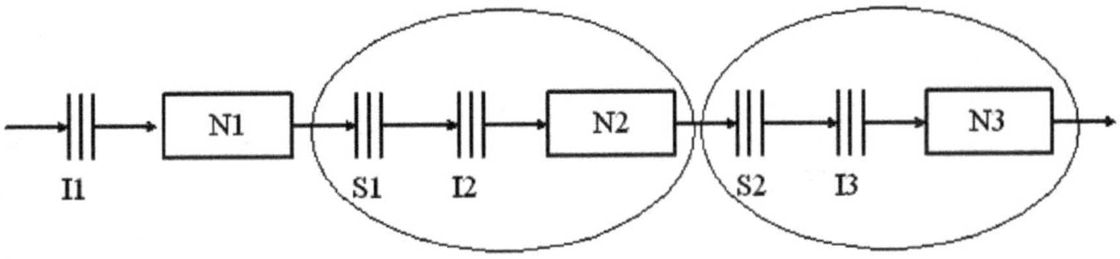

W_N), in which the given volume should be either in a warehouse close to their premises or directly in entry i.e. disposition warehouses. The above-mentioned products will be owned by the supplier up to the moment of their release from the disposition warehouse. It should be emphasized out once more that the products in disposition warehouses are owned by suppliers, which forces to each enterprises in the chain to make its sub-suppliers created I0, I1, I2 disposition warehouses in its concernment (see the Figure 5).

Now a case: if N1 and N2 are in the second type of SCM, i.e. N2 defines the level of storage products in the KANBAN (output warehouse) for N1 and levels of product in output warehouse are defined to the end of planning period (month).

The model of capacity planning company N1 must respect this information – regulation – limitation. The model of CPP utilized this KANBAN warehouse as regulation element it means that it is not enough order to capacity possibilities of the machine workshop utilization information about level of the product in the KANBAN buffer as virtual order which is assignment to capacity plan ($CP_J > CN_J$).

In the case, when ($CP_J < CN_J$) in the beginning of the planning period product from the KANBAN buffer are utilized for fulfilling the orders of the customer and by this way are decreased CN_J on the planning period.

Disadvantage to create the KANBAN buffer and its maintenance become advantage when model of CPP has assemble the regulation element describe on Figure 6.

T – planning period (month)

Heuristic Approach in Design of CPP Model

Heuristic approach of the synthesis LS assumes modelling of process principles as processing of information carried out by a man on various phases of his activities and while solving various tasks. This approach then bases on a principle of heuristic model creation. In the next, is described the example of the heuristic model synthesis for the CPP and production scheduling model (Malindžák, Kačmáry, Bednárová & Vrlíková, 2013).

Sequences of steps during creation of such heuristic model:

1. Definition of initial situation (problem definition).
2. Creation of possible variants for further situations (possible solutions).

Figure 5. The principle of demand chain application

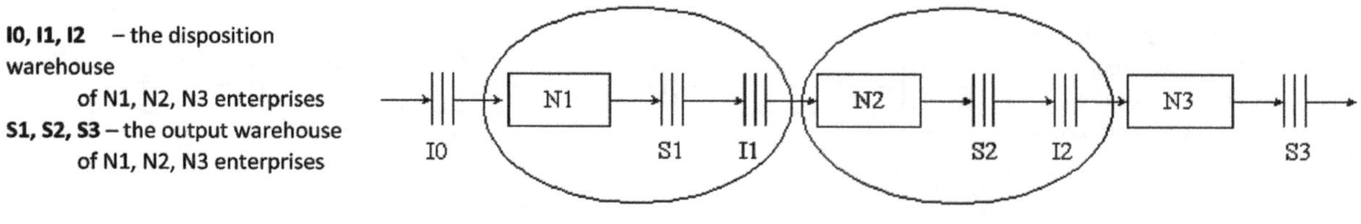

Figure 6. Utilization of KANBAN- warehouse for capacity regulation

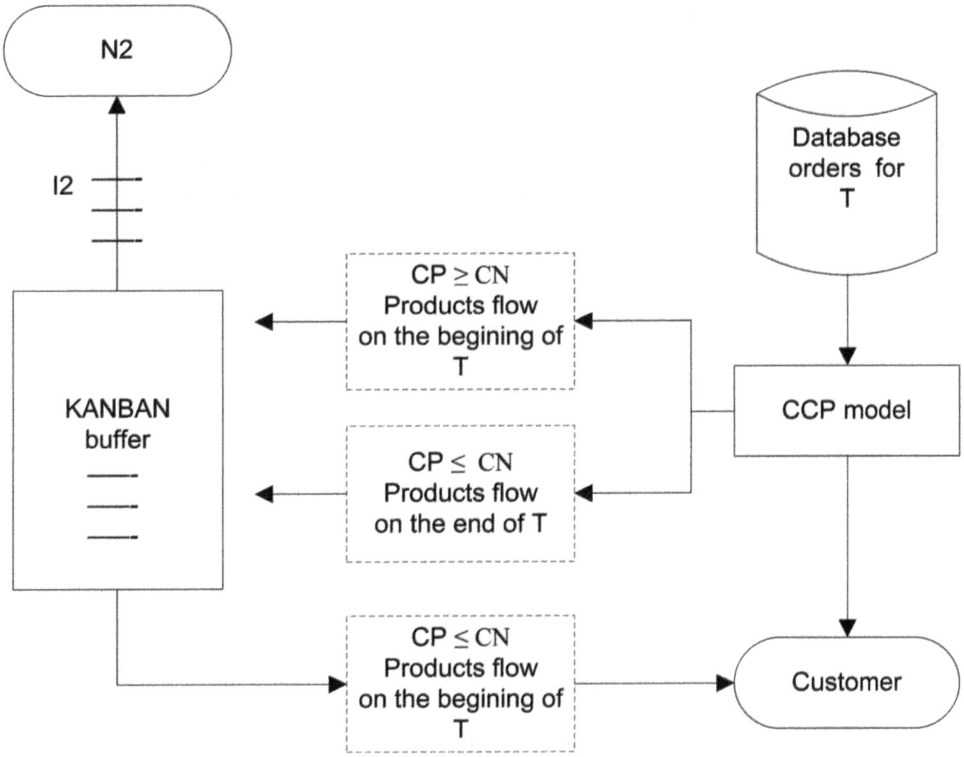

3. Rule creation.
4. Heuristic model synthesis.
5. Heuristic model verification.

The sequences of steps for creation of such heuristic model are illustrated in the Figure 7. Definition of rule set is performed as the result of analysis, technological processes, machines, equipment, organization and manufacturing process management, economy, capacity and optimality criterion.

Particular process, e.g. planning, has particular entry file of orders and by its analysis the rules were defined, which need to be fulfilled by the planning process.

The synthesis objective is to create an algorithm or model from these rules and from the definition of entry files structure.

Set of rules . $\{\overline{R}\} \in \{\overline{H}, \overline{TR}, \overline{ER}, \overline{O}, \overline{CO}\}$ comprise of following groups:

- Heuristic - $\overline{H}$
- Technological Rules - $\overline{TR}$
- Expert Rules - $\overline{ER}$

Figure 7. The schema of the creation heuristic model [4]

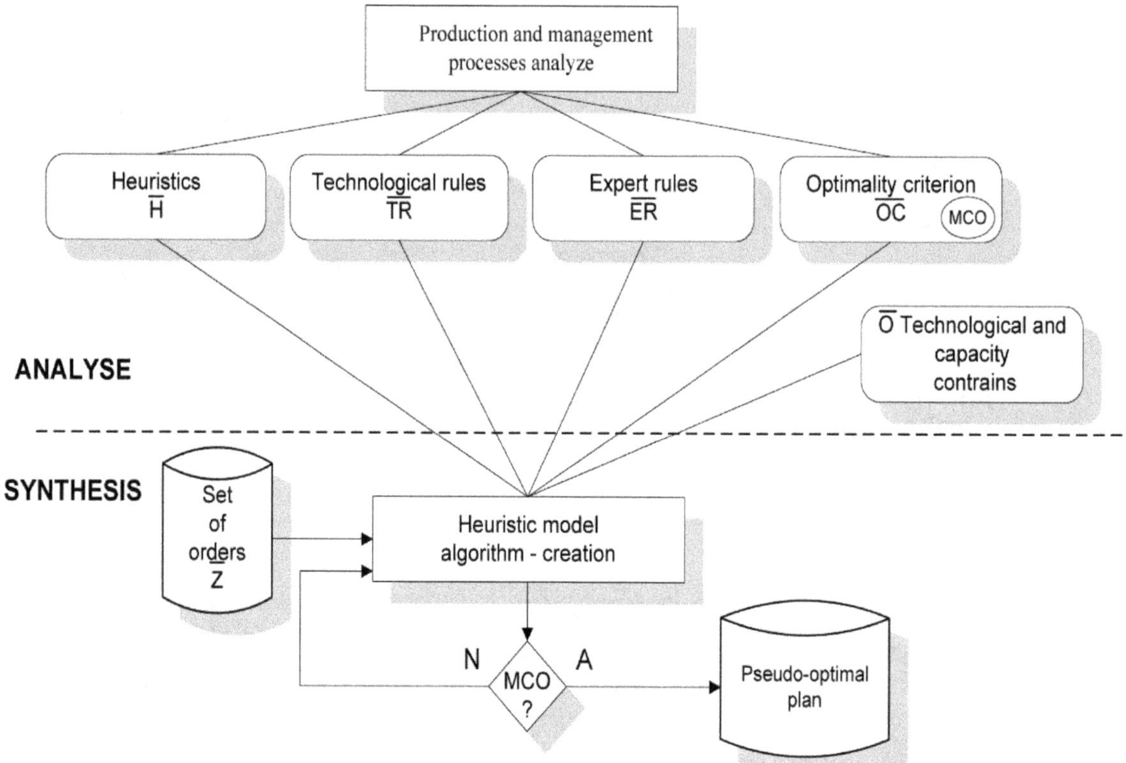

- Constraints - $\overline{O}$
- Optimality Criteria - $\overline{OC}$ (MCO – main criterion of optimization).

DESCRIBTION MANUFACTURING PROCESS AND MATERIAL FLOWS OF RS LIMITED

The enterprise RS Ltd. is producer of stator and rotors for industrial ventilation and air conditioning system and is conformed to German mother company. The basic production process starts at cutting of dynamo plates and their welding or riveting (PP1), its casting to aluminium alloys (PP2), finishing (PP3) and surface treatment (PP4), (see figure 8).

The file of recorded orders (FRO) is continuously actualised by adding new incoming orders, while planner can open the file anytime but on mentioned date he will receive the file through e-mail. By this step the actualisation of this file is finished and it is ready for planner to create new production plans.

Figure 8. Basic production processes and flows of RS Ltd. enterprise

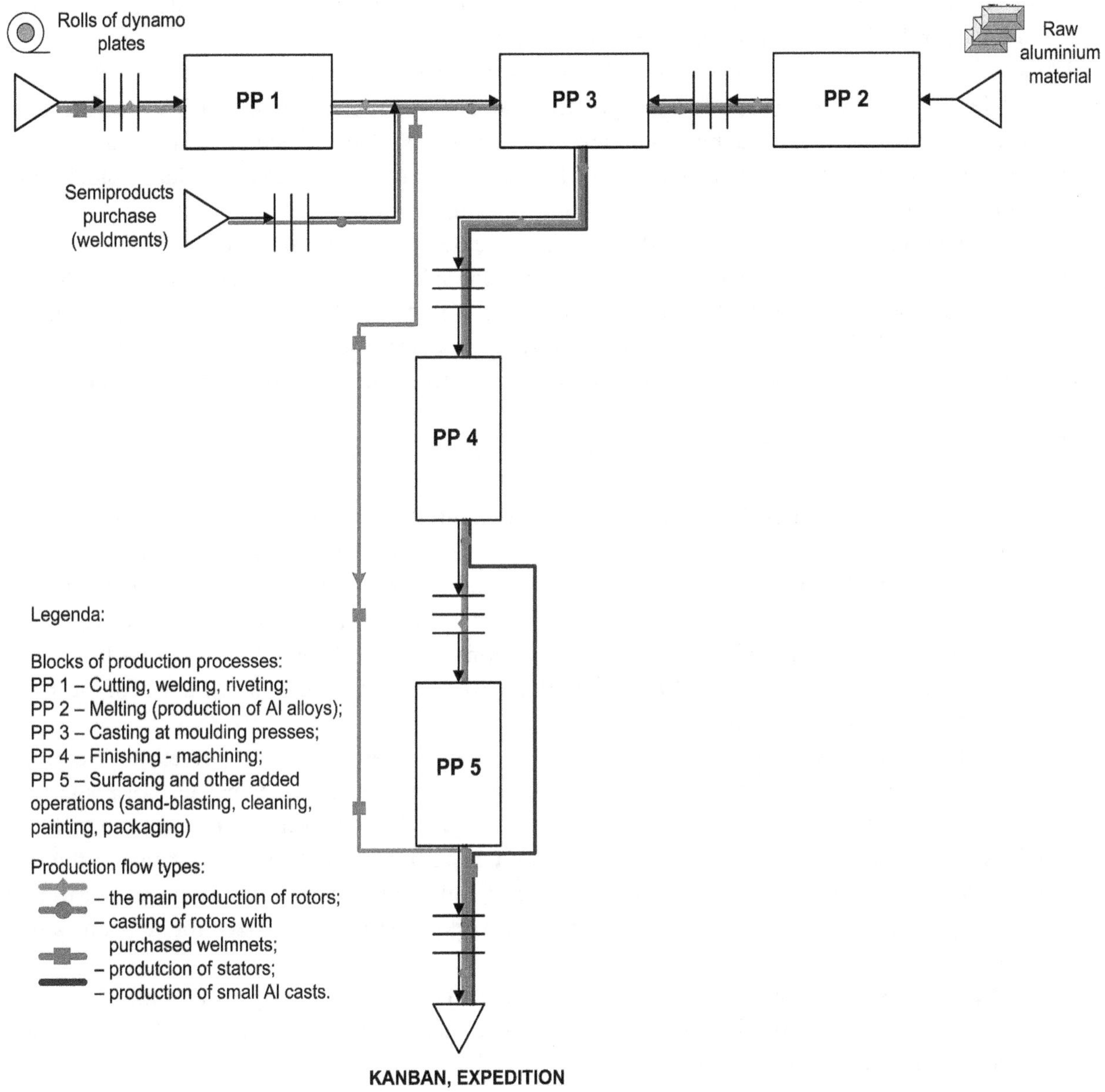

The plan is created separately for these divisions:

1. For cutting (CNC machines).
2. For casting.
3. For finishing (CNC machines), however this is not detailed plan given to a supervisor but it is a list of products and due dates and delivery dates.

The first set of information input is in-process production itself, which was planned in previous periods "N-1" for present period "N" (e.g. cut dynamo plates are not already casted in to aluminium body, casted are not finished and finished are not surfaced etc.)

The second set of information input is stock levels of finished products, so called "KANBAN", where the certain levels are need to be kept (levels represent quantity) by the end of a month. The decree, from the mother company which defined the certain levels of finished products, is actualised regularly, sometimes also when there is unexpected increase of orders of standard (often ordered) products. Stocks and then the missing items are planned first as a priority (Rosová, 2010).

Some of the semi-products (e.g. riveted stators and welded rotors) are not produced in this enterprise but they have to be ordered. This idea is used as a flexible item in capacity planning model. These purchased semi-products are also a reserve for smoothing of material flow especially between cutting and casting (Figure 8). The second task of purchased semi-products is a capacity balance of production process. The capacity needs (CN_J) calculation is done first with all incoming unproduced orders.

PP1 Cutting of Shreds, Riveting and Welding

- **Cutting:** Shreds are cut at the same, two cutting presses TSS 1 and 2. Presses are equivalent in speed and quality. Tact of the presses is 180-220 beats per minute. Cutting has enough reserve capacity. Scheduling strategy is to minimize the exchange of cutting forms, because replacing the forms takes about 8 hours (one whole shift). Replacing of the coil, the input item with which the presses are working, takes about 15 minutes. Minimizing of exchange of forms is achieved by creating families (family product - FP), i.e. product group, which varies only by a part of the same form, with the same strip width.
- **Riveting:** The shreds are weighed and sheets are stacked manually, before riveting. Weighing accuracy is ± 1 of a shred. It is not possible to automate number of plates because of respecting the tolerance weight of shreds (the sheet thickness is varied). However it can be automated to a height of shred column.The riveting machines are the two, by side (not parallel), because they do not work from the same buffer. The servicing is individual considering the service processes and their relationship to the riveting operation. Riveting workplaces are specialized for large and small stators only.
- **Welding:** It is similar operation as riveting, but the shred connection is got by welding of shred column at its perimeter. Welding rotor shreds is a bottleneck that is why it is possible to buy these weldments also from external suppliers.

PP2 – Production of Aluminium (Al) Alloys

- **Al Alloys:** Aluminium melting furnace produces 10 kinds of Al alloys for casting needs. The maximum batch is 450 kg / per one melting. The melting capacity is 350 kg per hour. The criterion for effectiveness is to minimize of alloy changeovers and so called flushing meltages and maximizing of melting batch (service times) (see Figure 9).

Figure 9. Time sequence of formation of aluminium alloys

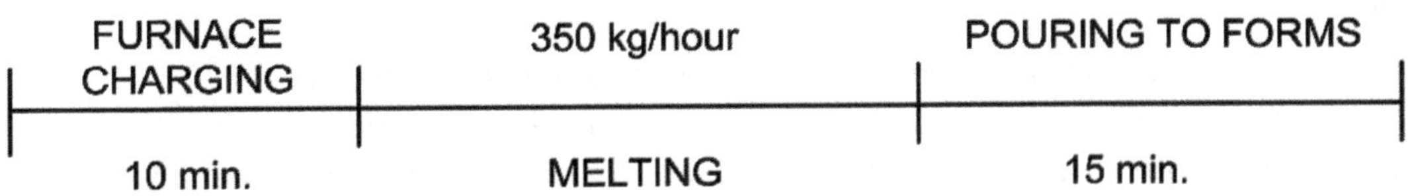

Minimizing of flushing meltages requires minimization of number of alloy modifications needed for casting presses. It means that the planning of meltages and their composition depends on the casting plan at injection moulding machines. These plans of casting and melting are created by the interaction, taking into account the criteria of both production points. Minimum number of meltage is2 per shift, maximum 5, flushing meltage takes at minimum 1 hour, its costs are inefficient costs (energy, metal loss, equipment depreciation, wages and loss of capacity). Because the melting furnace is the only one, it is a strategic device, by which the work of the whole company is influenced (see Figures 10 and 11).

Figure 10. Material flow of Al alloys creation

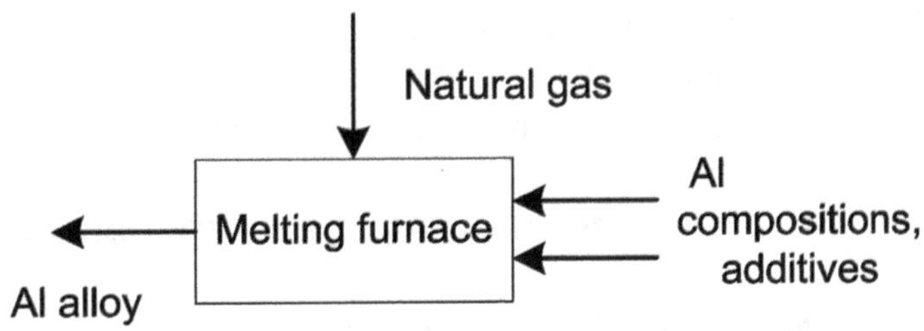

Figure 11. Material flow of casting process

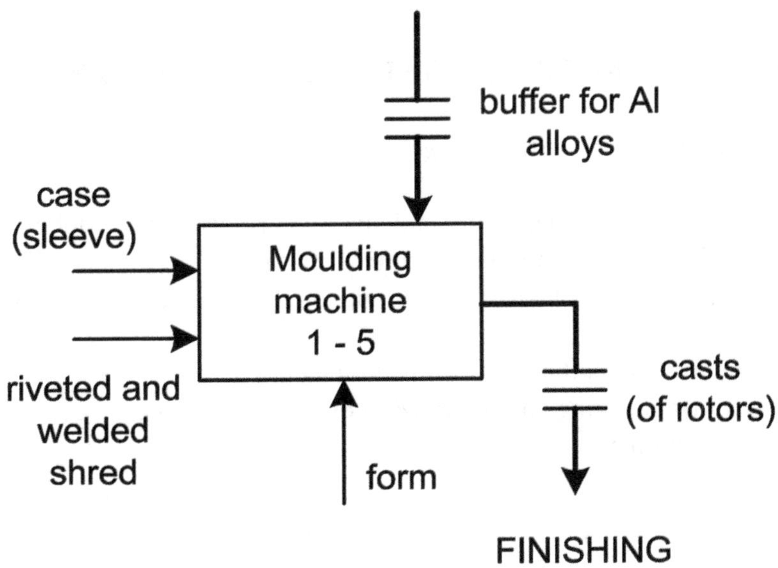

PP 3 Casting

Casting of rotors and casting is carried out on five injection moulding machines.

Casting, in terms of process control of logistic, is a point, which combines the two flows – first flow of welded and riveted parts and second flow of Al alloys, cases and forms. It is optimal to cumulate such a number of units that will require n-times batches (meltages) of the same alloy. Planning must also respect the size forms, contact force, speed, moulding quality. It means that the injection moulding machines are partly specialized in parts, i.e. not each assortment can be casted at each moulding machine with normal (the same) efficiency.

PP 4 Finishing, Surfacing

Currently, finishing of not controlled according to the operational plan of production, i.e., distribution of products to machines and workstations does not have its operational plan. Production manager assigns operational work to the machine workers according to the situation between the casting and surfacing.

The surfacing (post-processing) follows the finishing at the machining shop. The machining shop has 5 computer numerical control (CNC) centres. Centre 1 and 2 processes castings, the small orders are planned at centre 3 because of its fast reconfiguration, centres 4 and 5 processes larger orders due to their long reconfiguration (about 4 hours). There is also a criterion of smoothness change of processes at "large" centres, which can shorten the configuration (Malindžáková, 2011).

PP 5 Other added operations (sand-blasting, cleaning, painting, packaging)

These works are distributed to workers and machines operationally. There is not created operative production plan, but the planning for other aggregates and machines must calculate these operations (Malindžák et al, 2011).

PROPOSAL OF NEW CONCEPTION OF PRODUCTION LOGISTICS

The new proposal has to fulfil some specific targets:

- Completion of orders in required quantity and quality in priority to mother company.
- Flexibility and stability, minimization of operative changes.
- Increasing of technological and logistic discipline, keeping of a production plan and dates of delivery.
- Increasing of regularity (smoothness) – production tact.
- Consecutiveness order changing and minimizing of stores (buffers).
- Minimizing of production costs.

The proposal of the production logistic information system (SYVYLO – inner company name) is described in the Figure 12. The production logistics means managing, ensuring and providing of material flows in the RS Ltd. company, which are provided through these basic activities:

1. Annual planning of trade and production.
2. Order logistics:

Figure 12. The structure of production logistics system (SYVYLO) in RS Ltd.

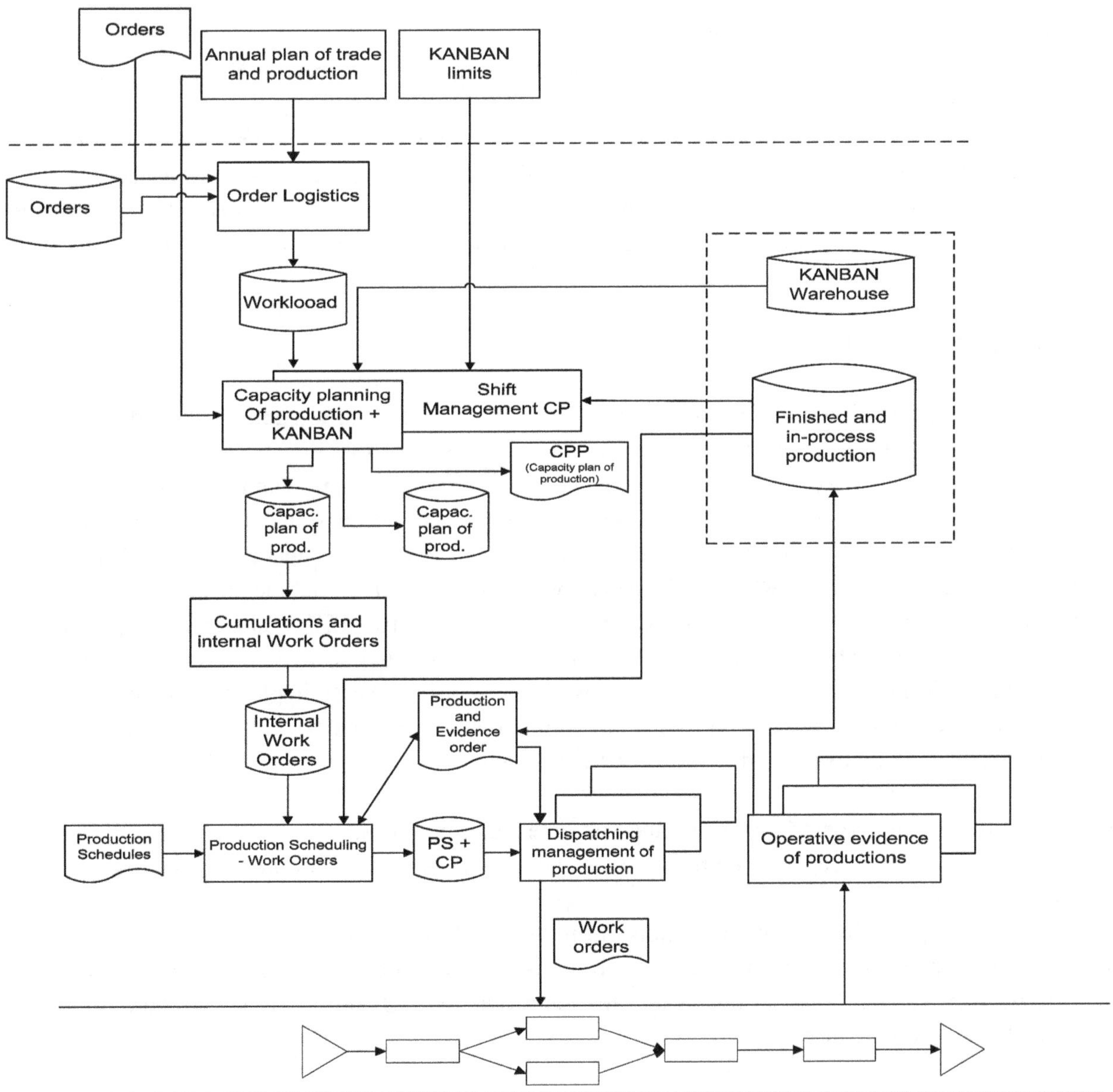

   a. Processing of workload (from incoming orders) into planning periods.
   b. Technological, material and economic order appraisal, material balance.
   c. Capacity planning
   d. Cumulations and creation of internal work orders.
3. Preparation of production scheduling for machines and equipment.
4. Shift managing of operative production plans.
5. Operative evidence of production.

SYVYLO is only the "core" of the production logistic system and does not include:

- Annual planning of trade and production.
- Warehousing management and material ordering, reality and material order assurance will be confirmed by a supplier.
- Preparation and recording of new technological descriptions for new products or changing in technological descriptions will be added into proposals manually through prepared interfaces.
- Operative evidence will not be automated (e.g. by bar codes) but will be entered through the terminals of information systems located in production area by production supervisors and also into KANBAN cards (Malindžák et al, 2011).

Proposal of Capacity Planning Model

Source of input information is the file of all orders. Each order has the certain quantity of ordered assortment and due dates. Present capacity planning consists of planning of all orders backwards from due date to first operation. There are calculated latest times of beginning of production and capacity need of each machine by this way, supposed that all income orders are put into production. Next, there is calculated in-process production and purchasing of semi-products, which can relieve capacity need at PP 1. In each production company, including RS Ltd., the capacity calculation have to include unexpected influences (lost times) i.e. breakdowns, increased time of setting etc., which is 10 – 15% of all times. That is why the machines are planned at about between 85 – 90% of their capacity, so this created a kind of time reserve.

The capacity planning comes out from known workload – orders to certain planning period. Its aim is to choose orders from the aggregated orders (file of incoming orders) and to put them into a production plan, into a certain planning period by keeping the following:

- To fulfil required ordered quantities and due-dates for customers.
- To give in balance capacity demands to machines, equipment, workplaces with their capacity possibilities.
- To keep the prescribed store level limits of products in KANBAN warehouse.
- To purchase material, semi-products or sub-deliveries based on capacity plan of production.
- To have the capacity plan as a basis for creation production scheduling, in which there is no need to take care about capacity.

Parameters and criteria for capacity planning:

1. The capacity plan is created for all company and it is divided to:
    a. U1 – cutting, welding, riveting (PP1 see Figure 8);
    b. U2 – casting and production of aluminium (Al) alloys (PP2 and PP3 see Figure 8);
    c. U3 – surfacing, finishing and other finalising (PP4 and PP5 see Figure 8).
2. Planning period will be at minimum 8 weeks. Planning point is between 10:00am – 12:00pm on Thursday (Figure 13 and 14).
3. There is used the principle of sliding planning, which is performed each Thursday at noon (12.00 o'clock).
4. There is a different obligation in weeks (1st week is definite, 2nd week is preliminary – set at approx. 80%, 3rd week – 8th week is forecasted). The exact authorisation of doing changes have to be defined, e.g. changes for 1st week – only general director can provide changes, for 2nd week – only delegated production supervisors or planner can provide changes, other weeks planner and other entitled persons can provide changes.
5. Planning by PULL system – it means from the end to beginning.
6. The products will have priorities "B" – ordinary, "S" – urgent, "SS" – super urgent.
7. Closed system of capacity planning (phase production) besides the products with priority "S" and "SS" (defined by general director) will be kept, i.e. what is produced at U1 in week N will be processed at U2 in week N+1 and at U3 in week N+2. It results to defined production time, which for "B" products is three weeks, for "S" products two weeks and for "SS" products is one week or orders are completed from KANBAN.
8. The capacity plan will be created at 90% of maximum capacity (10% is left for unexpected, unaware changes and interactions).
9. Initial variant of plan is created in the way that:
    a. Inputting "S" and "SS" orders into weeks according to DD_I to division U3.
    b. New "B" orders are assorted by: LRT – Longest Remaining Time, i.e. maximum time to their production is:

$$\sum_J OA_I * t_{I,J} = \max \text{, respectively according to } DD_I \text{ to division U3.} \quad (1)$$

Figure 13. Planning periods of defined divisions (U_1 - U_3)

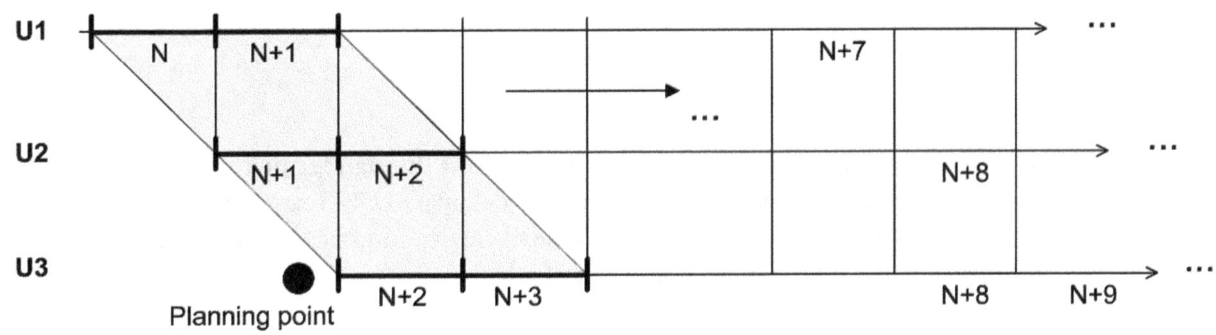

Figure 14. Algorithm of the capacity planning model

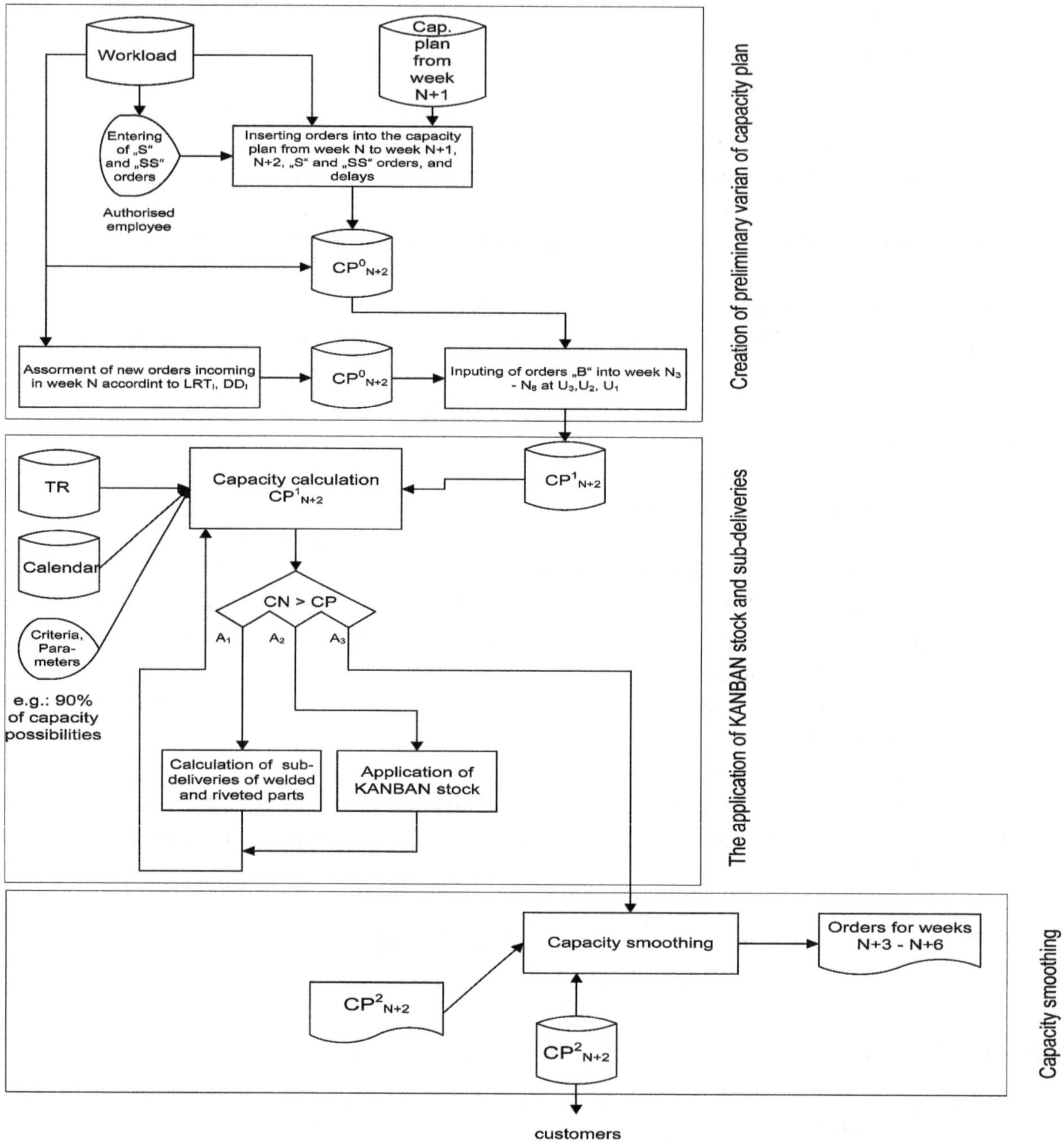

c. Orders are inserted to certain weeks from the end of U3 through U2 up to U1.
d. The capacity calculation is provided and there are calculated CN_J (capacity needs) in certain periods of weeks N+1 - N+8, but planning dead times, compulsory maintenance and in-process production have to be taken into calculation.

10. Calculation of bottle-neck.
11. Solution of bottle-neck through KANBAN stock levels, and point 10 is again repeated.
12. Calculation of sub-deliveries of welded and riveted parts and point 10 is again repeated.
13. The capacity smoothing is possible to reach:
    a. By using KAIZEN,
    b. By moving forward,
    c. By dividing one big order to 2, 3 smaller batches (internal work orders) and moving two, three batches a period back (Malindžák et al, 2011; Malindžák, Kačmáry & Moussttfa, 2012).

Model of the Production Scheduling

The dominant criterion is the continuity of changeovers in the casting process, as it was mentioned, because of the fixed relationship between the melting furnace and the casting injection moulding presses. This criterion also plays a primary role at the creation of production schedules and all orders are cumulated first at this moment, at that place, where formed groups use the same or related aluminium alloy. Other manufacturing operations are not fixed, they use a buffer and it allows some freedom to secondary and further cumulation. Mentioned freedom is given by the fact that cutting has sufficient capacity, capacity of welding can be added by purchasing of welded parts and in front of surfacing is also some sequence through the in-process production. Under these conditions there is defined progress of plans creation for each operation in the following order:

Schedule for press – moulding has to respect the continuity of chemical composition of AL alloys. The option of transition among press-mouldings without purification of presses are defined by "transition matrix" (Table 1). Followed by casting schedule it is created the production schedule for induction furnace. From the schedule of casting itis created the schedule of welding (riveting), or purchase rotors and cutting schedule, as well as schedule for machining (Figure 15).

Further cumulation of orders and generation of the sequence – the sequence of casting, production of alloys, cutting and welding, and finally the order casting are respected in particular plans.

The rules for creating of models of scheduling:

1. Creation of daily plans at moulding presses. There are some certain rules for all products which are produced on moulding presses: not that each product is suitable to produce at any machine.

Table 1. The transition matrix for furnace and presses

	DIN 226	DIN 231	DIN 230	D 106	S 106	D 165
DIN 226	Y	Y	N	Y	Y	Y
DIN 231	Y	Y	N	Y	Y	Y
DIN 230	Y	Y	Y	Y	Y	Y
D 106	N	N	N	Y	Y	Y
S 106	N	N	N	Y	Y	Y
D 165	N	N	N	Y	Y	Y

Figure 15. The sequence of the production scheduling models creation

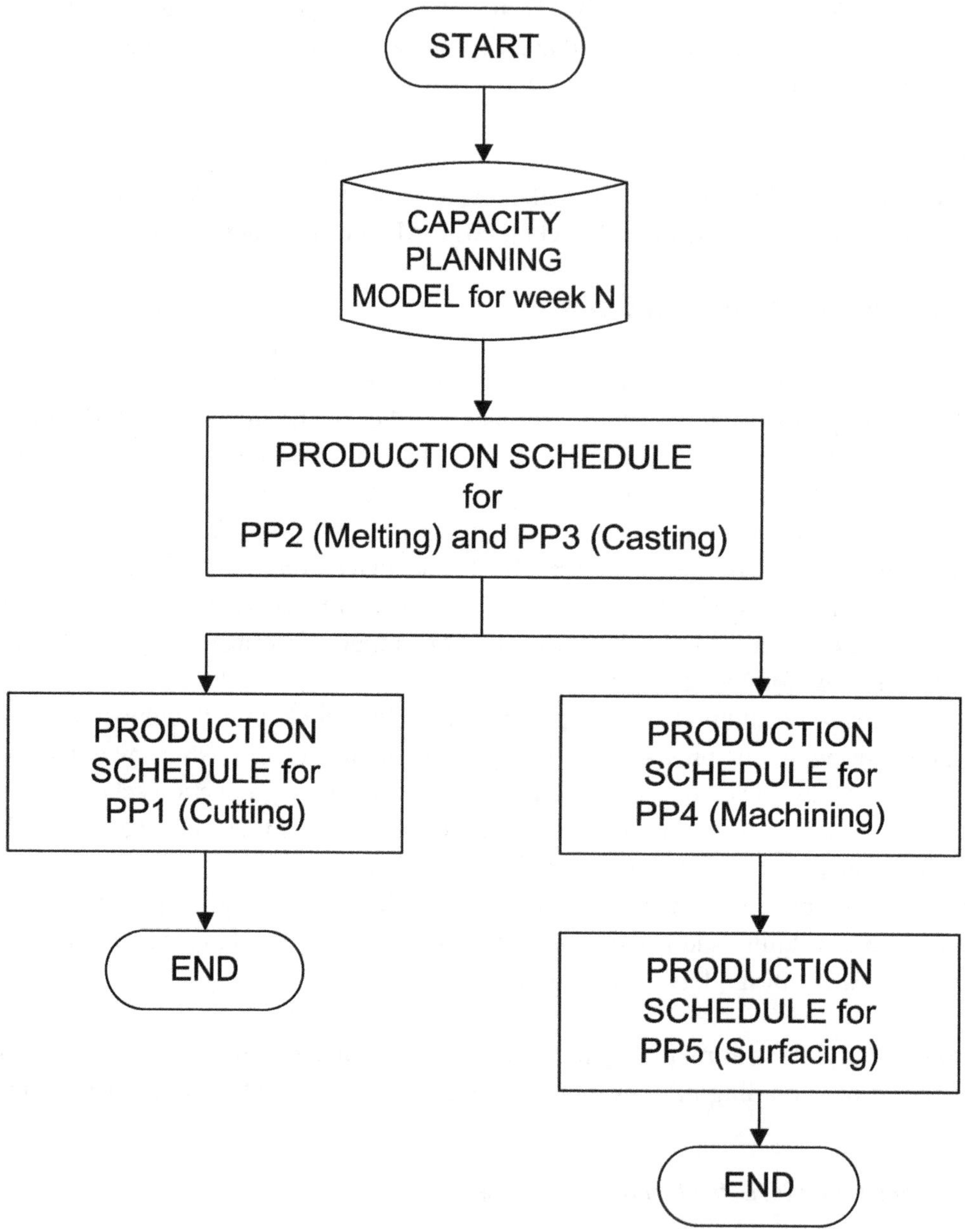

The existing technological rules determine for which of a product is suitable certain moulding press. Although ultimately any product can be produced at any moulding press, but with a different efficiency. The main product should always be produced on the main technology, so at the most appropriate moulding press. If it is not possible to provide in terms of overlapping operations, then a product can be produced at an alternative technology.

2. The daily plan for the production of alloys. The daily plan is given mainly in accordance with casting requirements for moulding presses and the alloy transition matrix. The biggest challenge is to ensure the continuous operation of the moulding machines, but which are fed from the only one furnace and thus one and the same alloy. Therefore it is necessary to cumulate the first contract for moulding presses.
3. The daily plan of cutting, welding and riveting. Each workplace is doubled, it means that there are two cutting machines, two welding machines and two riveting equipments. Plan for welding and welded parts is derived from plan of casting. Bought welded parts are ordered one week in advance for which a supplier is able to deliver them. The supplier is contractually bound to RS s.r.o.company with an assortment, which is managed by the Supply Chain Management. There are two riveting machines, where stators are riveted. The first is for small stators, the second for big stators, while there is a conjunction of middle dimensions that can riveted on both machines. The plan of cutting is derived from plans of welding and riveting on the basis of cumulation of orders.
4. The daily plan of surfacing. There are five machining CNC centres available. The first two centres are intended purely for castings and other three machining centres are designed for rotors that come directly from the casting. The newest CNC machine is determined for orders with a small number of products, because it is reconfigured fast. It is planned to include this CNC machine to produce the largest rotors in the future. Here again, any rotor can be processed on each machine, but there is optimal assignment of the rotors to the individual machines, i.e. specialization. To minimize setup times (configuration) of a machine it has to be minimized the change of rotors size. The larger is dimension jump, the longer time is needed for adjustment of machines. Another important fact to be calculated when daily plans reparation for surfacing that there is only one operator for two CNC machines in one shift.

The daily plans are not created for subsequent operations such as washing, painting, packaging etc., there are only calculations of the total time of these operations and the finish time of the production is moved forward about this calculated time.

5. The daily plan for finishing. Finishing in the current planning is not included i.e. finishing does not receive plan. There is only plan for casting and then subsequent operations of finishing are derived (known is only assortment, due date and quantity). Finishing is essentially used to as a time fulfilling, i.e. as a flexible reserve for the replenishment of stock levels.

Production scheduling determines the order of products, volumes to days and changeovers for particular machines based on capacity plan for the week N and week N+1.

The Procedure Starts at the Section U2 – Casting: The Model of Casting and Melting

1. The arrangement of orders by chemical composition, according to the number of products, the possibility of transition by transition matrix (Table 1).
2. Derivation of orders for components requiring a maximum of two batches.
3. The order with the purest alloy is selected and it is assigned to the moulding press – to the main technology for the certain product, a second batch, if it exists, is assigned to the alternate technology, if there is any.

4. The other product of the same alloy is taken, if exists, and it is assigned to the main technology, if not occupied (and the procedure is repeated from the step 3).
5. The other product with different alloy is taken, on which is the best transition and it is assigned to previously (in step 3, 4) not occupied moulding press.
6. This cycle is repeated until all products are assignedfrom KP^2_{N+1}.
7. When assigning a product to the moulding press, it will reduce the capacity of the operating time required for a given order. If the capacity of the moulding press runs out or there is no such order, which the capacity uses, the moulding press is not further taken.
8. The plan of melting and the order of moulding press supplies are calculated on the basis of this procedure.
9. The sequence of orders (and number of pieces of products) for individual moulding presses are created, from which daily and shift plans are calculated.

The Scheduling Model for the Section U1

Orders casted in week N+1 are produced at U1in the week N. It is determined the order of cutting at the cutting machines L1 and L2 based on the capacity plan for a week N for U1 and schedule for the section U2 for the week N+1 as follows:

The volume of orders cumulated in FP (family product) and individual orders are split into approximately two equal volumes:

- 2 by ½ of the number of pieces (whole orders) for riveting,
- 2 by ½ of the number of pieces (whole orders) for welding.

The rules at cutting are:

To cut large size shreds at L1 for the riveting workplace N1 – specialized in larger stators and weldments – workplace Z1;

To cut small size shreds at L2 for the riveting workplace N2 and welding workplace Z2;

To determine order of cutting according to FP and the width of the strip from the most wide to the narrowest or vice versa.

The order of riveting, as it is the last operation, is not important. The order of welding is determined from the order of pressing. The order of the welding Z1 and Z2 is according to the sequence of the cutting.

The Model of Scheduling for the Section U3

Orders casted in week N+1 are processed on the section U3 in the week N+2 as followed:

To divide cumulated product to:

O1, O2 – casts and division to O1 a O2 according to the main technology and equally usage O1 a O2,
O3 – small rotors (approximately 1/3 of week volume),

O4 – middle sized rotors (approximately 1/3 of week volume),
O5 – big rotors (approximately 1/3 of week volume).

The order of production to CNC machining centres is determined according to the criteria: from the minimum dimension to the maximum in even weeks and from maximum to minimum in odd weeks. Thereby it is achieved minimizing the number of changeovers. Or it is determined to assign those products (orders), which are the earliest due date (DD) to have enough time for finishing (Malindžák et al, 2011; Malindžák, Kačmáry & Mousssttfa, 2012).

CONCLUSION AND DISCUSSION

This real and implemented system for the company of RS Ltd. is not only the model for capacity planning itself. It was also created the model cumulating of orders and production scheduling with the evidence of work in progress and finished products for the each machine. The system, in this state, is already similar to the professional information systems for production management, with the difference, that there are integrated all the features, tools and experience they really need by the logistic workers (experts) and they use all the system properties, which were taken from their production practice. Description of such a system would have far exceeded the scope of this chapter and further description of this model is omitted.

The proposed model of capacity planning itself is a key activity of the whole system, because in this model it is created a file of possible workload that is possible to process, in the certain week with the given capacity potential of machines.

The strong side of this system is that the creation of models of CPP and production scheduling gives a big opportunity for SMEs to have the fully adapted system for their environment in relatively short time (the project lasted 6 months form the task definition, through analysis and final implementation) and with low costs. This system is not a black box for the company, because it was created on the base of their ideas, rules, principles, procedures and they can still adjust and set all the system features by themselves. However the weak side of the system is in difficult application of the system to another company because this system is a tailored solution for the given company.

REFERENCES

Bindzár, P., Kačmáry, P., & Lavrin, A. (2009). Projekt KASSETTS ako pomoc malým a stredným podnikom k riešeniu dopravného problému. *Transport & Logistics*, *6*, 28–31.

Christopher, M. (1998). *Logistics and supply chain management: Strategies for reducing costs and improving services*. Financial Times Publishing, London.

Johns, D. T., & Harding, H. A. (1989). *Operations management – A personal skills handbook*. Southampton: Gower.

Johnson, J. C., & Wood, D. F. (1996). *Contemporary logistics*. Upper Saddle River, New Jersey: Prentice Hall.

Khouri, S., & Kačmáry, P. (2009). Optimalizácia tokov informácií v modelovej firme prostredníctvom zavedenia ERP systému. *Transport & Logistics*, *6*, 125–127.

Lenort, R., Staš, D., & Samolejová, A. (2012). Heuristic Algorithm for Planning and Scheduling of Forged Pieces Heat Treatment. *Metalurgija*, *51*(2), 225–228.

Levin, R. I., Rubin, D. S., Stinson, J. P., & Gardner, E. S. r. (1989). *Quantitative approaches to Management* (7th ed.). New York: McGraw Hill Inc.

Malindžák, D. (1983). *Application of heuristic methods in manufacturing operative planning in Považské Strojárne, Považská Bystrica*. TU Košice.

Malindžák, D. (1983). Determination of the optimum sequence of heuristics. In production Management tasks. *Transaction TU Košice*.

Malindžák, D. (1997). *Production Logistics I*. Košice: Štroffek.

Malindžák, D., ... (1993). Modely pre automatizáciu a koordináciu operatívnych plánov TŠP 1700 a narážacích pecí, Záverečná správa projektu, Košice.

Malindžák, D. et al.. (2011). *Návrh systemu výrobnej logistiky a layout firmy Rosenberg Slovakia spol. s r.o.: analýza a návrh koncepcie. Project realisation*. TU Košice.

Malindžák, D., & Kačmáry, P. et al.. (2013). *Technical project: Systémová analýza a návrh algoritmov výrobného rozvrhovania výroby základný chfólií Chemosvit fólie a.s., TU Košice*. F-BERG.

Malindžák, D., Kačmáry,P., Bednárová,L., & Vrlíková, J. (2013). *The Production Capacity Planning Model in Term of Supply Chain Management*. Metal, Tanger, Brno.

Malindžák, D., Kačmáry, P., & Moussttfa, A. A. (2012). Models for capacity planning in the condition for SMEs connected to supply chain management. *Proceedings of Carpathian Logistic Congress, Tanger, Jeseník*.

Malindžák, D., Lenort, R., & Mervart, J. (2011). The logistics principles for fast flexible strategy design of the company in Crisis Time. *Managing global transition*, 9(2).

Malindžák, D., & Takala, J. (2005). Projektovanie logistických systémov: Teória a prax. Express Publicit s.r.o., Košice.

Malindžáková, M. (2011). Significance evaluation of environmental aspects. *Communications*, *13*(3), 1335–4205.

Michlowicz, E. (2002). *Podslawy logistyki przemyslowej UWND*. AGH Krakow.

Rayward-Smith, V. J., Osman, I. H., Reeves, C. R., & Smith, G. D. (1996). *Modern heuristic search methods*. John Wiley &Sons Ltd.

Rosová, A. (2010). Sústava ukazovateľov distribučnej logistiky, logistiky dopravy a materiálové hotoku ako jeden z nástrojov controllingu v logistike podniku. *Acta Montanistica Slovaca*, *15*(1), 67–72.

Schmenner, R. W. (1990). *Production / Operations Management: Concepts and Situations*. New York: Macmillan Publishing Company.

Schroder, R. (1989). *Operational management*. McGraw Hill, New York.

Takala, J., Malindžák, D., & Straka, M. et al.. (2007). *Manufacturing Strategy – Applying the Logistics Models* (p. 206). Vaasan Yliopisto.

Zimmermann, J. J., & Sorergein, M. O. (1974). Quantitation models for Production Management. PH, New Jersey.

Section 3

Resilience and Risk within Engineer-to-Order Supply-Chain Management Businesses

Chapter 8
Management of Risk and Resilience within Collaborative Business Network

Ahm Shamsuzzoha
Sultan Qaboos University, Oman

ABSTRACT

Global business communities are facing tremendous challenges from market places with respect to reduce cost and offer true customized products or services to the end customers. To cope such challenges companies are nowadays considering forming a business network with the objective to achieve several business benefits. However, to execute such business network is not risks free but always facing some problems for its continuation successfully. In such situation, it is necessary to formulate risk mitigation plan and strengthen the resilience within business network. The objective of this article is therefore to identifying and sharing risks within the collaborative business network and proposing necessary mitigating plan and resilience for it. In this research, a framework is also highlighted that provides a structural approach for identifying and assessing potential risks and resilience in business networks and their possible impacts on different levels of collaboration. The study is concluded with future research directions.

INTRODUCTION

Today's shorter product life cycle and increasing demand for customized product exerts extra pressure to the manufacturing companies to stay competitive in the markets. Due to cost pressure and achieving competitive benefits, companies are adopting globalization and business network strategies (Peck, 2006; Oshri et al., 2015). The basic concept behind collaborative business environment is that an individual company cannot successfully compete by itself. Therefore, many companies seek coordinate cross-com-

DOI: 10.4018/978-1-5225-0021-6.ch008

pany activities and work together to achieve superior performance (Varrichio et al., 2012). Companies initiates such collaborative arrangement in order to share risks and rewards. Business collaboration or networking, whether it is temporary or permanent offers higher performance and security than would be achieved by operating individually.

To be successful in today's business environment, companies must deal with intertwined layers of information, raw materials, analytical data, customer communication and service while maintaining countless secure relationships with third-party organizations such as suppliers, outsources and government regulations (Rohrbeck et al., 2013). Although the economic and organizational impacts for collaborative business are high, their vulnerabilities remain largely unexplored. Such vulnerabilities depends largely on transitional alliances, global markets for investment capital, goods and services, that can affect customers preferences, legal issues, achieved benefits and ultimately partners value (Stadler, 2015). These vulnerabilities are considered as independence risks, which are exposed across the business network that are beyond an individual partner's direct control (Parise & Casher, 2003).

Major manufacturing companies are aware of establishing business collaborations with their network partners in order to achieve associated advantages. Especially, small and medium-sized enterprises (SMEs) are participating in collaborative network with the objective to enhance their competitiveness by increasing their agility, responsiveness and adaptability. Such strategic benefits contribute them to deal with the rapid market evolutions and the globalization of business. However, there are challenges for the participating companies, especially for SMEs, which are derived from the shortage of resources and capabilities and the limitations associated to cultural barriers. As the business collaboration consists of individual and autonomous partners, each one defines its own business strategies that often may result in conflict situations among enterprises of the same business network.

Different business strategies among collaborative partners creates diversity in business goals that negatively affects overall business network's strategy. The lack of coherence and coordination among business strategies leads to misalignment in the network goals. If the conflict among business strategies that are aroused in the business network are not mitigate on time it could lead, in the end to the business collaboration. The overall business performance with higher success in the collaboration is directly related with the activation of a proper combination of strategies of each partner belonging to the network (Kleindorfer & Saad, 2005). The objective of business alliance is to provide partners appropriate mechanisms to remove the strategies misalignment problem, with the view to establish long-term collaborative relationships.

This research deals with the fundamental concepts of business collaboration, its evolution and life cycle along with associated benefits, risks and resilience, which are in combination rarely discussed in the literature. This paper discusses the role of collaborative networks as a facilitator for resilience and risk management. The focus of this research mainly to identify different type of risks, causes of the risks and their mitigation plans as evolves during business collaboration. A risk mapping and resilience management framework is also highlighted within the scope of this research.

BACKGROUND

Perspectives of Risks within Collaborative Business Network

Within the business network there exists various types of risks during its execution. The network risk sometimes is interpreted as unreliable and uncertain resources creating business processes interruptions. As in most literature, it is agreed that risk is associated with negative consequences of impact (Christopher and Lee, 2004; Paulson, 2005; Wagner and Bode, 2006). However, it is difficult to define the expectation of risk. Furthermore, the expectation of risk often cannot be described by probability (Tang and Musa, 2011). In such situation, the business network's risk can therefore be defined as the events with small probability but may occur abruptly.

The risks within the business network can be of two types such as internal and external. The internal risks are related to processes and their control mechanisms, while external risk are associated with demand and supply chain network and external environment to the network. To mitigate both types of risks network partners need to adopt an end-to-end perspective of the flows of product and related information from the raw materials source to the final delivery to the potential customers and often beyond the delivery such as after-sales service. The risks within the business network can be minimized through the identification of potential sources of risk and implementation of appropriate strategies through a coordinated approach among partners, to reduce business vulnerability (Schonsleben, 2000; Nigro & Abbate, 2011).

Risks in business network can arise from many sources, including external sources such as a natural disaster and internal sources such as a failure to integrate all functions in a business network (Thun & Hoenig, 2011). Such disruptions happen rapidly and without warning. These kinds of disruptions can also result from attempts to create a more efficient, cost-conscious business environment (Shamsuzzoha et al., 2013; McNeil et al., 2015). In many business network, logistics activities such as raw materials supply, manufacturing, and even inventory management are outsourced to suppliers that are located across the globe (Camarinha-Matos, 2009; 2014). Such a business scenario has created a supply chain-dependent environment, where any abnormalities can have a much more pronounced effect either upstream or downstream in the supply chain.

Assessment of Resilience in Collaborative Business Network

One main objective of risk management is to increasing the resilience of the network. From this perspective, the emphasis is on managing and mitigating risks and, thereby, fostering the resilience of the business network. According to Christopher and Peck (2004), the resilience can be defined as "the ability of a system to return to its original state or move to a new, more desirable state after being disturbed". The business network resilience is therefore deals with the notion of flexibility and adaptability within the collaborative business processes. Furthermore, partner organizations need to an awareness to consider the resilience of their processes as part of their overall approach to risk and business continuity management (Walters, 2011).

There are various types of operational risks found in literature within a business network, which can be summed up in a table as shown in Table 1. These risks are needed to be properly managed or mitigated through adopting appropriate resilient plan (Sutton et al., 2008). Some of the common risks, their possible causes and associated resilient plans are presented in Table 1. From Table 1, it is noticed that four common type of operational risks are found in business network namely demand uncertainty, delivery delay, cost management and profitability and resource scarcity.

COLLABORATIVE BUSINESS NETWORKS: LIFE CYCLE PHASES

To be successful in today's turbulent and competitive market environment, companies need to acquire new business models, business strategies, governance principle, processes and technological capabilities (Ermilova & Afsarmanesh 2006). There are growing challenges towards manufacturing enterprises such as increased product variety, product's complexity and quality demands, reduced product life cycle and decreased revenue margins. In order to tackle such challenges companies, especially small and medium

Table 1. Business network's operational risks, possible causes and resilient plans

No.	Risk Type	Possible Cause	Resilient Plan
1	Demand uncertainty	1. Economic situation 2. Poor supply chain 3. Increase competition 4. Product architecture 5. Decrease market share 6. Wrong information management	1. Information visibility and sharing 2. Increased planning capacity 3. Coping with both usual and unusual demand 4. Improve customer service 5. 10. Adoption of appropriate forecasting
2	Delivery delay to end customer	1. Demand fluctuation 2. Operational problems 3. Poor supply chain 4. Failure in process adaptability 5. 10. Poor planning and scheduling	1. Matching with demand and logistics capability 2. Appropriate performance measures 3. Up-to-date Inventory management 4. 8. Shorter order cycle
3	Cost management and profitability	1. Changes of fixed and variable cost 2. More investment 3. Inappropriate cost structure 4. Failure to cost control 5. 10. Non-identified cost drives	1. Increase market share 2. Assets utilization 3. Increase cash flow 4. Manage promotion and pricing 5. Improve overall performance 6. 12. Pay for performance
4	Resource scarcity	1. Demand fluctuation 2. Inefficient resource pool 3. Prioritize supplier 4. Change of investment 5. 10. Mismatch resource alignment	1. Maintain inventory replenishment 2. Promote resource pool 3. Collaborative planning and scheduling 4. Integrated policies 5. 10. Resource development

enterprises (SMEs) need to join efforts with others through collaboration. This collaboration offers companies to bring the potential of dynamically adjusting the needs in terms of dynamic inter-organizational models, distributed business process management, integration and coordination.

In case of collaborative business networks, partners' companies need to form and execute a virtual business network, which is connected by the ICT-enabled technology such as Internet or web-enabled platform. Such virtual business network is formed and executed through an independent communication infrastructure, which is supported by ICT tools. The collaboration starts after identifying business opportunity, which is ended after achieving it successfully. The complete life cycle of virtual collocation can be explored as follows:

Virtual collaboration (VC) consists of several phases in its life cycle as displayed in Figure 1. From Figure 1 it is noticed that VC consists of five phases namely infrastructure, management, creation, execution and dissolution. The VC infrastructure phase concerns with information sharing and security, process interoperability, reference model and methodology, etc., while VC management phase deals

Figure 1. Different life cycle phases within virtual collaboration

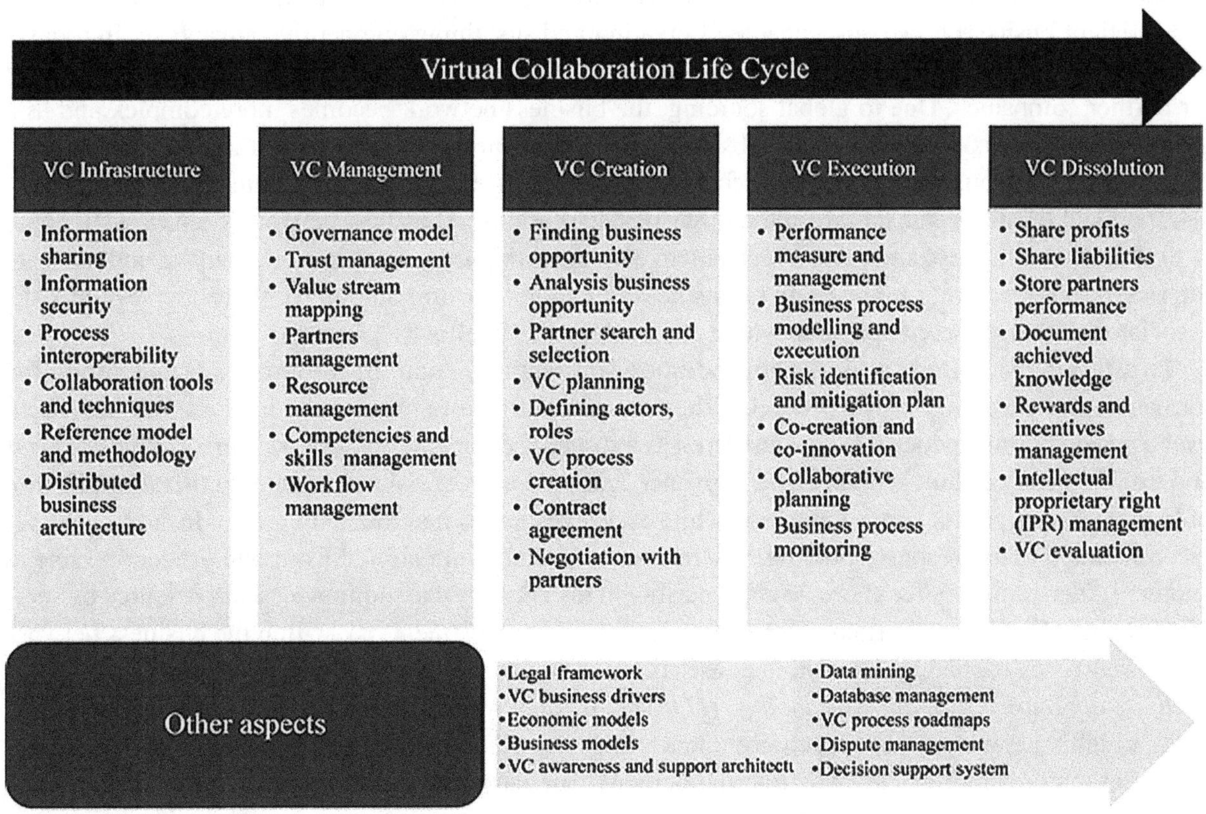

with governance model, partners management, resource management, competence and skills management etc. Identifying and analyzing business opportunity, partner search and selection process, contract agreement, negotiation with partners, etc., are confined with VC creation phase.

The VC execution phase is responsible for business process monitoring within the collaborative processes, performance measure and management, risk identification and mitigation plan etc. Final phase of the VC life cycle accounts for dissolving the collaboration. In the dissolution phase profits and liabilities are shared among partners and partners performances are stored, rewards and incentives are managed and IPR is management, etc. In addition to five phases of the VC life cycle, there are other issues like legal framework, business and economic models, process roadmaps, dispute management, etc., are included within these lifecycle phases.

POTENTIAL RISKS WITHIN BUSINESS NETWORKS

Different Types of Risks within a Business Network

Brokers who initiates the business networks strive to achieve the ideals of fully integrated, effective and efficient business processes, capable of creating and sustaining competitive advantage. In today's uncertain and turbulent markets, business network vulnerability has become an issue of significance for partner companies. Due to global sourcing, the business network becomes more complex and as a result, associated risks are increases substantially (Christopher et al., 2011). In the age of unpredictable and changing business world in the form of natural disasters, industrial dispute, government rules, geographical location, etc., have resulted in serious disruptions to business network's activities (Chopra & Sodhi, 2004). In these situations, the challenge to network partners are to plan, manage and mitigate the risk through creating more resilient business collaboration. In addition to, there are several risks associated with networked business, which can be defined as follows:

Investment risk: In any business collaboration substantial amount of resources are invested to the formation and execution of the network. Often the needs for resources can be increased or decreased over a period of time, which can be considered as a strategic choice of network partners. Both use of less and excess resources hurts financial performance and increase the risks of resource utilization. In case of low resources, production is hampered while excess resources maximize the waste. In such case, it is recommended to use resource pooling, where excess resources are shifted from one production line to another if there is a resource shortage and a partner in the network can minimize excess resource by serving distributed partners locations. There are other forms of investment risk within the business network such as currency exchange rate risk, increases of prices by the suppliers of the collaborative partners.

Risks of secure information technology (IT): The threat of information technology security is more in today's highly networked business environments. The larger the size of the business network, the greater the threat that a failure can anywhere in the network can cause failure everywhere. Any failure within the network can be devastated. The failure of information security can be occurred due to various forms and formats of Internet bugs that can cause fast-spreading infection shut down to emails, intranet, causing huge financial damage and integrity within a network. Although, IT risk is considered as one of the important issue facing business network, most business network often does not have a tested and up-to-date risk management method. As the formation and execution of business network is fully dependent on IT systems, they are also becoming more vulnerable to IT threats. It threats can be of different types such as data security risks, computer viruses, patent protection risk, legal risks etc.

It is needed to behave network partners to perform their risks analysis correctly following determining the risk level and take necessary measures accordingly. The greater use of automated information channels has the higher potential to secure information risk than manual information processing. Such risk of information security can be avoided or may be minimized through robust backup systems and well designed, well-communicated recovery processes that duplicate all the essential information.

Risk of inventory management system: The excess storage for both raw materials and finished products increase the risk of inventory management system. Inventory risk effects on three factors such as product obsolesces, product value and uncertainty of product's demand and supply. In case of excess inventory, there exists always risk of falling products prices that hurts financial performance. The inventory risks can be minimized through (i) inventory pooling, where components or finished products inventory can be distributed among the partner organizations, (ii) adapt component commonality principle in order to reduce differentiated inventory and (iii) postpone the production until confirmed order is received. From the beginning of the business network the partners need to come in a common consensus of component commonality and production postpone with the view to reduce inventory cost as well as risk. Partner companies can also minimize inventory risk by working with highly responsive supply chain, especially for complex, high-value and shorter life cycle products.

Intellectual proprietary right (IPR) risk: This risk is related when partner organizations outsource to the same manufacturers as used by their business competitors. In order to maintain competitive power, profitability and same business model this risk often has drastic and long-term implications within the collaborative network. Such kind of risk can be mitigated by keeping most of its core production processes in-house or at least under full control of the network partners. Another way to protect the IPR risk is to provide after sales service of the collaborative products by the selected partner(s) within the business network in order to prevent any possibility to share proprietary information with competitors.

Risks Structure within a Business Network

The objective of utilizing risk structure in a business network is to gain functional advantages from utilizing logical reasoning in the implementation. Thus, the way risk concept is defined has a huge impact on the possible advantages gain. Such advantages can be attained through the classification of risk concept by its other sub-concepts. Fundamentally, the pursuit outcome from inference is to have an idea of the risk's type and span of impact on the business network, if there is any.

The overall risk infrastructure in a business network can be seen in Figure 2. Figure 2 presents the central sub-concepts used to define the risk concept. From Figure 2, it is observed that business network facet has its own substructure, which is dependent on each other. Firstly, risk structure has three main phases such as deviation, effect range and reaction. Secondly, each phase has sub-phases as seen in Figure 2. In addition, the properties that define the relationship between these and risk concept are also depicted in Figure 2.

The deviation concept within the risk structure represents in a simple form the problem, inadequacy or shortcoming associated to a risk (Wagner and Bode, 2009). This concept relates to deviation attribute and has deviation type. The effect range concept of the business network risk has effect type, has source partner and ranges to business collaboration level. The concept reaction within the risk has also risk type depending on the damage level as depicted in Figure 2 too.

Working with Risks within a Business Network

Risks in a business network occur because we never know exactly what might be happen in the coming days. Although, in any business network necessary forecasting is done and do every possible analysis but there is always uncertainty about future events. This uncertainty creates risks, which might come in a huge variety of different forms (Weick, & Sutcliffe, 2007). They can appear at any segment in a business network from initial production process to final customers. In a business network, each partner's need to manage risk within both its company as well as its collaborative companies too.

Figure 2. Overall risk infrastructure in a business network

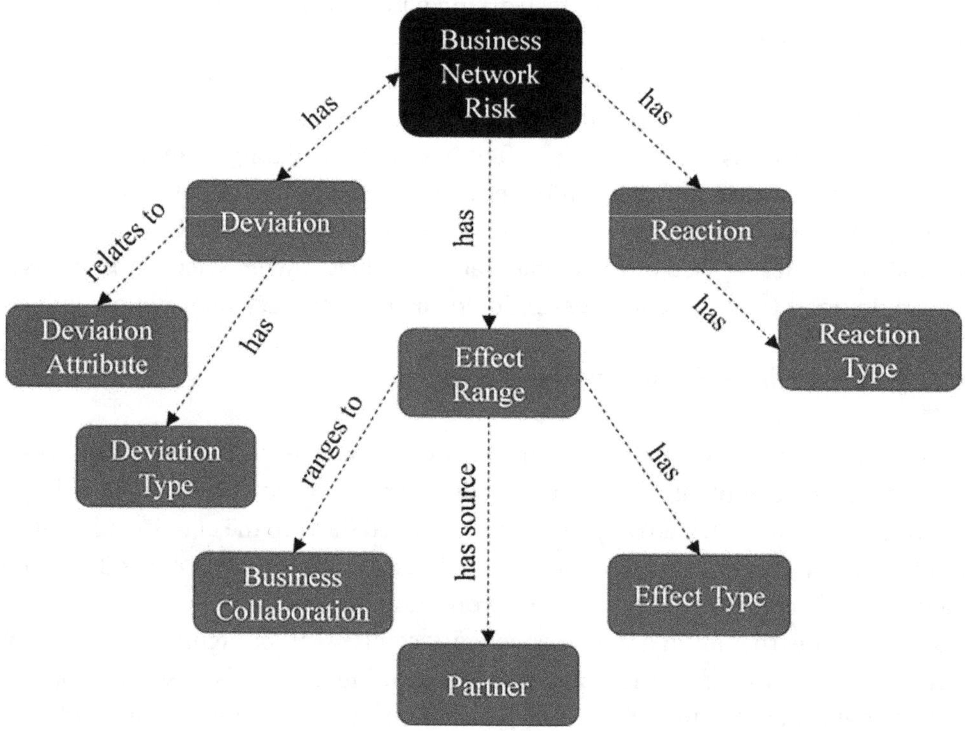

Three core elements of network's risk management are: (a) to design of a resilient business network, (b) procedure to manage risk and (c) agility. Other risk management strategies can be expressed as; identifying significant risks – which means finding events that may occur at any time, analyzing risks consequences – to find the likelihood of events and possible damage and designing appropriate mitigation plans – to define alternatives and assess their relative merits (Walters, C.D.J. (2011). Figure 3 highlights the fundamental process of risk management in business network.

PRINCIPLES OF MANAGING RISKS WITHIN BUSINESS NETWORK

Business communities are nowadays facing challengers in an unprecedented ways. Many of these challenges have been the scope to severely affect the continuity of a business network. It is clear that business network risk is currently greater now than ever before. One of the key source of such risk is the increasing turbulent business environment. Other causes of risks are evolved from reducing both the production cost, adoption of lean practices and increasing the overall network's efficiency (Christopher, 2004). These unavoidable risks can be mitigated through a systematic approach to business network risk management. Following a number of fundamental principles, underpin the creation of a more resilient business network.

Understanding the Business Network

The basic condition to create business network resilience is an understanding of the network that connects the business to its partner organizations up to the end customers. There needs to develop a mapping tool that can help to identify the weakest links or critical paths within the network. These critical points are usually characterized as bottlenecks, where there is a limit of capacity and where alternative options might not be available. Such bottlenecks often would place a heavy strain on the rest of the working systems.

Business Strategy within the Business Network

Overall business strategy might be needed to change in order to avoid possible risks in the business network. For instance, single sourcing, where one supplier is responsible to supply specific component

Figure 3. The fundamental process of business network's risk management

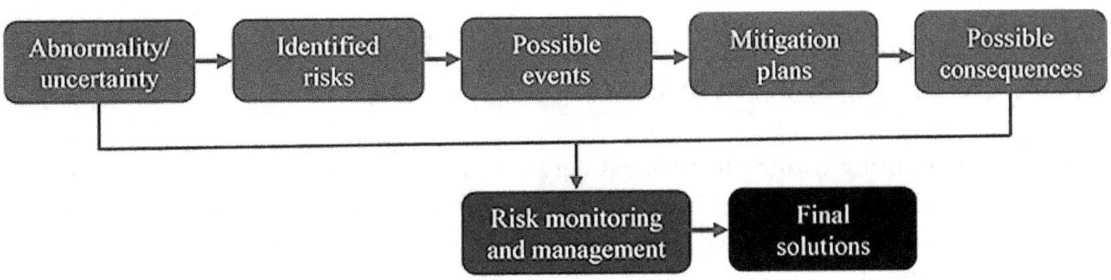

or item might be dangerous in case of supply disruption, although it might be advantageous from a cost and quality management perspective. This single sourcing strategy could be dangerous in terms of business network's resilience. In order to avoid potential risk factors it is desirable to have a lead supplier, where possible, alternative sources should be available.

Trust and Collaboration among the Business Network

It will be apparent that since business network vulnerability is by definition a network wide terminology, management of risk has to be network wide too. Tight and trusty collaboration across the business partners can help to mitigate risk. In recent days, there is a growing concern to form and execute a collaborative business environment, especially in the area of highly customized product development with shorter life cycle. Such collaboration offers added benefits through collaborative planning, forecasting and replenishment initiatives.

Agility within the Business Network

Agility is considered one of the powerful ways of achieving resilience in the business network, which offers more rapid response to changed business conditions. In agile business environment, the time required to respond to new situations is dramatically reduced. The focus of agility is to compress the time to attend abnormality. The heart of agile business strategy is to seek for opportunities for eliminating non-value activities in the business network and better focus to response unexpected events. There are two components within agility such as visibility and speed. Through visibility business network reduces uncertainty and enables demand driven business network. It also reduces risk through shared information, both upstream and downstream of the network's operational activities. In case of speed, partner companies within the network require shorter end-to-end pipelines which themselves are dependent on sourcing decisions as well as internal process improvement.

Developing Risk Management Culture within the Business Network

The development of risk management culture within the business network is also considered as an important issue to identify and manage unexpected risk. Network partners could recognize the concept of Total Quality Management (TQM) in the network in order to engender a culture, which contributed quality concern of everyone in the network. To promote the culture of risk management, an active leadership is necessary within the business network.

Developing Real-time Business Process Monitoring Tool within the Business Network

Business process monitoring within the business network describes the processes and technologies that provide real-time situation awareness in the form of analyzing key performance indicators, which is based on event-driven source of data (Shamsuzzoha et al., 2013). Through the implementation of advanced ICT tool, collaborative business processes can be effectively monitored and managed to avoid associated risks. The purpose of business process monitoring tools, which are mainly sensor-based technologies are to improve the speed and effectiveness of business operations by keeping track of what is occurring currently and raising awareness of issues as soon as they can be detected. Such tools and technologies are nowadays recognized as so called 'smart objects', which application can emit alerts about a process problem, display an event and offer drill-down features.

CREATING RESILIENT BUSINESS NETWORK: FOUNDATION TO CAPTURE RISKS

Resilience in the business network can be indeed made possible by the creation of a risk management culture in the partner organizations. A basic pre-requisite for improved network resilience is an understanding of the network that connects the business processes to its suppliers and their suppliers and to its end customers. The resilience of collaborative business network is related to change that represents the ability of a network to cope with severe disturbances or disruptions and back to its original state. The objective of resilience within business network is to make fast and effective response plan to unexpected and even disruptive changes.

There is a strong relationship among business network resilience with risk management, collaboration, information visibility and agility, which can be displayed as in Figure 4. From Figure 4, it is noticed that each of the branch of business network resilient has its sub phases that are tightly integrated. For instance, business network risk management phase is consisted of sub-phases as of risk factors, risk strategy, risk register, risk design, risk assessment and risk monitoring. In business collaboration phases the sub-phases are trust management, collaborative planning, decision-making and liability, whereas information visibility phase is consisted of information security, real time, velocity and accuracy. The agility phase, which is the center of the entire business network's resilience model, is consisted of visibility, responsive, acceleration and adaptive sub-phases. Each of the sub-phases within the resilient model has its own functionally that can be used to create a business network more resilient.

In case of an individual company, the resilience plan is more independent than the plan within a business network. It also varies from one business network to another one but the generic phases are remained same with some additional sub-phases. Risks management strategy in any business network also highly dependent on its resilience strategy. Better resilience plan contributes to eliminate or minimize the risks as occurred during execution of the business network.

The establishment of business network resilience must involves all the participating partner companies within the network. With their active involvement, a new risk management approach can be developed which will improve network governance and enhance decision making within a network. After setting the risk mitigation strategy, the network partners should agree on a timetable for long-term resilience goals. Overtime, enhanced business network's intelligence and faster information sharing need to be developed to promote greater situation awareness within the partner organizations. In today's economy, business have always faced risks, however, not all risks can be anticipated but they can be managed by the partners working together to create a resilient network. The expectations of the partners in business network are nowadays higher than ever, and business collaborations that are more resilient will experience more rewards from market segments with respect to better customers and partners' loyalty.

Figure 4. Creating the resilient business network

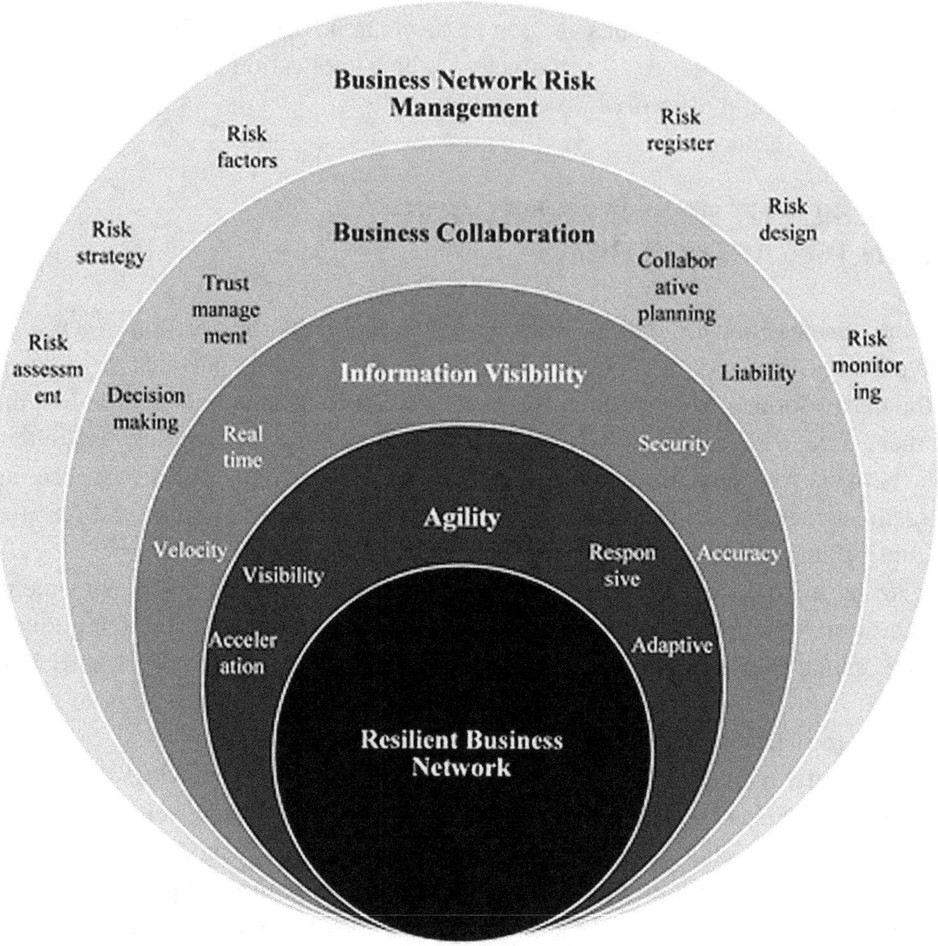

CONCEPTUAL FRAMEWORK OF BUSINESS NETWORK RESILIENCE AND RISK MANAGEMENT

The proposed conceptual framework as presented in this research highlights the relationship between business network reliance with risk with respect to business network's ICT solutions point of view. This framework will contribute to minimize the overall risks within a business network. The conceptual framework is presented in Figure 5. From Figure 5, it is noticed that there is a substantial interdependencies of the business network's ICT solutions with network's resilience and risk management.

The network's ICT solutions are dependent on information management, which monitor the business processes. This business process-monitoring component is directly interfaced with supply chain management (SCM), enterprise resource planning (ERP) and manufacturing execution systems (MES) components. The SCM within the network ensures the supply chain activities in order to create net value,

Figure 5. Conceptual framework of the relationship between business network resilience and risk management

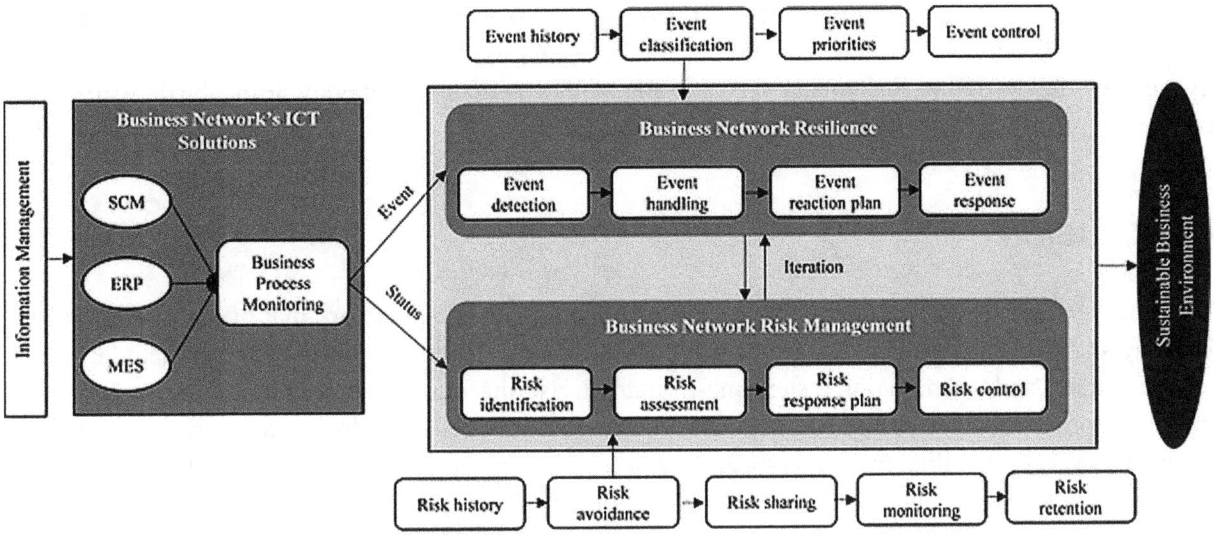

building a competitive infrastructure and synchronizing supply with demand within the collaborative processes. The ERP provides an integrated view of collaborative business processes; often in real-time – typically a business network can use to collect, store, manage and interpret data from various business activities such as planning, manufacturing, inventory management etc. The MES provide the right process information at the right time to improve production output. It works in real-time to enable the control of the various collaborative production processes.

The business network resilience initiates when an event is created from business process monitoring component as displayed in Figure 5. The resilience process starts with detecting an event followed by event handling, event reaction plan until finally response an event. In the event detection phase, critical event is identified and required steps are taken to handle it properly. Necessary event reaction plan is also prepared well ahead in order to response an event efficiently. In addition to adopt such steps to response an event finally, there are also other sub-processes such as event history, event classification, event priorities and finally event control which are also directly interfaced with the network resilience to support event monitoring and management process.

In order to know the risk level within a business network the business process-monitoring component within the network's ICT solutions evokes a status message to the network level. This status message initiates the network's risk management, which starts after risk identification followed by risk assessment, risk response plan until the end process risk control. In the risk identification phase, all the possible risks are identified based on the risk levels, which are then assessed according to their possible damage levels. Depending on the damage levels, all the identified risks are responded according to the prescheduled plan. Finally, the risks are control based on the predefined plan. In addition to such controlling of network's risks, other sub-processes such as risk history, risk analysis, risk sharing, risk monitoring and risk retention are also support the business network's risk management process.

Within the presented integrated framework, the business network's resilience and risks management plan are iterated between each other in order to manage both domains effectively. More resilience the network is more risk free it is too. Finally, if the overall relationships among the business network's ICT solutions, resilience and risk management plan are coordinated effectively it will contribute to maintaining sustainable business environment. The sustainable business environment within a business network can create value for customers, partners and the environment.

CONCLUSION

Dynamic relationships between companies are resulted the business network that offers innovativeness in today's competitive market. The principal benefits of networking as identified so far include: access to new markets and technologies, faster time-to-market, risk sharing, safe guard property rights, resource pooling, complementary skills, etc. The benefits of business collaboration or network is not only to achieve survival capability, but it also offer exploitation of new business opportunities. Development of today's so-called Future Internet facilitates to spreading collaborative networks in many domains. Due to such wider applications, the business networks are facing new challenges and risks from the surrounding unstable environments. In such unstable situations, the collaborative business network paradigm can support organizations resilience that contributes to manage risks or abnormalities.

The objective of risks management in business network is to minimize potential loss by identifying, prioritizing and safeguarding business processes that need protection, with goal of the partner organizations being able to reduce risks exposure and save valuable resources in the event of abnormalities or events. A properly prepared risks plan should detail a timely and complete response to a specific risk or a cluster of risks. This risk plan to be resilient or prepared to respond to and store operations after an unnoticed disruption occurs. There can be three ways to manage network risks, firstly, to address primarily the probability of network risks (e.g. avoidance), secondly, risk actions focusing on the risk effects (e.g. risk sharing or continuity plans) and, thirdly, any actions which increase the knowledge about business risks.

The resilient and risk management framework as presented within the scope of this research can use to recover from a disaster that cause a disruption to business networks operations. The potential impacts of a business network disruption and the risk exposure created by a disruption make it critical to identify the appropriate managerial preparation and response for planning and responding to these types of situations in a way that allows the network to maintain a certain performance level. The emergence of flexibility within the network is considered as an important strategic capability, with potential to enhance networks ability to minimize the impacts of a disruption.

The concept of networks resilience can be represented as a multidimensional phenomenon. It is relatively new concept within the business network and it requires a clear understanding and conceptualization. Network's resilience can be defined as the adaptive capability of the network to prepare for unexpected events, respond to abnormalities and recover from them by maintaining continuity of operations at the desired level. In any process abnormalities, adaptive capability is noted as a critical component of resilient ecosystems. In resilience business network, all collaborative processes maintain the same control over structure and function when an abnormality or disruption occurs.

This paper responds to the need to enhance common understanding of business network, its resilience and associated risks management plan. Although, the works is presented theoretically but future research can be done to present such ideas in real life collaborative networks. Other possibilities to make the business network more agile and flexible can be studied in the future research work. Additionally, a Web-based collaborative risk management tool can be designed and developed to support the real-time risk response plan and to make the business network more resilient.

REFERENCES

Camarinha-Matos, L. M. (2009). Collaborative networked organizations: Status and trends in manufacturing'. *Annual Reviews in Control, 33*(2), 199–208. doi:10.1016/j.arcontrol.2009.05.006

Camarinha-Matos, L. M. (2014). Collaborative networks: a mechanism for enterprise agility and resilience. In K. Mertins, F. Benaben, R. Poler, & J.-P. Bourriers (Eds.), *Enterprise Interoperability VI: Interoperability for Agility, Resilience and Plasticity of Collaborations*. Switzerland: Springer International Publishing. doi:10.1007/978-3-319-04948-9_1

Chopra, S. & Sodhi, M. (2004). Managing risks to avoid supply-chain breakdown. *MITSloan Management Review*, 2004(Fall).

Christopher, M. (2004). Creating resilient supply chains. *Logistics Europe*, 2004, 14-21.

Christopher, M., & Lee, H. (2004). Mitigating supply chain risk through improved confidence. *International Journal of Physical Distribution & Logistics Management, 34*(5), 388–396. doi:10.1108/09600030410545436

Christopher, M., Mena, C., Khan, O., & Yurt, O. (2011). Approaches to managing global sourcing risk. *Supply Chain Management: An International Journal, 16*(2), 67–81. doi:10.1108/13598541111115338

Christopher, M., & Peck, H. (2004). Building the resilient supply chain. *The International Journal of Logistics Management, 15*(2), 1–14. doi:10.1108/09574090410700275

Das, T. K., & Teng, B.-S. (1999). Managing risks in strategic alliances. *The Academy of Management Perspectives, 13*(4), 50–62. doi:10.5465/AME.1999.2570554

Ermilova, E., & Afsarmanesh, H. (2006). *Competency and profiling management in virtual organization breeding environments. Network-centric collaboration and supporting frameworks* (pp. 131–142). Boston: Springer. doi:10.1007/978-0-387-38269-2_14

Kleindorfer, P. R., & Saad, G. H. (2005). Managing disruption risks in supply chains. *Production and Operations Management, 14*(1), 53–65. doi:10.1111/j.1937-5956.2005.tb00009.x

Lo Nigro, G., & Abbate, L. (2011). Risk assessment and profit sharing in business networks. *International Journal of Production Economics, 131*(1), 234–241. doi:10.1016/j.ijpe.2009.08.014

McNeil, A. J., Frey, R., & Embrechts, P. (2015). *Quantitative Risk Management: Concepts, Techniques and Tools*. New Jersey, USA: Princeton University Press.

Oshri, I., Kotlarsky, J., & Willcocks, L. P. (2015). *The Handbook of Global Outsourcing and Offshoring: The Definitive Guide to Strategy and Operations* (3rd ed.). UK: Palgrave-Macmillan. doi:10.1057/9781137437440

Parise, S., & Casher, A. (2003). Alliance portfolios: Designing and managing your network of business-partner relationships. *The Academy of Management Perspectives*, *17*(4), 25–39.

Paulson, U. (2005). Developing a supply chain flow risk model. *Proceedings of the NOFOMA 2005 Conference*, Copenhagen, Denmark (pp. 9-10).

Peck, H. (2006). Reconciling supply chain vulnerability, risk and supply chain management. *International Journal of Logistics Research and Applications*, *9*(2), 127–142. doi:10.1080/13675560600673578

Rohrbeck, R., Konnertz, L., & Knab, S. (2013). Collaborative business modelling for systemic and sustainability innovations. *International Journal of Technology Management*, *63*(1/2), 4–23. doi:10.1504/IJTM.2013.055577

Schonsleben, P. (2000). With agility and adequate partnership strategies towards effective logistics networks. *Computers in Industry*, *42*(1), 33–42. doi:10.1016/S0166-3615(99)00059-7

Shamsuzzoha, A., Kankaanpaa, T., Carneiro, L. M., Almeida, R., Chiodi, A., & Fornasiero, R. (2013). Dynamic and collaborative business networks in the fashion industry. *International Journal of Computer Integrated Manufacturing*, *26*(1-2), 125–139. doi:10.1080/0951192X.2012.681916

Stadler, H. (2015). Supply chain management: an overview. In H. Stadler, C. Kilger, & H. Meyr (Ed.), Supply Chain Management and Advanced Planning (pp. 3-28). Springer Berlin Heidelberg. doi:10.1007/978-3-642-55309-7_1

Sutton, S. G., Khazanchi, D., Hampton, C., & Arnold, V. (2008). Risk analysis in extended enterprise environments: Identification of critical risk factors in B2B e-commerce relationships. *Journal of the Association for Information Systems*, *9*(3-4), 151–174.

Tang, O., & Nurmaya Musa, S. (2011). Identifying risk issues and research advancements in supply chain risk management. *International Journal of Production Economics*, *133*(1), 25–34. doi:10.1016/j.ijpe.2010.06.013

Thun, J.-H., & Hoenig, D. (2011). An empirical analysis of supply chain risk management in the German automotive industry. *International Journal of Production Economics*, *131*(1), 242–249. doi:10.1016/j.ijpe.2009.10.010

Varrichio, P., Diogenes, D., Jorge, A., & Garnica, L. (2012). Collaborative networks and sustainable business: A case study in the Brazilian system of innovation. *Procedia: Social and Behavioral Sciences*, *52*, 90–99. doi:10.1016/j.sbspro.2012.09.445

Wagner, S. M., & Bode, C. (2006). An empirical investigation into supply chain vulnerability. *Journal of Purchasing and Supply Management, 12*(6), 301–312. doi:10.1016/j.pursup.2007.01.004

Wagner, S. M., & Bode, C. (2009). Dominant risks and risk management practices in supply chains. In G. Zsidisin, & B. Ritchie (Eds.), Supply Chain Risks: A Handbook of Assessment, Management, and Performance (Vol. 129, pp. 271-290). doi:10.1007/978-0-387-79934-6_17

Walters, C. D. J. (2011). *Supply Chain Risk Management: Vulnerability and Resilience in Logistics* (2nd ed.). UK: Kogan Page Limited.

Weick, K. E., & Sutcliffe, K. M. (2007). *Managing the Unexpected: Resilient Performance in an Age of Uncertainty*. San Francisco, USA: Jossey-Bass Publisher.

Section 4

Strategies in Supply-Chain Management and Engineering-to-Order Processes

Chapter 9
Supply Chain Design Approaches for Dual Demand Management Strategies

Can Celikbilek
Ohio University, USA

Gürsel A. Süer
Ohio University, USA

ABSTRACT

Supply chain management involves efficiently integrating suppliers, manufacturers, warehouses, stores, and customers. To survive in a highly competitive business environment, manufacturing, resource planning and scheduling and distribution operations are the significant drivers that need to be optimized within supply chain management. In this chapter, we design the supply chain system considering dual demand management strategies simultaneously for the same company, both engineer-to-order (ETO) and make-to-order (MTO). This research has been inspired from the window manufacturer which manufactures and distributes vinyl windows to meet new construction and replacement/remodeling sector demand in the State of Ohio. The company manufacturers windows based on make-to-order strategy for new construction projects and at the same time builds replacement windows to individual customer specifications in very small quantities to be used for replacement in homes. In this study, a total of 174 individual customers and six big contractors are considered throughout the State of Ohio. This paper proposes to separate products based on demand management strategy and develop different supply chain networks for each group. The idea is to design bigger facilities for high volume (make-to-order products) as transportation cost per unit is reduced due to economies of scale whereas to place smaller and more facilities for low volume engineer-to-order products to be closer to the clients where it may not feasible to carry a only few products over long distances. All in all, this study provides nested models to integrate both design and operational aspects of supply chain system in the presence high-volume and low-volume of window products. Moreover, all location, design and manufacturing operations are performed by considering

DOI: 10.4018/978-1-5225-0021-6.ch009

new mathematical models (mixed-integer and integer mathematical models) and heuristics in engineer-to-order demand management environment. Normally-distributed, probabilistic demand environment is considered in our design and operational phase of the study. Preliminary results show that, each design has its own strategic advantage and outcome and the ultimate objective has been accomplished in our design in this study. Briefly, four manufacturing facilities are established to meet the demand of replacement/remodeling sector and two manufacturing facilities are situated to meet the demand of new construction sector. The results revealed that, 29 layered-cells and a total of 200 machines are opened and utilized for replacement/remodeling sector. Additionally, 15 layered-cells and a total of 104 machines are needed to cover the entire demand of new construction sector. Also for the new construction sector, three distribution centers are needed to facilitate the products over the region. In terms of daily cell loading and scheduling phase, the results are almost doubled in replacement/remodeling sector demand compared to new construction sector demand volume due to having more cells and machines in the new construction design strategy. The supply chain work involves location and number of manufacturing facilities, number and location of distribution centers, detailed design of manufacturing systems and performing scheduling to confirm the validity of the manufacturing system design.

1. INTRODUCTION

Supply chain management (SCM) is described as a "set of approaches utilized to efficiently integrate suppliers, manufacturers, warehouses, and stores, so that merchandise is produced and distributed at the right quantities, to the right locations, and at the right time, in order to minimize the systemwide costs while satisfying service level requirements" (Simchi-Levi, 2009).

Simchi-Levi (2009) emphasizes that the objective of supply chain management is to be efficient and cost effective across the entire system. The overview of a supply chain system is illustrated in Figure 1.

Chopra (2007) indicated supply chain network design decisions that include the assignment of facility role, location of manufacturing, storage, or transportation-related facilities and the allocation of capacity and markets to each facility. In this dynamic supply chain, facility location decisions have significant impact on the performance of supply chain (Simchi-Levi, D., 2009). Thus, multiple factors such as distance, transportation cost, land cost, plant capacities and customer demands are considered for the facility location and capacity allocation model in this study.

1.1. Challenges of Supply Chain Network Design

To survive in a highly competitive business environment, manufacturing and distribution operations are the two significant drivers that need to be optimized within supply chain management. To achieve operational excellence in a supply chain, it is critical to integrate these two operations jointly and schedule them accordingly in a coordinated fashion. However, due to the highly dynamic market, it is not that easy for the companies. Due to fierce competition and globalization, overall demand keeps increasing and demand uncertainties become a prominent issue. In order to compensate that, companies start to build more safety stock and hold excess capacity that negatively affects the costs. In both manufacturing

Figure 1. An overview of Supply Chain Network

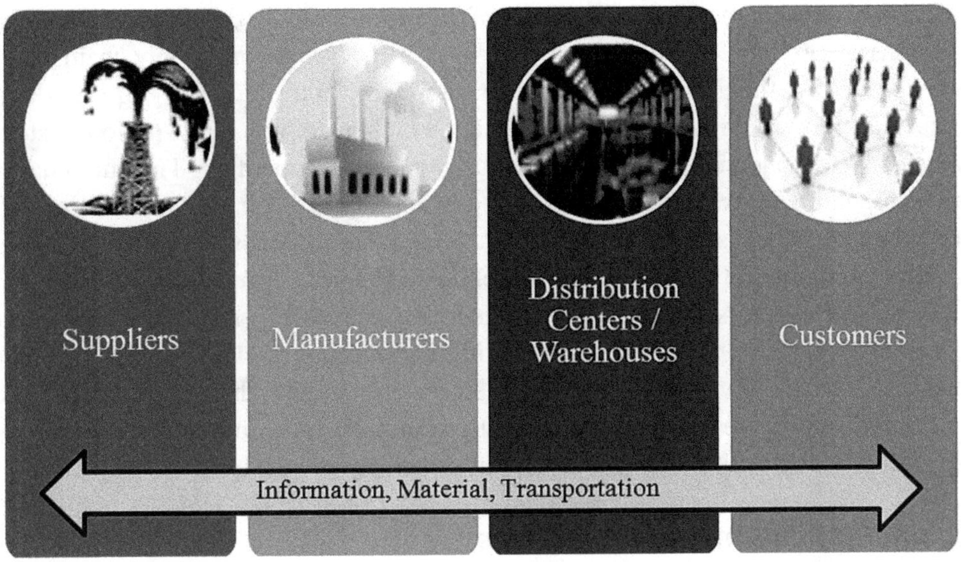

and distribution operations, facility location decisions are always a strategic challenge. By having few centralized facilities, a company may benefit from economies of scale and increase efficiency; however, they might reduce responsiveness to their customers. On the other hand, having smaller distributed facilities may improve customer responsiveness but increases the complexity of the product and process flow. Thus, in this research we will be analyzing and optimizing facility location decisions by applying quantitative models considering various capacity, transportation and cost parameters.

Additionally, resource planning and scheduling determines how to allocate different products, resources and services to suppliers, manufacturers, distribution centers and customers over a period of time. It involves operational planning level decisions such as determining how many products should be supplied, manufactured, scheduled and delivered sequentially in a timely manner. Diversity of products and widespread geographical locations also triggers companies to focus on their manufacturing processes and adjust their production plans. In this chapter, we will be focusing on resource allocation and design in manufacturing processes and those with distribution operations in supply chain hierarchy.

1.2. Integrated Supply Chain Design

This study is motivated by the need for integrated supply chain network models and the currently limited available research. A few research works have been discussed in the literature related with integrated supply chain network design that jointly considers different aspects of supply chain network which considers strategic, tactical and operational level of decisions. Recently, Huang (2012) approached the supply chain design problem by considering two major decisions within supply chain management; resource management and sourcing strategies. Briefly, those strategies determine "which products go where at what amount and what time" in order to fulfill the customers' requirements. Mainly, facility location/

allocation decisions, cellular manufacturing system design, purchasing decisions and supplier profiles are investigated in order to demonstrate the ability of capturing supplier dynamics and making the right sourcing decisions accordingly. On the other hand, Ates (2013) tackled the supply chain network design problem by incorporating the global competition in supply chain design. The framework of the study is to incorporate layered cellular manufacturing design and then employ multi-period strategic analysis where various competition games are studied considering price competition. The outcomes of different strategies are presented and evaluated in terms of profitability and market share. More recently, Xia (2013) proposed an integrated supply chain network design that incorporates facility location problem, inventory control problem and vehicle routing problem. Inventory level decisions are performed for each opened distribution centers and this study also incorporates mixed-product transshipments between distribution centers with multiple truck size selection and routing delivery to retailers. Supply chain network design is evaluated by considering direct shipment and also through distribution centers for the same set of products. However, the manufacturing system design component is not discussed in the study which primarily affects the structure of supply chain network design.

1.3. Research Motivation and Objectives

In this study, we design the supply chain system considering dual demand management strategies simultaneously for the same company, both engineer-to-order and make-to-order. This research is inspired from the window manufacturer which manufactures and distributes vinyl windows to meet new construction and replacement/remodeling sector demand. The company manufacturers windows based on make-to-order strategy for new construction projects and at the same time builds replacement windows to individual customer specifications in very small quantities to be used for replacement in homes.

Accordingly in this research, supply chain design will be performed by considering vinyl windows' new construction sector demand volumes and replacement/remodeling sector demand volumes. The research aims to conduct a comprehensive joint manufacturing and distribution system design, and is expected to make the following contributions:

- Provide nested models to integrate both design and operational aspects of supply chain network.
- Design a supply chain system in the presence of a mixture of high-volume and low-volume products.
- Manage all location and manufacturing operations for products with high and low volume.
- Propose new mathematical models, heuristics, and layered cellular manufacturing system design for integrated production-distribution system in engineering to order demand management environment.

This research covers the multiple aspects of the supply chain design with different demand volumes and demand management strategies. The overall flow of the research is shown in Figure 2.

Figure 2. Overall flow of the research

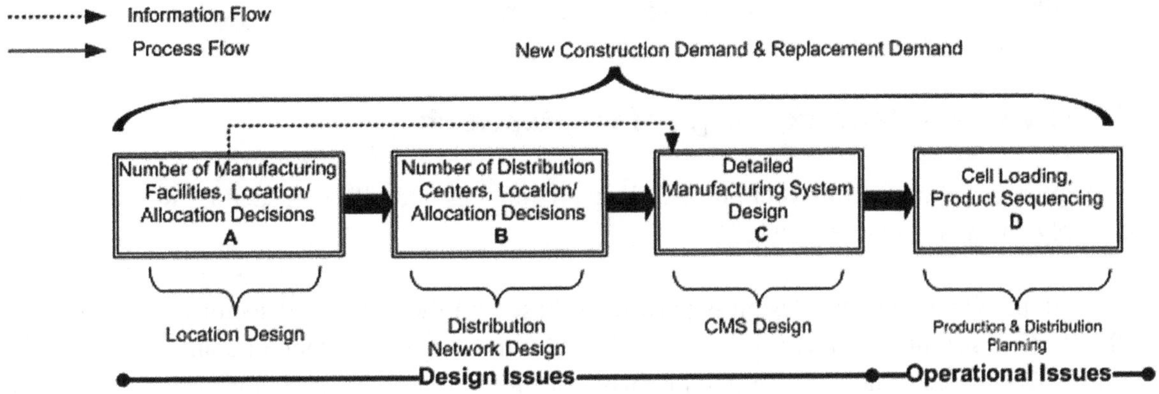

2. RESEARCH PROBLEM

The company that was considered as an inspiration has a product spectrum that includes vinyl windows, vinyl siding, aluminum coil and steel siding accessories. The company has multiple manufacturing facilities, multiple distribution centers/distributors and multiple customers all across the state of Ohio. According to the company's market strategy, approximately 65% of company's net sales are generated from the remodeling market and 35% of net sales are generated from the new construction market. The study only considers the vinyl windows and its 24 different product types.

2.1. Research Scope

In this chapter, we consider two different multi-product supply chain network designs. There has been limited available research discussing multi-product multi-level supply chain network design problem. Under a multi-product supply chain system, distribution centers (D/Cs) play a significant role. Distribution centers receive products from multiple manufacturing facilities and then deliver those mixed products to retailers or customers through consolidation of shipments by using vehicles. The advantage of network design with distribution centers/consolidation centers can be significant, especially when the manufacturing facilities are far from retailers or customers. The advantage of rapid replenishment, shorter transportation lead-time and faster delivery time can be considered as an advantage. However, investment cost/fixed cost, labor & utility cost, overhead and storage/inventory carrying cost can be considered as the drawbacks of having distribution centers. On the other hand, lower investment cost/fixed cost, labor & utility cost, overhead, storage/inventory carrying costs can be considered as the advantages of point-to-point network design. Another important consideration is to take the advantage of economies of scale. Economies of scale can be considered as the decrease of unit transportation cost when the shipment quantity increases. In point-to-point network design, since the facility is further from customers, the company is able to take the advantage of special pricing for products and services (such as bulk pricing, volume discounts). Figure 3 illustrates both distribution center network design and point-to-point network design.

2.2. Supply Chain Network Design of the Research

In this chapter, based on the different product demand volumes and different sectors, we will have different supply chain network designs.

1. The supply chain network design is performed based on the individual sector demand volumes. Similarly, distribution centers will be considered for meeting the demand of contractors in the new construction sector and point-to-point network design will be considered for meeting the demand of the replacement/remodeling customers (shown in Figure 4).

Figure 3. Distribution Center Design vs. Point-to-Point Network Design

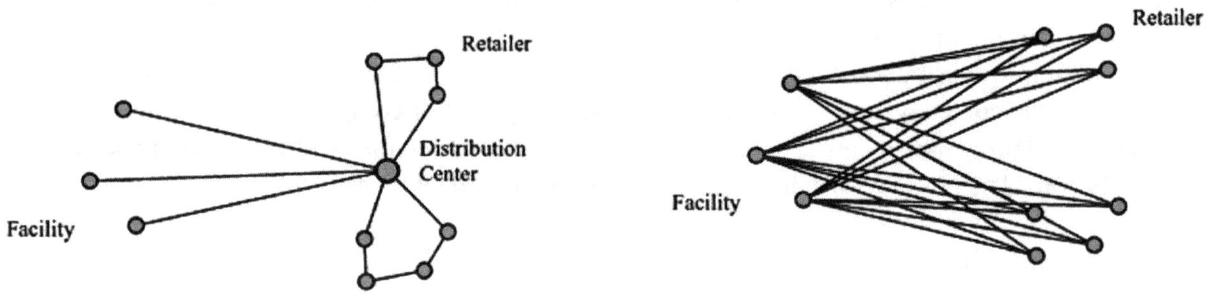

Figure 4. Illustration of two different supply chain designs

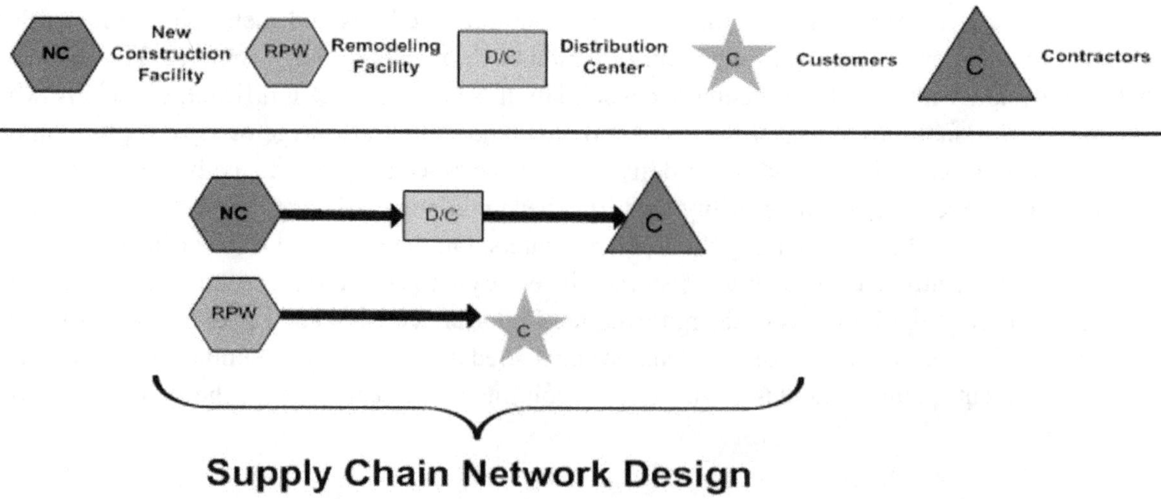

2.3. Justification of the Research

Highly competitive industry standards require companies to manage supply chain operations in a concurrent manner. Well designed, planned and implemented supply chain operations are necessary for the companies to be able to compete with their peers. Therefore, there is a need to concurrently design, manage, schedule and control the manufacturing and distribution operations of a company. The main emphasis of this chapter is to design a nested supply chain network in which the facility location/allocation decisions, cellular manufacturing system design and network system design are highly related to each other. Briefly, hierarchical joint input-output supply chain system design and analysis is conducted for both high volume and low volume products.

In the design phase, first of all, the number, location and allocation of manufacturing plants are determined for the specified region with a minimum cost based on the annual demand. Next, the number, location of D/C's and the served customer zones are determined based on the annual demand with a minimum total cost. Later, cellular manufacturing system design is conducted for the opened manufacturing plants based on the annual demand in the design phase. The design phase is conducted for individual (replacement and new construction) window demand volumes. In the operational phase, cell loading and product sequencing is performed.

The main objectives are summarized as follows:

1. To determine and design the locations of manufacturing facilities (M/Cs) and allocation to customers for a specified region based on annual demand
2. To determine and design the locations and number of distribution centers (D/Cs) as well as identify the customer zones of each distribution center based on annual demand, also a new heuristic approach will be proposed for this phase.
3. To design a CMS for each opened manufacturing facilities based on annual demand
    a. Identification of part families for alternative designs,
    b. Determine the cell configurations (Demand Coverage Probabilities & Expected Cell Utilizations),
    c. Determine the cell types (dedicated, shared or remainder cells) by using a mathematical model
    d. Group scheduling (cell loading) by using a mathematical model
4. To propose a cell loading and product sequencing mathematical model based on weekly/daily demand,
5. To explore the critical features of between high-volume (make-to-order) supply chain system design and low-volume (engineer-to-order) supply chain system design.

An overall representation of the considered system can be illustrated as in Figure 5, basically we have manufacturing center(s), distribution center(s), contractors, and customers throughout the Ohio region.

Figure 5. Overall Illustration of a typical network structure

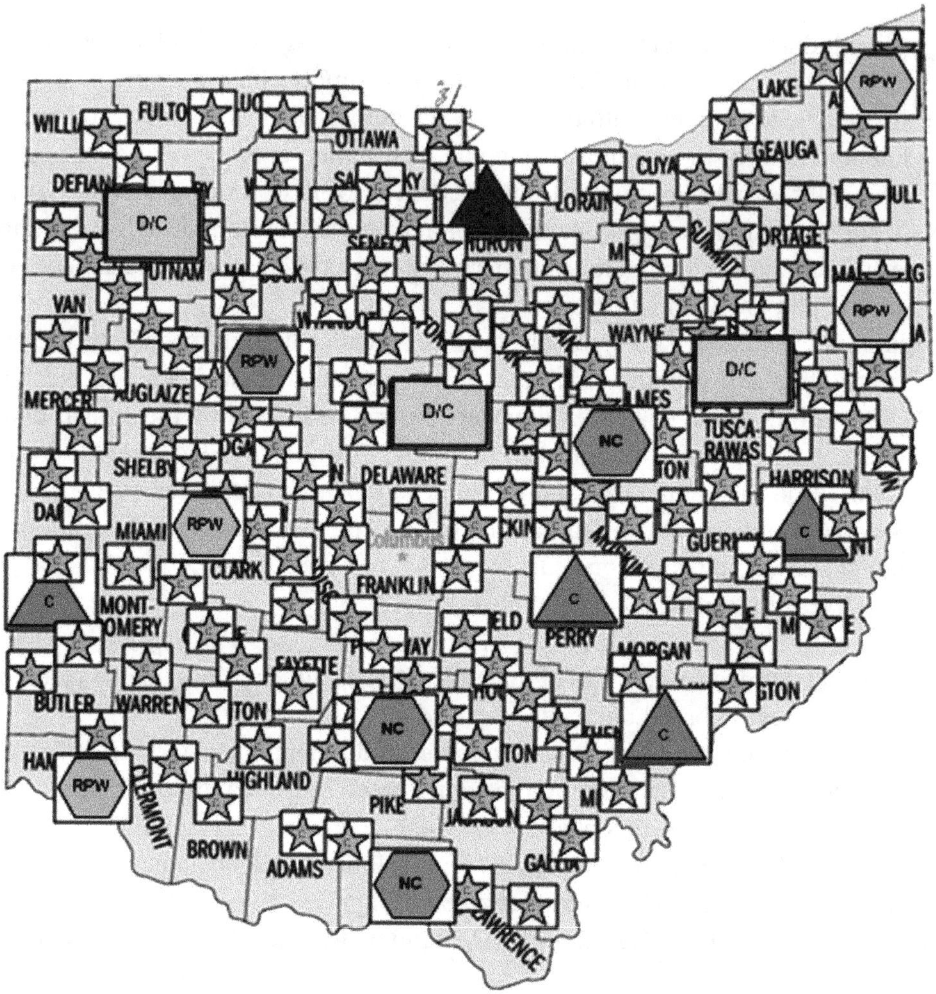

3. LITERATURE REVIEW

This section briefly discusses the literature on supply chain design, cellular manufacturing design and cell loading literature.

3.1. Supply Chain Design

These types of problems are considered as tactical level problems in the literature. Firstly, Cohen and Lee (1989) studied a four-stage model where demand of each product is deterministic and a penalty cost is associated if the demand is not satisfied. The study covered three echelons (plant, D/C and market region) within a supply chain environment. International issues such as taxes in different countries and offset requirements (minimum number of orders that are needed to enter into the country) are considered in the model. The model is formulated as mixed-integer nonlinear program and inspired from a PC manufacturer. A heuristic approach is also developed for the problem.

Chen and Wang (1997) studied three-echelon (multiple suppliers, multiple plants and multiple customers) problem that is also motivated by the steel production and distribution sector. The problem is formulated as a linear programming (LP) problem and the demand values of products are considered as deterministic. The objective of the model was to find the maximum profit for the company. Sabri and Beamon (2000) considered a four-echelon supply chain network in which the demand values are treated in a deterministic fashion. Fixed costs associated with D/C's and transportation costs between D/Cs and customer zones were considered. Two objectives were considered for this study and these were; i) total cost and ii) volume flexibility. The strategic sub-model of the problem was formulated as a multi-objective mixed-integer programming and also with two operational sub-models (links suppliers and factories) are formulated as a non-linear mathematical model. An overall iterative procedure is proposed that combines strategic and operational decisions.

Dhaenens-Filipo and Finke (2001) studied a three echelon supply chain problem that considers multiple products, multiple plants, multiple warehouses and multiple customers. The demand is considered as deterministic and should be met in a just-in-time fashion. There is a fixed setup cost for producing a product on each production line and transportation cost is considered as a linear function. The model is formulated as a MIP and solved by CPLEX. Amiri (2006) studied the production and distribution network problem by considering plants, warehouses and customers in order to minimize the total costs of distribution network for a single product. Amiri (2006) developed a MIP model as well as an heuristic approach for solving larger instances of the problem. Moreover, Russell (2008) studied multi-product production and distribution problem as a newspaper logistics problem. This paper introduced a tabu-search meta-heuristics to aid the coordination and synchronization of the production and delivery of newspaper in bulk delivery locations. The vehicle routing component of the problem considered the VRP with time-windows with different zoning constraints. Pan and Nagi (2013) studied supply chain network design problem in agile manufacturing scenario with multiple echelons in multiple periods in which multiple customers have heavy demand. Production, inventory and transportation component is formulated to minimize the total transportation cost including fixed alliance costs between two companies, production, raw material holding, finished products holding, and transportation costs by considering production and transportation capacity limits. A langrangian heuristic is proposed. The main drawback of this paper is to consider single product which doesn't reflect the real supply chain structure of the companies.

3.2. Cellular Manufacturing System Design

Initially, several types of integer programming formulations have been proposed over the last three decades by a number of researchers: Purcheck (1974) studied the group technology problem and utilized linear programming. Kusiak (1987) proposed a p-median formulation and showed the improvement in the quality of cell formation. Shtub (1989) introduced Generalized Assignment Problem (GAP) which is used for simple cell-formation problem. Rajamani, Singh and Aneja (1990) implemented three integer programming models for CMS design and assessed the impact of alternative process plans. Wei and Gaither (1990) applied an optimal 0-1 integer programming model to provide an analysis for the CMS design problem in order to minimize the cost of manufacturing exceptional products outside of the cellular manufacturing system by considering the machine capacity constraints. Kamrani, Parsaei and Leep (1995) proposed a mathematical modeling approach and tested the performance of the proposed model with discrete event simulation in four phases. Heragu and Chen (1998) developed a mathematical modeling approach which encompasses benders' decomposition to CMS design problem. Continuingly, Chen (1998) introduced an integer mathematical model to minimize the material handling, machine costs and cell reconfiguration cost for a multi-period planning horizon. Akturk and Turkcan (2000) developed an integrated algorithm to solve the product-family and machine-cell formation problem by concurrently considering the within-cell layout problem. Won and Lee (2004) proposed a modified p-median approach for efficient cell formation with the objective of maximizing the sum of similarities between machines in the same cell. In real life applications Won and Lee's (2004) modified p-median problem is faster and more capable of solving different problem sizes. Solimanpur, Vrat and Shankar (2004) proposed a multi-objective integer programming model for the design of cellular manufacturing systems with independent cells.

Wang, Tang, Gong and Chen (2008) presented a dynamic cell formation (DCF) problem by considering multiple objectives. They proposed a nonlinear multi-objective mathematical model for DCF problem by assigning weights to three conflicting objectives such as machine relocation cost, utilization rate of machine capacity and total number of inter-cell movements. Chan, Lau, Chan and Lo (2008) developed two-phase approach to tackle the cell formation problem by considering both intra-cell and inter-cell part movements. In the first phase of their approach, mathematical model is developed with multi-objective function to identify the machine cells and part families. Then in the second phase, another mathematical model is presented with a single objective function which minimizes the total intra-cell movements and inter-cell part movements. Due to the complex structure of the aforementioned models, a meta-heuristic solution is provided. Moghaddam, Bourani, Amin and Siadat (2012) approached the CMS design problem by considering four objectives simultaneously. These were; (i) minimizing the total fixed and variable costs; (ii) minimizing the cost of intercellular movements; (iii) maximizing the utilization of machines in the system; and (iv) minimizing deviations among the levels of cell utilization. Firstly, these objectives were weighted accordingly and then a new mathematical model was proposed. Again, for the large-scale applications scatter search algorithm is proposed.

In the last decade, widely applied meta-heuristics such as Genetic Algorithms (GA), Simulated Annealing (SA), Tabu Search (TS) have been utilized in the cellular manufacturing system. The research conducted by Süer (1999), Süer, Pena and Vazques (2003), Chen and Cao (2004), Jayaswal and Adil (2004), Solimanpur, Vrat and Shankar (2004), and Egilmez, Süer and Ozguner (2012) highly utilized the meta-heuristic approaches in CMS design.

The papers that have been discussed so far considers deterministic CMS design problem. Considering the changing business conditions and dynamics, uncertainty in demand, supply and processes are inevitable. In the literature of CMS design, Seiffodini (1996) tackle the uncertainty of the product mix for a single period and developed a probabilistic model for machine cell formation in order to minimize the expected inter-cell material carrying costs of the system. Ghezavati et al. (2010) applied queuing theory to the CMS design problem by considering the uncertainty issues. The objective of the proposed theory is to minimize the idleness cost of the machines and the total cost of resource underutilization. Additionally, a frontier research is proposed by Süer, Huang and Maddisetty (2010) in which cellular manufacturing system deals with the uncertainty due to product demand and processing times. A benchmark approach named as layered cellular manufacturing system, consisted of three cell types namely dedicated cell (DC), shared cell (SC) and remainder cell (RC). The product-family formation and product cell assignments are performed by considering the probability theory. Probabilistic product demand and probabilistic processing times are utilized to perform quantitative analysis. Their research is fortified by applying discrete-event simulation and then capability of the designed system under uncertainty is observed.

Another interesting study is conducted by Egilmez et al. (2012), in this research a mathematical model is proposed to solve the stochastic cellular manufacturing system design. In this study, production rates and product demands are treated as a stochastic fashion. The main objective is to minimize the number of manufacturing cells by grouping products to families based on the processing similarity for different risk levels. The results revealed that, the number of required manufacturing cells and part families' decreases when the risk level increases. The cell performances are observed by using discrete event simulation for each CMS design. All in all, stochastic cellular manufacturing system design with 10% risk level provided the best results in terms of cell utilization, work-in-process (WIP), total waiting time and total number of waiting.

The methodology proposed by Süer et al. (2010) is in this chapter for the purpose of designing manufacturing system by using the vinyl windows industry demands for both new construction windows under make-to-order strategy and replacement/remodeling demand volumes under engineer-to-order strategy. A normal distribution is applied to the vinyl windows demand and processing times to observe the effects of probabilistic parameters in the system. The details of the manufacturing system design and analysis will be explained in further sections.

3.3. Cell Loading Literature

Süer, Vazquez, and Cortes (2005) proposed a genetic algorithms and local optimizer to solve cell loading and job sequencing to minimize the total number of jobs. The results indicated that, number of cells and crossover strategy significantly affected the number of tardy jobs found. Süer, Cosner & Patten (2009) discussed mathematical models to minimize makespan, machine requirements and manpower transfers in a cellular manufacturing environment. Süer, Arikan, and Babayigit (2009) studied cell loading problem by considering manpower restrictions and developed a fuzzy bi-objective cell loading problem in labor-intensive cellular manufacturing environments in order to minimize the total manpower levels. Also, Süer et al. (2009) proposed heuristic procedures and mathematical models for the cell loading and scheduling problem in a shoe manufacturing company. Süer and Mese (2011) developed a mathematical model and genetic algorithms approach for cell loading when family splitting is allowed and setup times are family oriented. Also, jobs within the families have their individual due dates. Akturk (2011) studied an integrated cell loading, lot-sizing and cell scheduling problem and proposed a multi-objective mathematical model and a multi-objective search heuristic in order to minimize the total cost of production, setup, and inventory holding and regular as well as overtime capacity costs by considering production and balance constraints.

Recently, Egilmez and Süer (2013) studied the design and control aspects of cellular manufacturing systems in the presence of uncertainty. First, stochastic CMS design phase is handled and then stochastic cellular control phase is introduced. Two stochastic non-linear models are implemented for cell formation and cell scheduling problem and then Monte-Carlo simulation is utilized to assess the impact of uncertainty in design phase on the control phase. Various design risk and control risk scenarios are experimented to evaluate the impact of uncertainty on the cell formation and the cell scheduling performance. The results indicated that lower risks in design phase trigger the scheduler to take higher risks in the control phase in a cellular manufacturing environment. Süer et al. (2014) studied cell loading and family scheduling problem in order to minimize the maximum tardiness of jobs. Each job has an individual due date in a family and family splitting is allowed among cells. Mathematical modeling, genetic algorithms (GA) and heuristic approaches are introduced for finding the optimal solution. The experimental results indicated that GA is capable of finding optimal solutions in larger problem sizes and heuristics were easy to implement but they could not find the optimal solution compared to mathematical model. It was indicated that family splitting is a good strategy for finding the optimal solution.

4. PROPOSED SOLUTION METHODOLOGIES

In this section, the proposed mathematical models are presented for determining manufacturing facilities, determining distribution centers and cellular manufacturing system design and scheduling.

4.1. Determining Manufacturing Facilities

The capacitated plant location model considers capacity, total the capacity, total cost of transportation and customer demand while deciding which manufacturing plant will be opened. The following formulation describes the model in detail as such.

Parameters

n: number of plant locations
m: number of customers/ demand points
Di: Demand of customer
s_j: capacity of plant j
c_{ij}: total cost of transportation from plant j to serve customer i

Decision Variables

x_{ij}: fraction of demand of customer i supplied from plant j.
f_j: is the fixed cost of plant j when it is open
y_j: 1, if plant j is open, 0 otherwise.

Minimize

$$\sum_{i=1}^{m}\sum_{j=1}^{n} c_{ij}x_{ij} + \sum_{i=1}^{n} f_j y_j \tag{1}$$

Subject to

$$\sum_{j=1}^{n} x_{ij} = 1, \quad i = 1..m; \tag{2}$$

$$\sum_{i=1}^{m} d_i x_{ij} \le s_j y_j \quad j = 1..n; \tag{3}$$

$$0 \le x_{ij} \le y_j \le 1, \quad i = 1..m; j = 1,..n; \tag{4}$$

The objective is to minimize the total cost locating facilities as shown in Equation (1). The constraint (2) guarantees that the demand of every client is satisfied and constraint (3) guarantees that each opened plant does not supply more than its capacity, and the clients are supplied only from opened plants.

4.2. Determining Distribution Centers (DCs)

The system design consists of set of identified origins (j), hubs/consolidation terminals (k) and destinations (l). The set of commodities i that move through the network are represented by the set of I. The amount of commodities i to be transported from origin terminal i to destination terminal k is denoted by d_{il}. Decision variable z_k is a binary variable and equal to 1 if a consolidation terminal is located at site k. On the other hand, y_{kl} is a binary variable equal to 1 if D/C k is linked to customer l. Finally, variable x_{ijkl} denotes the amount of flow for commodity i with origin j, destination l, passing through terminal k. The following formulation solves the network design problem, where c_{ijk}, c_{ikl} stands for the unit transportation cost between manufacturing plant to DCs and DCs to customer zones, respectively.

Notation

Indices:

i: product index
j: mfg. plant index
k: hub/ D/C index
l: customer index

Parameters:

I: number of products
J: number of mfg. plants
K: number of distribution centers
L: number of customers
$\mathbf{d_{il}}$**:** Demand of product i in customer zone l
$\mathbf{sp_{il}}$**:** Sales price of product in customer zone l
$\mathbf{c_{ijk}}$**:** Unit cost of transportation for product i from any plant j to any D/C k
$\mathbf{c_{ikl}}$**:** Unit cost of transportation for product i from any D/C to any destination l
$\mathbf{f_k}$**:** Fixed cost of D/C k
$\mathbf{S_{ij}}$**:** Plant capacity for product i at plant j,
$\mathbf{Q_k}$**:** D/C Capacity

Decision variables:

$\mathbf{X_{ijk}}$**:** demand flow of product i from plant j to D/C k
$\mathbf{y_{kl}}$**:** 1 if D/C k is linked to customer zone l, 0 otherwise.
$\mathbf{z_k}$**:** 1 if D/C k is open, 0 otherwise.

Maximize Profit

$$\left(\sum_{i=1}^{I}\sum_{k=1}^{K}\sum_{l=1}^{L} sp_{il} * d_{il} * y_{kl}\right) - \left(\sum_{i=1}^{I}\sum_{j=1}^{J}\sum_{k=1}^{K} c_{ijk} * x_{ijk} + \sum_{i=1}^{I}\sum_{k=1}^{K}\sum_{l=1}^{L} c_{ikl} * d_{il} * y_{kl} + \sum_{k=1}^{K} f_k * z_k\right) \quad (5)$$

Subject to

$$\sum_{k=1}^{K} x_{ijk} <= S_{ij}, \quad \forall i, \forall j \quad (6)$$

$$\sum_{i=1}^{I}\sum_{j=1}^{J} x_{ijk} <= Q_k * z_k, \quad \forall k, \tag{7}$$

$$\sum_{j=1}^{J} x_{ijk} \geq \sum_{l=1}^{L} d_{il} * y_{kl}, \quad \forall i, \ \forall k, \tag{8}$$

$$\sum_{k=1}^{K} y_{kl} = 1, \quad \forall l, \tag{9}$$

$$y_{kl} \leq z_k \ \forall l, \ \forall k, \tag{10}$$

$$y_{kl} \in \{0,1\} \ \forall k, \ \forall l, \tag{11}$$

Equation (5) is the objective function which is maximizing the total net profit. The total transportation and total fixed cost of the system is deducted from the total revenue which is obtained by multiplying the product demand with the sales price. Equation (6) ensures that manufacturing plant capacity is not exceed. Equation (7) guarantees that DC or hub capacity is not exceeded. Equation (8) guarantee demand constraint is satisfied. Equation (9) ensures that each customer zone must be served by a single DC. Equation (10) ensures to open D/C if it serves to the customer l. Equation (11) is the binary constraint.

4.3. Cellular Manufacturing System Design and Scheduling

The cell requirements may vary significantly under a highly fluctuating demand situation, which results in poor utilization of resources (Süer et al., 2010). Thus, Süer et al., (2010) brought some flexibility into the cellular manufacturing system by introducing dedicated, cell and remainder cells. In the proposed layered cellular manufacturing system, dedicated cells can process only one product family, shared cell can handle two product families and remainder cells can process more than two product families at the same time.

4.3.1. Identifying Product Families

The manufacturing systems considered in this chapter are designed by considering product families instead of individual products. The reason of using product families is because all products within a family have the same manufacturing operations. The similarity matrix is needed to be constructed based on the similarity coefficient. In this regard, Jaccard's similarity coefficient proposed by McAuley (1972) is utilized. This similarity coefficient is modified to consider similarity of parts instead of machines which is shown in Equation (12).

$$MS_{ab} = \frac{\text{Number of machines processing parts 'a' and 'b'}}{\text{Number of machines processing either parts 'a' or 'b'}} \tag{12}$$

After the product families are formed, the mean and standard deviation of each product family is converted to capacity requirements in order to compare with the yearly hours available and to calculate the utilization of a cell. The mean capacity requirements and standard deviation for each and every part family is calculated by using the equations (13) and (14), respectively. Assuming that, n is denoted as the number of products in the product family and Di is considered as the demand for product *'i'* in the product family, the mean demand of each product and standard deviation are calculated as follows;

$$E\left(D_1 + D_2 + D_3 + \ldots + D_n\right) = \mu_1 + \mu_2 + \mu_3 + \ldots \mu_n \tag{13}$$

$$\sigma\left(D_1 + D_2 + D_3 + \ldots + D_n\right) = \sqrt{\sigma_1^2 + \sigma_2^2 + \sigma_3^2 + \ldots + \sigma_n^2} \tag{14}$$

4.3.2. Proposed Heuristic for Identifying Number of Machines

The mean capacity requirements and standard deviations are calculated for the corresponding product families. Then, each product families consist of different products which yields different processing times in different machines. Therefore, the production rate of each product family is dependent on the processing times. The slowest machine within a product family is called the bottleneck machine. The bottleneck machine directly affects the mean capacity requirements which may yield poor capacity requirements. Therefore, there is a dire need to control the bottleneck machine processing times through machine duplication. By the help of machine duplication, the system may overcome the difficulties resulting from bottleneck machine. The proposed machine duplication process is summarized as follows;

Step 1: Identify each product machine matrix and processing times in each product family,
Step 2: Calculate each product's bottleneck machine (slowest machine for each product),
Step3: Calculate the mean capacity requirements for each product ((mean demand * bottleneck machine)/60),
Step 4: Sum each product's mean capacity requirements to obtain mean capacity requirement for product family,
Step 5: Find out the bottleneck machine in the product family,

Step 6: Duplicate that machine for the product family,
Step 7: Recalculate the mean capacity requirements for the product family,
Step 8: If the change in total mean capacity is higher than 5%, then that machine needs to be duplicated for the entire product family,
Step 9: Update and repeat step 8 for each bottleneck machine and observe the increase in mean capacity requirements, Check whether the change of total MCR $\geq 5\%$.
Step 10: If one machine is already duplicated (MCR change is ≥ 5%) and the second machine's duplication changes the total MCR less 5%, then if those machines are duplicated simultaneously and 5% increment is observed for the total MCR, then duplicate the second machine as well.
Step 11: Repeat step 10 for all machines and observe whether total MCR is greater or equal to 5%,
Step 12: Record which machines are duplicated and calculate total MCR based on the duplicated machines,
Step 13: Stop.

4.3.3. Determining Cell Configurations

Once, the mean capacity requirements and standard deviation requirements are determined for product families, the probability of covering demand for different number of cells is determined. As a matter of fact, the number of cells required increases when the demand for product family increases and consequently increases the probability of covering the demand. Therefore, number of cells needed is highly dependent on the demand volumes. For each additional cell, the expected utilization is calculated by considering the utilization percentage of the previous cell, the probability of the previous cell covering the demand, utilization percentage of the additional cell and the probability of the additional cell covering the demand (Süer et al., 2010). Consequently, expected utilization of the cell is calculated by using the following formula.

$$E(C = X) =$$

$$P(CR > X)*PU_1 + P(X - 1 \leq CR \leq X)*PU_2 + P(CR < X - 1)*PU_3 \quad (15)$$

$E(C = X)$, is the expected utilization for the Xth cell assigned to a product family. $P(CR > X)$ is the probability that the number of cells required (CR) is greater than X. PU_1, is the percent utilization of cell X when CR > X, PU=1. ..is the probability that the number of cells required falls within X-1 and X. PU_2 is the percent utilization of cell X when X-1≤ CR ≤ X. $P(CR < X - 1)$, is the probability that the number of cells required is less than X-1. PU_3 is the percent utilization of cell X when CR < X-1, $PU_3 = 0$. PU_2 is calculated by the following equation (16) and solved in MATLAB by using the mean and standard deviation of each product family and each cell.

$$PU_2 = \int_{2000(X-1)}^{2000X} \frac{y*f(y)}{2000*A} dy - (X-1) \quad (16)$$

where, f(y) is the probability density function for the number of cells required. y is the random variable that represents the number of cells required. A is denoted as the probability that cells required is between X-1 and X, $P(X-1 \leq CR \leq X)$. The probability density function *f(y)* for the normal distribution is calculated in the equation.

4.3.4. Determining Number of Cells and Types

Once, the expected cell utilization and demand coverage probability values are obtained, those values are used in the mathematical model in order to determine cell configuration and cell types. The model minimizes the number of opened cells to cover fluctuated demand. The outcome of the model determines the number and type of cells (dedicated, shared and remainder) to be opened. The model is explained below.

Notation

Indices

i: Family index
j: Coverage segment index
k: Cell index

Parameters

p: number of families
q: number of coverage segments
r: number of cells
b_{ij}: Expected cell utilization for part family i and coverage segment j
c_{ij}: Probability of covering demand for part family i and coverage segment j
m: maximum allowable expected cell utilization for a cell
n: minimum demand coverage probability for each family

Decision variables

X_{ijk} = 1, if part family i is assigned to coverage segment j and cell k, 0 otherwise
Y_k = 1, if cell k is opened, 0 otherwise

Objective function

$$\mathrm{Min}\,\mathrm{Z} = \sum_{i=1}^{n} \mathrm{Y}_{k} \tag{17}$$

Subject to:

$$\sum_{j=1}^{n} \mathrm{X}_{ijk} \leq \mathrm{Y}_{k} \quad \text{for} \begin{cases} i = 1,2,\ldots,p \\ k = 1,2,\ldots,r \end{cases} \tag{18}$$

$$\sum_{j=1}^{q}\sum_{i=1}^{p} \mathrm{b}_{ij} * \mathrm{X}_{ijk} \leq \mathrm{Y}_{k} * \mathrm{m} \ \text{for } \mathrm{k} = 1,2,\ \ldots,\ \mathrm{r} \tag{19}$$

$$\sum_{j=1}^{q} \mathrm{c}_{ij} \left[\sum_{k=1}^{r} \mathrm{X}_{ijk}\right] \geq \mathrm{n} \quad \text{for } \mathrm{i} = 1,2,\ldots,\mathrm{p} \tag{20}$$

$$\sum_{k=1}^{r} \mathrm{X}_{ijk} \leq 1 \quad \text{for} \begin{cases} i = 1,2,\ldots,p \\ k = 1,2,\ldots,r \end{cases} \tag{21}$$

$$\sum_{j=1}^{n} \mathrm{X}_{i,j-1,k} \geq \sum_{j=1}^{n} \mathrm{X}_{ijk} \quad \text{for} \begin{cases} i = 1,2,\ldots,p \\ k = 1,2,\ldots,r \end{cases} \tag{22}$$

$$\mathrm{X}_{ijk} \in \left[0,1\right],\ \mathrm{Y}_{k} \in \left[0,1\right] \tag{23}$$

The aim of this model is to minimize the number of opened cells as given in Equation (17). Equation (18) enforces that number of coverage segments from the same part family cannot exceed 1. If the cell is not opened, the coverage segment is not assigned to that cell. Equation (19) prevents that expected cell utilization never exceeds the upper limit for cell utilization. Equation (20) enforces that the demand coverage probability for each part family to be higher than the minimum value. Equation (21) guarantees that, each coverage segment of a product family can be assigned to only one cell. Also, equation (22) enforces that the coverage segments of a product family are assigned to the cells, respectively. The rest are the binary constraints.

4.3.5. Cellular Manufacturing Scheduling/Cell Loading

Once the number and the types of cells are identified based on the annual demand, then products are scheduled on a daily basis to those cells in order to meet the customer due date in a timely fashion. The mean daily demand and standard deviation for each and every product is calculated by using the equations 24 and 25, respectively. Assuming that *'b'* is scalar (number of days in year) and X_i is considered as the demand for product *'i'*, the mean daily demand of each product and standard deviation are calculated according to the property of variance rule;

$$E\left[\mathrm{bX}\right] = \mathrm{b}E\left[\mathrm{X}\right] \tag{24}$$

$$\mathrm{Var}\left[\mathrm{bX}\right] = \mathrm{b}^2\mathrm{Var}\left[\mathrm{X}\right] \tag{25}$$

$$\sigma = \sqrt{\frac{\sigma_{\mathrm{X}}^2}{\mathrm{b}^2}} \tag{26}$$

Those normally distributed daily demand values are used in the proposed cell loading mathematical model. The proposed mathematical model both considers alternative manufacturing due dates and a customer due date. At first, all manufacturing due dates are sorted in the increasing order (including all alternative manufacturing due dates as well) and then the mathematical model is solved.

Notation

Indices:

V: product index
Z: cell index

Parameters:

N: number of products
s: number of cells

$d_{[k]}$: k^{th} smallest manufacturing due date
W_k: set of products with due date less than or equal to d_k
r_{vz}: revenue for product v when built in cell z
mc_{vz}: manufacturing cost for product v when built in cell z
st_{vz}: setup time for job v when built in cell z with a due date u
t_{vz}: in-cell time of job v in cell z

Decision variables:

x_{vz}: 1 if job v is assigned to ce z
y_{vz}: 1 if setup is performed for job v assigned to cell z, 0 otherwise.

Objective Function:
Maximize the Total Profit:

$$\sum_{v=1}^{n}\sum_{z=1}^{s} r_{vu} * x_{vz} - \sum_{v=1}^{n}\sum_{z=1}^{s} mc_{vu} * x_{vz} \tag{27}$$

Subject to:

$$\sum_{z=1}^{s} x_{vz} \leq 1, \quad v = 1 \ldots n \tag{28}$$

$$\sum_{v \in Wk} \left(t_{vz} * x_{vz}\right) \leq d_{[k]}, \quad z = 1 \ldots s \quad \& \quad k = 1 \ldots N \tag{29}$$

Equation (27) is the ultimate objective function which maximizes the total net profit. The total net profit is obtained by subtracting the total manufacturing cost from the total revenue of products. Equation (28) guarantees that each product is assigned to only one cell. Equation (29) enforces that each product should be manufacturing in a cell before its corresponding due date.

In this study, probabilistic demand is used for determining processing times for each product in each cell on a daily basis. The processing time is calculated dividing demand of each product by the corresponding cell output rate. The sales price of each product is known in advance. The total revenue is calculated multiplying demand with unit sales price. The total net profit of each product on each cell is obtained by subtracting manufacturing cost from the revenue. The manufacturing cost is obtained by multiplying hourly cost of labor with the number of workers in machines and unit processing time. In this research, the number of workers is assumed to be equal to the number of machines.

All in all, this model determines the daily scheduling of products in cells on a timely basis by considering the transportation lead-time such that the customer due date is met.

5. EXPERIMENTATION AND RESULTS OF SUPPLY CHAIN DESIGN STRATEGIES

This section discusses the detailed experimentation of supply chain strategies for different demand volumes. The detailed results for each of the individual demand volumes (new construction and replacement) are presented. This section discusses the experimentation performed based on the individual demand volume strategy (new construction and replacement window industry demand). The State of Ohio is considered for an analysis. Six big contractors and 174 customers all around the region are considered as shown in Figure 6.

5.1. Results of New Construction Demand Volume Strategy

This section discusses the results of new construction demand volume (make-to-order) strategy. The number, location and allocation of manufacturing and distribution centers are determined. Then, the manufacturing system design for each established manufacturing centers is performed.

5.1.1. Location of Manufacturing Facilities

The considered region has 17 potential manufacturing plants which will satisfy the demands of 6 contractors. Each contractor and customer annually needs to satisfy 24 different window products. The contractors constitute 34.7% of the total demand units (441,580) whereas the customers constitute 65.3% (832,420) of the total demand units. The standard mileage rate is considered as $0.56/mile, which is used to calculate the transportation cost among potential plants. This value is used based on the Internal Revenue Service (IRS) standard mileage rates for employees, self-employed individuals, or other taxpayers to use in computing the deductible costs of operating an automobile for business, charitable, medical, or moving expense purposes. The manufacturing facility is considered as 600,000 sq. ft of large manufacturing facility and the capacity of each manufacturing facility is considered as 700,000 units. The capacitated facility location model is used to find how many manufacturing facilities are needed to satisfy demand locations (see Figure 7).

According to the results, two new construction manufacturing facilities (North Ridgeville-OH and Tipp City-OH) are established to meet each individual demand. The total cost of establishing those two manufacturing facilities is $2,000,128.

5.1.2. Location of Distribution Centers

After determining how many manufacturing facilities are opened, now based on the manufacturing facilities, the location, the number of distribution centers as well as the serving contractors are should be determined. It is important to note that, only contractors are supplied from DCs. The product flow from manufacturing centers to distribution centers (after opened) and to customers should be also decided. 14 potential distribution centers are considered within the region. Therefore for this purpose,

Figure 6. Representation of Contractors and Customers Locations

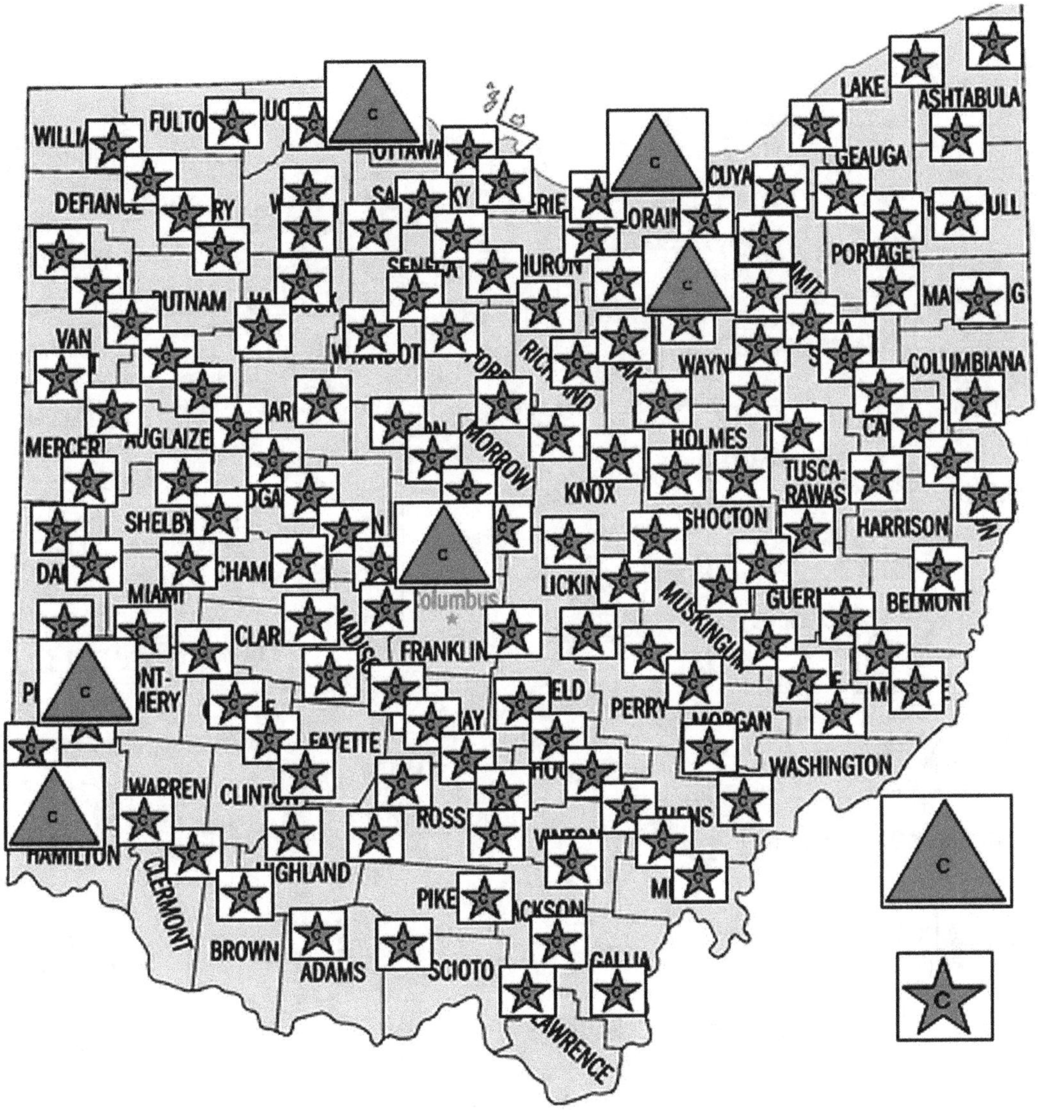

multi-commodity production and distribution network (MCPDN) model is used. There are two aspects of this model, the first one considers the transportation costs between two manufacturing centers and potential distribution centers for each product and transportation costs from potential distribution centers to known customers for each product type. The capacity of each distribution center is considered as 200,000 units. The fixed cost is determined according to the 50,000 sq.ft facility and the real fixed costs are calculated accordingly.

The results indicated that three distribution centers are opened and those are briefly located in the cities of Mount Vernon, Springboro and Troy. The results are presented in Table 1 and Table 2.

Figure 7. Established Manufacturing Facilities in North Ridgeville, OH and Tipp City, OH for New Construction Demand Volume

Table 1. Opened distribution centers and volumes

Name	DC	Served Contractors	Total Units (all products)
MountVernon	7	Akron, Cleveland, Toledo	**196,600**
Sprinsboro	10	Cincinnati	**62,430**
Troy	11	Columbus, Dayton	**182,550**
Total Demand			**441,580**

Table 2. Summary of results

	Non-Split Model	Split Model	Net Profit Savings
Total REVENUE	$ 65,843,907.00	$ 65,843,907.00	
SHIPPLANTDC	$ 56,606.02	$ 52,767.60	
SHIPDCCUSTOMER	$ 86,377.20	$ 283,970.00	
Fixed Cost	$ 14,058,500.00	$ 14,058,500.00	
Total Cost	$ 14,401,483.00	$ 14,395,237.90	
Total Net Profit	$ 51,442,423.00	$ 51,448,669.00	$ 6,245.3

Figure 8. Designed representation of new construction demand volume

5.1.3. Manufacturing System Design

The cellular manufacturing system studied in this chapter consists of 24 products and 8 machines. The processing times are approximately determined based on the different product and machine capabilities. The maximum of the processing times is considered to be the bottleneck machine processing time. It is assumed that products move in units and set-up times for the products are assumed to be zero and the total number of hours a cell works is taken to be 2000 hours per year based on 50 weeks and 40 hours/ week. The remaining weeks are allocated for covering unexpected events. Annual demands and associated standard deviation (25% of demand) of products are used in this analysis for the situated facilities in North Ridgeville and Tipp City. The first phase as aforementioned is to identify the part families based on the processing similarity of parts on machines. Table 3 shows the routes, processing times

Table 3. Routes, processing times (min) and demand of products for North Ridgeville- OH Plant

	Machines								Demand		
Product	**1**	**2**	**3**	**4**	**5**	**6**	**7**	**8**	**mean**	**SD**	**Variance**
P1	1.80		2.40		2.22	1.50			8171	2042.73	4172725.4
P2	2.58	3.36			2.40		3.00		8257	2064.18	4260818.4
P3	3.68		1.50		2.40		1.50	2.70	7851	1962.78	3852485.7
P4	2.58	3.60			2.40		2.70		7910	1977.4	3910110.8
P5	3.00	2.58		1.80				2.28	7914	1978.48	3914363.3
P6	2.16	3.00		2.40		1.20			7970	1992.43	3969757.4
P7	1.68	3.66		2.22				2.70	7798	1949.55	3800745.2
P8	2.22	2.58			2.40	2.40			7860	1964.93	3860930.3
P9	2.22	2.10	2.88		1.62				7954	1988.5	3954132.3
P10	2.70		1.50	2.58		2.70			7855	1963.85	3856706.8
P11	1.98	3.18	3.54		1.98		1.50	1.98	8074	2018.5	4074342.3
P12	3.18		2.88		1.50		1.50	2.88	7855	1963.85	3856706.8
P13	1.80	0.00	2.70	3.30		2.40		2.22	7951	1987.78	3951249.5
P14	0.90	3.36	2.64	2.70		2.40			7944	1985.98	3944096.7
P15	1.38		1.98		2.22		1.98	1.62	7927	1981.7	3927134.9
P16	2.10	1.50	3.30			2.28	1.38		7960	1989.9	3959702.0
P17	1.50	0.00	2.40	2.40		1.50		2.22	7890	1972.4	3890361.8
P18	2.58	3.60			2.28		2.88		7930	1982.4	3929909.8
P19	2.22	3.24		3.30				2.58	7970	1992.4	3969657.8
P20	2.46		3.18		2.64			3.00	7861	1965.25	3862207.6
P21	1.98		3.78		1.50			1.50	7945	1986.35	3945586.3
P22	2.52	1.68		3.00				2.88	7917	1979.2	3917232.6
P23	3.78	3.66		3.00				2.58	7887	1971.7	3887600.9
P24	2.64	1.50	3.30	2.22			1.50		7935	1983.83	3935561.6

Total Demand 190,584

(min) and demand of products for North Ridgeville, OH facility. ALINK clustering algorithm (Seifoddini and Wolfe, 1986) is used in this analysis. Once again, ALINK generates part families considering the average of similarity coefficient. The identified product families for North Ridgeville-OH facility are given in Table 4.

Once the product families are identified, it is necessary to determine number of machines to process those products in that product family. The proposed heuristics for determining number machines in cellular manufacturing system in section 4 is used to determine number of machines for each product family. The resulting machines for each product family are given in Table 5.

Depending on the number of machines in each product family, the mean capacity requirements (MCR) and standard deviations for each product family is determined. The resulting MCR and standard deviation for each product family is given Table 6.

Once the MCR and standard deviations are determined, the demand coverage probability is determined as in section 4. The demand coverage probability for each product family is determined and given in Table 7.

Table 4. Identified Product Families

	Product Family	**2000hours**
ALINK	**ST**	**0.5**
100%UTILIZATION		
Product Family	**Products**	
PF1	p1,p2,p4,p8,p9,p18	
PF2	p3,p11,p12,p15,p20,p21	
PF3	p5,p7,p19,p22,p23	
PF4	p6,p10,p13,p14,p16,p17,p24	

Table 5. Number of Machines in Product Families

Machines	**PF1**	**PF2**	**PF3**	**PF4**
m1	1	1	1	1
m2	2	1	2	1
m3	1	2	0	2
m4	0	0	1	1
m5	1	1	0	0
m6	1	0	0	1
m7	1	1	0	1
m8	0	1	1	1
Total machines	**7**	**7**	**5**	**8**

Table 6. Mean Capacity Requirements (MCR) and Standard Deviation for Product Families

PF	Capacity Requirement(hours)		
	MCR	Varc	SDc
PF1	2172.4	49588.7	222.7
PF2	2271.7	55941.9	236.5
PF3	2077.7	54713.6	233.9
PF4	2601.3	61617.1	248.2

Table 7. Demand Coverage Probability

Family	1st Cell	2nd Cell
1	0.21941	1.00000
2	0.12532	1.00000
3	0.36996	1.00000
4	0.00771	1.00000

Having determined the demand coverage probabilities, it is now time to calculate expected cell utilization for cells. It is assumed that the number of available hours in a cell is 2000 hours, which implies that if two cells are needed 4000 hours are available to tackle with the demand. As seen from the results; by assigning one cells to product family 1, the probability that the cell can cover the demand is 21.9%. If an additional cell is allocated-the second cell-, the demand would be fully covered by 100%. Continuing with that, to be able to cover the complete demand, two cells are needed. Once those demand coverage probabilities for each product family is determined, the percent utilizations of different cells on different product families are calculated and shown in Table 8. Once again, I eliminated 3rd cell.

The results indicated the percent utilizations of cells to be able to cover the entire demand. The solutions indicated for the product family 1 that, the first and seconds cells are utilized by 98.5% and 44% to be able to cover the entire demand. Having determined all the expected cell utilizations and demand coverage probabilities, cell configurations and cell types (dedicated, shared and remainder cells) are

Table 8. Expected Cell Utilizations for Product Families

Family	1st Cell	2nd Cell
1	0.985	0.440
2	0.988	0.509
3	0.970	0.367
4	0.996	0.647

identified based on the mathematical model in section 4. Based on the designed optimal results, North Ridgeville- OH facility requires 7 cells that consist of one shared and 6 dedicated cells with a total of 47 machines to be able to cover the entire demand (see Table 9).

Similar steps are performed for the Tipp City-OH facility as well and summary of the number and types of cells and their corresponding expected cell utilizations for Tipp City-OH facility with a total of 57 machines are presented in Table 10.

5.1.4. Daily Cell Loading and Scheduling Decisions

The manufacturing system considered in this study has five different operations. The number of machines used within each cell varies based on the type and speed. The production rates are defined in vinyl windows/hour. For each machine type, the production rate is different. North Ridgeville-OH facility has 7 cells and Tipp City-OH facility has 8 cells after the design process. It is now find the daily demand and schedules for each cell configuration. Here is the mean daily demand and number of machines and their production rates in cells for North Ridgeville-OH facility (see Table 11 and Table 12).

Table 9. Number and Types of Cells and their Corresponding expected Cell utilizations North Ridgeville, OH

4 PF		C1	C2	C3	C4	C5	C6	C7
PF 1	ithcell/ECU	1/0.985			2/0.440			
PF 2	ithcell/ECU				2/0.509		1/0.988	
PF 3	ithcell/ECU		2/0.367	1/0.97				
PF 4	ithcell/ECU					2/0.647		1/0.996
Total ECU		98.5%	36.7%	97.0%	94.80%	64.70%	98.80%	99.60%
Machines		**7**	**5**	**5**	**7**	**8**	**7**	**8**
Cell Types		DC	DC	DC	SC	DC	DC	DC

DC: Dedicated Cell SC: Shared Cell RC: Remainder Cell

Table 10. Number and Types of Cells and their Corresponding expected Cell utilizations Tipp City, OH

4 PF		C1	C2	C3	C4	C5	C6	C7	C8
PF 1	ithcell/ECU			1/1.00	2/0.713				
PF 2	ithcell/ECU						1/1.00		2/0.744
PF 3	ithcell/ECU	1/0.997				2/0.676			
PF 4	ithcell/ECU		2/0.856			3/0.026		1/1.00	
Total ECU		99.7%	85.6%	100.0%	71.30%	70.20%	100.00%	100.00%	74.40%
Machines		**5**	**8**	**7**	**7**	**8**	**7**	**8**	**7**
Cell Types		DC	DC	DC	DC	SC	DC	DC	DC

Table 11. Mean daily demand for all products: North Ridgeville, OH facility

Product 1	Product 2	Product 3	Product 4	Product 5	Product 6	Product 7	Product 8	Product 9	Product 10	Product 11	Product 12
24	51	29	44	38	37	29	35	29	31	43	29
Product 13	**Product 14**	**Product 15**	**Product 16**	**Product 17**	**Product 18**	**Product 19**	**Product 20**	**Product 21**	**Product 22**	**Product 23**	**Product 24**
32	39	38	36	25	36	33	17	29	36	39	38

Table 12. Number of machines and their production rates in cells: North Ridgeville, OH facility

Production Rate (windows/hour)	Operation 1	Operation 2		Operation 3		Operation 4		Operation 5	Output (windows/hour)	Number of MCs
	Type I	Type I	Type II	Type I	Type II	Type I	Type II	Type I		
	10	20	15	17	13	12	18	23		
Cell 1	1	1	0	0	1	2	1	1	42	7
Cell 2	1	1	1	0	1	0	0	1	35	5
Cell 3	1	1	1	0	1	0	0	1	35	5
Cell 4	1	1	0	0	1	2	1	1	42	7
Cell 5	1	1	0	2	1	1	1	1	47	8
Cell 6	1	1	0	0	1	2	1	1	42	7
Cell 7	1	1	0	2	1	1	1	1	47	8

In this study, we use mean daily demand values for determining processing times for products. The processing time is calculated by dividing the demand value of a product by the corresponding cell output rate. Therefore, total processing time of a product depends on which cell it is produced. The sales price of each product is known. The net profit of each product on a cell is obtained by subtracting manufacturing cost from the sales revenue. Then, the manufacturing cost is obtained by multiplying the hourly cost of labor with the number of workers and unit processing time (hour). In this study, the number of workers is assumed to be equal to the number of machines in that cell. The median wage rate in Ohio Labor Market Information is used for calculating the labor cost. Similar work is conducted for the Tipp City, OH facility as well and both results are presented in Figure 9 and Figure 10, respectively.

The cell loading and scheduling mathematical model explained in section 4 is used to obtain the above Gantt charts. The daily optimal net profit for North Ridgeville-OH manufacturing facility is $57,860.74. Based on mean daily demand of each product and manufacturing system capacity, a product might be assigned to different cells with different manufacturing due dates. Based on the information provided and depending on the cell loading mathematical model discussed in section 4, a daily optimal solution is found for 8 cell manufacturing environment and the total net optimal profit is found as $71,061.91.

Figure 9. Cell loading for North Ridgeville-OH

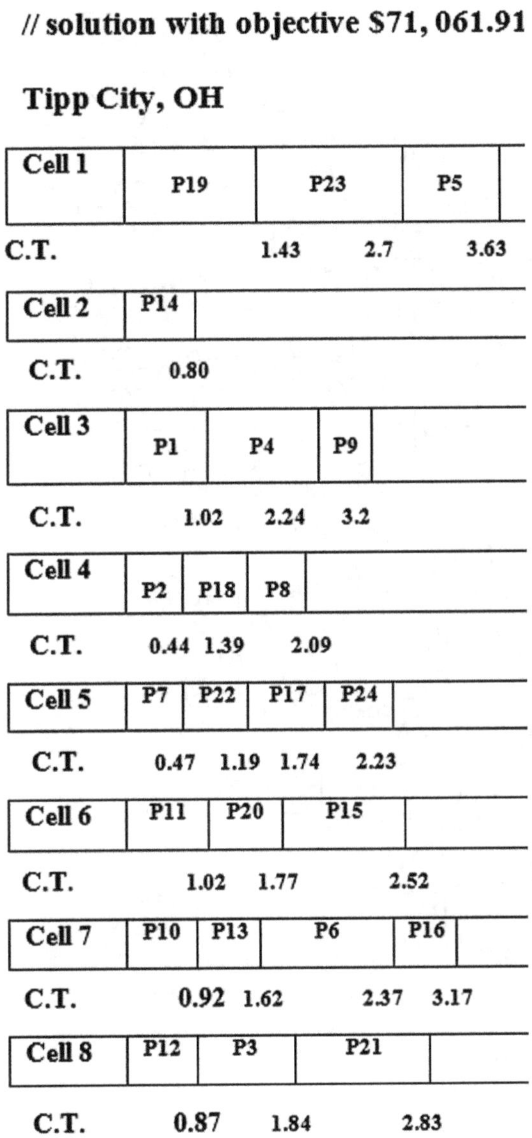

Figure 10. Tipp City- OH facility- Daily

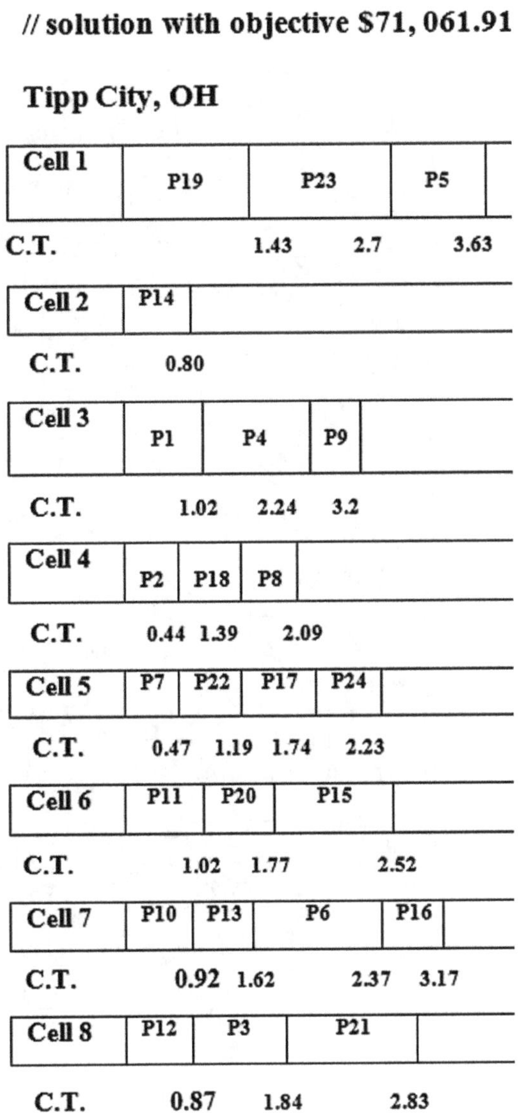

5.2. Results of Make-to-Order and Engineer-to-Order Strategies

Similar methodology is conducted for replacement window supply chain strategy. This section briefly summarizes the results of replacement and new construction demand volume strategies. The final supply chain network illustration for each replacement and new construction demand volume is shown in Figure 11 and Figure 12, respectively.

Figure 11. Replacement window SC Strategy

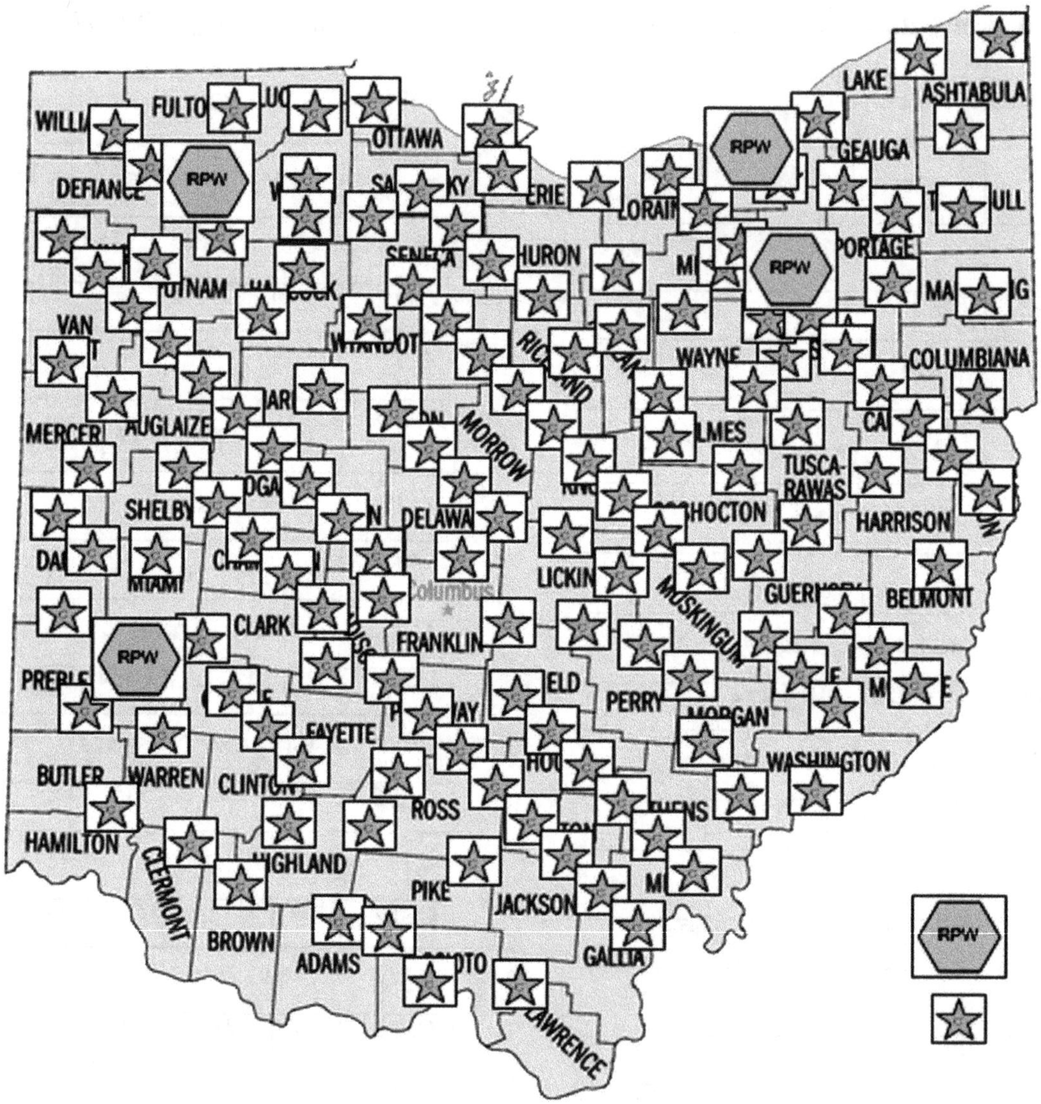

For each design, the manufacturing facility is considered as 250,000 sq. ft of manufacturing facility and the capacity of each manufacturing facility is considered as 250,000 units. For new construction design, each distribution centers is considered as 200,000 sq. ft of distribution center facility and the capacity of each distribution center is considered as 200,000 units. The results indicated that, four replacement window (Highland Heights-OH, Montgomery-OH, Napoleon-OH and Portage Lakes-OH) and two new construction manufacturing facilities (North Ridgeville-OH and Tipp City-OH) are established to meet each individual demand. For new construction design alternative, three distribution centers (Mount Vernon-OH, Springboro-OH and Troy-OH) are situated and the demand is satisfied through these distribution centers. The detailed comparison for all supply chain design scenarios will be given in the next section.

Figure 12. New construction SC Strategy

5.3. Comparison of Supply Chain Design Alternatives

This section briefly summarizes the replacement and new construction design supply chain (SC) design alternatives. Based on the company's responsiveness and design strategy, each supply chain design strategy provides a unique outcome and solution preference to company's strategic priorities. Table 13 shows the results of two supply chain design alternatives.

Table 13. Comparison supply chain design alternatives

Parameters		Supply Chain Design Scenarios	
		Replacement SC	New Construction SC
# of Mfg. Facilities		4	2
Total Cost of Mfg. Facilities ($)		$6,004,318	$2,000,128
Total # of Mfg. Cells	DC	25	13
	SC	4	2
	RC	-	-
Total # of Machines		200	104
Total # of Distribution Centers		-	3
Total Net Profit of Distribution Centers ($)	Split	-	$51,448,669
	Non-split	-	$51,442,423
Total Net Profit of Daily Mean Schedule ($)		$245,857.2	$128,922.64

In replacement demand volume strategy, four manufacturing facilities are opened. The total of four shared and 25 dedicated manufacturing cells are opened. In new construction supply chain strategy, two manufacturing facilities are established and a total of 15 manufacturing cells (13 dedicated, 2 shared) are opened. Also in new construction SC strategy, three distribution centers are established to meet the contractor demand throughout the Ohio.

6. CONCLUSION AND DISCUSSIONS

This study presents integrated models to capture both design and operational aspects of supply chain system in the presence high-volume and low-volume of window products. Two different supply chain designs are performed; one for make-to-order window products for high-volume contractors, and another one for engineer-to-order window products for low-volume customers. This research is inspired from the window manufacturer which manufactures and distributes vinyl windows to meet new construction and replacement/remodeling sector demand. Supply chain design was performed by considering vinyl windows' new construction sector demand volumes and replacement/remodeling sector demand volumes. New mathematical models, heuristics, and layered cellular manufacturing system design are proposed for integrated production-distribution system in make-to-order and engineer-to-order demand management environments.

For both replacement and new construction designs, the locations of manufacturing facilities (M/Cs) and allocation to customers for a specified region based on annual demand are determined. For only the new construction sector, the locations and number of distribution centers (D/Cs) are determined as well as customer zones of each distribution center based on annual demand are identified. Layered CMS is implemented for each opened manufacturing facilities based on annual demand. Part families, cell configurations, cell types and scheduling steps are performed. The results revealed that, each design has its own strategic advantage and outcome. Briefly, few large facilities are established to meet high-volume demand (make-to-order products) of contractors, and smaller but more facilities are situated for low-volume demand (engineer-to-order products) of customers such that it becomes feasible to carry few products over short distances.

All in all, four manufacturing facilities are required to meet the demand of replacement/remodeling sector (ETO) and two manufacturing facilities are situated to meet the demand of new construction sector (MTO). The results pointed out that, 29 layered-cells and a total of 200 machines are required and utilized to meet replacement/remodeling sector. Additionally, 15 layered-cells and a total of 104 machines are required to meet the entire demand of new construction sector. Also for the new construction sector, three distribution centers are needed to facilitate the products over the region. In terms of daily cell loading and scheduling phase, the results are almost doubled to meet ETO demand management strategy compared to MTO demand management strategy due to having more cells and machines in the ETO design strategy.

As a future work, it would be nice to see the effect of a combined demand volume strategy. Furthermore, incorporating a transportation aspect (vehicle routing) of a supply chain component to observe the overall tradeoff among different supply chain strategies in terms of delivery leadtime and delivery cost would be very important as well. Another possible extension is to perform simulation for different daily schedules over a period of time.

REFERENCES

Akturk, M. S. (2011). Joint cell loading and scheduling approach to cellular manufacturing systems. *International Journal of Production Research*, *49*(21), 6321–6341. doi:10.1080/00207543.2010.532165

Akturk, M. S., & Turkcan, A. (2000). Cellular manufacturing system design using a holonistic approach. *International Journal of Production Research*, *38*(10), 21. doi:10.1080/00207540050028124

Albadawi, Z., Bashir, H. A., & Chen, M. (2005). A mathematical approach for the formation of manufacturing cells. *Computers & Industrial Engineering*, *48*(1), 3–21. doi:10.1016/j.cie.2004.06.008

Altiok, T. (1996). *Performance Analysis of Manufacturing Systems*. Springer.

Amiri, A. (2006). Designing a distribution network in a supply chain system: Formulation and efficient solution procedure. *European Journal of Operational Research*, *171*(2), 567–576. doi:10.1016/j.ejor.2004.09.018

Askin, R. G. (2013). Contributions to the design and analysis of cellular manufacturing systems. *International Journal of Production Research*, *51*(23-24), 6778–6787. doi:10.1080/00207543.2013.825745

Askin, R. G., & Estrada, S. (1999). An Investigation of Cellular Manufacturing Practices. In S. Irani (Ed.), *Handbook of Cellular Manufacturing* (pp. 25–34). New York: Wiley. doi:10.1002/9780470172476.ch1

Askin, R. M., & Standridge, C. R. (1993). *Modeling and analysis of manufacturing systems* (1st ed.). New York: Wiley.

Ates, O. K. (2013). Global Supply Chain and Competitive Business Strategies: A Case Study of Blood Sugar Monitoring Industry [Doctoral dissertation]. Ohio University.

Chan, F.T.S., Lau, K.W., Chan, L.Y., & Lo, V.H.Y. (2008). Cell formation problem with consideration of both intracellular and intercellular movements. *International Journal of Production Research*, *46*(10), 2589–2620. doi:10.1080/00207540500478843

Chandra, P., & Fisher, M. L. (1994). Coordination of Production and Distribution Planning. *European Journal of Operational Research*, *72*(3), 503–517. doi:10.1016/0377-2217(94)90419-7

Chen, M., & Cao, D. (2004). Coordinating production planning in cellular manufacturing environment using Tabu search. *Computers & Industrial Engineering*, *46*(3), 571–588. doi:10.1016/j.cie.2004.02.002

Chen, M., & Wang, W. (1997). A Linear Programming Model for Integrated Steel Production and Distribution Planning. *International Journal of Operations & Production Management*, *17*(6), 592–610. doi:10.1108/01443579710167276

Choobineh, F. (1988). A framework for the design of cellular manufacturing systems. *International Journal of Production Research*, *26*(7), 1161–1172. doi:10.1080/00207548808947932

Chopra, S., & Meindl, P. (2007). *Supply chain management. Strategy, planning & operation*. New Jersey, Pearson Prentice Hall.

Cohen, M.A. and H.L. Lee, (1989). Resource Deployment Analysis of Global Manufacturing and Distribution Networks. *Journal of Manufacturing and Operations Management*, 1989, 81-104.

Dhaenens-Flipo, C., & Finke, G. (2001). An Integrated Model for an Industrial Production- Distribution Problem. *IIE Transactions*, *33*(9), 705–715. doi:10.1080/07408170108936867

Egilmez, G., & Süer, G. (2013). The impact of risk on the integrated cellular design and control. *International Journal of Production Research.*

Egilmez, G., Süer, G. A., & Huang, J. (2012). Stochastic cellular manufacturing system design subject to maximum acceptable risk level. *Computers & Industrial Engineering*, *63*(4), 842–854. doi:10.1016/j.cie.2012.05.006

Ghezavati, V., & Saidi-Mehrabad, M. (2010). Designing integrated cellular manufacturing systems with scheduling considering stochastic processing time. *International Journal of Advanced Manufacturing Technology*, *48*(5-8), 701–717. doi:10.1007/s00170-009-2322-2

Huang, J., & Süer, G. A. (2012). Long-Term Sourcing Model for Multi-Objective Supplier Selection Under Uncertainty. *Proceedings of 42nd International Conference on Computers & Industrial Engineering*, Cape Town, South Africa.

Hyer, N., & Vemmerlov, U. (2002). *Reorganizing the Factory - Competing Through Cellular Manufacturing. Usa.* Productivity Press.

Jayaswal, S., & Adil, G. K. (2004). Efficient algorithm for cell formation with sequence data, machine replications and alternative process routing. *International Journal of Production Research*, *42*(12), 2419–2433. doi:10.1080/00207540310001652914

Jeon, G., Leep, H. R., & Parsaei, H. R. (1998). A cellular manufacturing system based on new similarity coefficient which considers alternative routes during machine failure. *Computers & Industrial Engineering*, *34*(1), 21–36. doi:10.1016/S0360-8352(97)00148-4

Kusiak, A. (1987). The generalized group technology concept. *International Journal of Production Research*, *25*(4), 561–569. doi:10.1080/00207548708919861

Pan, F., & Nagi, R. (2013). Multi-echelon supply chain network design in agile manufacturing. *Omega*, *41*(6), 969–983. doi:10.1016/j.omega.2012.12.004

Pirkul, H., & Jayaraman, V. (1998). A multi-commodity, multi-plant, capacitated facility location problem: Formulation and efficient heuristic solution. *Computers & Operations Research*, *25*(10), 869–878. doi:10.1016/S0305-0548(97)00096-8

Russell, R., Chiang, W.-C., & Zepeda, D. (2008). Integrating multi-product production and distribution in newspaper logistics. *Computers & Operations Research*, *35*(5), 1576–1588. doi:10.1016/j.cor.2006.09.002

Sabri, E., & Beamon, B. M. (2000). A Multi-Objective Approach to Simultaneous Strategic and Operational Planning in Supply Chain Design. *Omega*, *28*(5), 581–598. doi:10.1016/S0305-0483(99)00080-8

Seifoddini, H. (1989). Duplication process in machine cells formation in group technology. *IIE Transactions*, *21*(4), 382–388. doi:10.1080/07408178908966245

Seifoddini, H. (1989). A note on the similarity coefficient method and the problem of improper machine assignment in group technology applications. *International Journal of Production Research*, *27*(7), 1161–1165. doi:10.1080/00207548908942614

Seifoddini, H. K. (1989). Single linkage versus average linkage clustering in machine cells formation applications. *Computers & Industrial Engineering*, *16*(3), 419–426. doi:10.1016/0360-8352(89)90160-5

Selim, H. M., Askin, R. G., & Vakharia, A. J. (1998). Cell formation in group technology: Review, evaluation and directions for future research. *Computers & Industrial Engineering*, *34*(1), 3–20. doi:10.1016/S0360-8352(97)00147-2

Shtubt, A. (1989). Modelling group technology cell formation as a generalized assignment problem. *International Journal of Production Research*, *27*(5), 775–782. doi:10.1080/00207548908942586

Simchi-Levi, D. (2009). *Designing and managing the supply chain concepts strategies and case studies*. Tata McGraw-Hill Education.

Skorin-Kapov, D., & Skorin-Kapov, J. (1994). On tabu search for the location of interacting hub facilities. *European Journal of Operational Research*, *73*(3), 502–509. doi:10.1016/0377-2217(94)90245-3

Solimanpur, M., Vrat, P., & Shankar, R. (2004). Ant colony optimization algorithm to the inter-cell layout problem in cellular manufacturing. *European Journal of Operational Research*, *157*(3), 592–606. doi:10.1016/S0377-2217(03)00248-0

Süer, G., & Bera, I. S. (1998). Multi-period cell loading and cell size determination. *Computers & Industrial Engineering*, *35*(1-2), 85–88. doi:10.1016/S0360-8352(98)00026-6

Süer, G., & Dagli, C. (2005). Intra-cell manpower transfers and cell loading in labor-intensive manufacturing cells. *Computers & Industrial Engineering*, *48*(3), 643–655. doi:10.1016/j.cie.2003.03.006

Süer, G., Pico, F., & Santiago, A. (1997). Identical machine scheduling to minimize the number of tardy jobs when lot-splitting is allowed. *Computers & Industrial Engineering*, *33*(1-2), 277–280. doi:10.1016/S0360-8352(97)00092-2

Süer, G., Saiz, M., & Gonzalez, W. (1999a). Evaluation of manufacturing cell loading rules for independent cells. *International Journal of Production Research*, *37*(15), 3445–3468. doi:10.1080/002075499190130

Süer, G., Saiz, M., & Gonzalez, W. (1999b). Evaluation of manufacturing cell loading rules for independent cells. *International Journal of Production Research*, *37*(15), 3445–3468. doi:10.1080/002075499190130

Süer, G., Vazquez, R., & Cortes, M. (2005). A hybrid approach of genetic algorithms and local optimizers in cell loading. *Computers & Industrial Engineering*, *48*(3), 625–641. doi:10.1016/j.cie.2003.03.005

Süer, G. A. (1996). Optimal Operator Assignment and Cell Loading in Labor Intensive Manufacturing Cells. *Computers & Industrial Engineering*, *31*(1/2), 155–158. doi:10.1016/0360-8352(96)00101-5

Süer, G. A. (1999). Cell Loading in Connected Cells. *Proceedings of the 26th international conference on computers and industrial engineering,* New Orleans (pp. 438-441).

Süer, G. A., Arikan, F., & Babayiğit, C. (2008). Bi-objective cell loading problem with non-zero setup times with fuzzy aspiration levels in labour intensive manufacturing cells. *International Journal of Production Research, 46*(2), 371–404. doi:10.1080/00207540601138460

Süer, G. A., Arikan, F., & Babayigit, C. (2009). Effects of different fuzzy operators on fuzzy bi-objective cell loading problem in labor-intensive manufacturing cells. *Computers & Industrial Engineering, 56*(2), 476–488.

Süer, G. A., Ates, O. K., & Mese, E. M. (2014). Cell loading and family scheduling for jobs with individual due dates to minimise maximum tardiness. *International Journal of Production Research.*

Süer, G. A., Cosner, J., & Patten, A. (2009). Models for cell loading and product sequencing in labor-intensive cells. [Elsevier Ltd.]. *Computers & Industrial Engineering, 56*(1), 97–105. doi:10.1016/j.cie.2008.04.002

Süer, G. A., Cosner, J., & Patten, A. (2009). Models for cell loading and product sequencing in labor-intensive cells. *Computers & Industrial Engineering, 56*(1), 97–105. doi:10.1016/j.cie.2008.04.002

Süer, G. A., & Dagli, C. (2005). Intra-cell manpower transfers and cell loading in labor-intensive manufacturing cells. *Computers & Industrial Engineering, 48*(3), 643–655. doi:10.1016/j.cie.2003.03.006

Süer, G. A., Huang, J., & Maddisetty, S. (2010). Design of dedicated, shared and remainder cells in a probabilistic demand environment. *International Journal of Production Research, 48*(19), 5613–5646. doi:10.1080/00207540903117865

Süer, G. A., & Mese, E. M. (2011). Cell Loading and Family Scheduling for Jobs with Individual Dates. In V. Modrak, & R. Sudhakara Pandian (Eds.), Operations Management Research and Cellular Manufacturing, Innovative Methods and Approaches. Hershey, PA, USA: IGI Global.

Süer, G. A., & Ortega, M. (1996). *Flexibility Considerations in Designing Manufacturing Cells: A Case Study. Group Technology and Cellular Manufacturing - Methodologies and Applications* (pp. 97–127). Amsterdam, The Netherlands: Gordon and Branch Science.

Süer, G. A., Pena, Y., & Vazquez, R. (2003). Cellular design to minimise investment costs by using evolutionary programming. *International Journal of Manufacturing Technology and Management, 5*(1/2), 118–132. doi:10.1504/IJMTM.2003.002536

Süer, G. A., & Saiz, M. (1993). Cell loading in cellular manufacturing systems. *Computers & Industrial Engineering, 25*(1-4), 247–250. doi:10.1016/0360-8352(93)90267-2

Süer, G. A., Saiz, M., Dagli, C., & Gonzalez, W. (1995). Manufacturing cell loading rules and algorithms for connected cells. *Manufacturing Research and Technology, 24*, 97–127. doi:10.1016/S1572-4417(06)80038-0

Süer, G. A., Saiz, M., Dagli, C., & Gonzalez, W. (1995). Manufacturing cell loading rules and algorithms for connected cells. In A. Kamrani, H. Parsaei, & D. Liles (Eds.), *Planning, Design and Analysis of Cellular Manufacturing Systems*. Amsterdam: Elsevier. doi:10.1016/S1572-4417(06)80038-0

Süer, G. A., Saiz, M., Dagli, C., & Gonzalez, W. (1995). *Planning, Design, and Analysis of Cellular Manufacturing Systems* (Vol. 24). Elsevier.

Süer, G. A., Saiz, M., & Gonzalez, W. (1999). Evaluation of manufacturing cell loading rules for independent cells. *International Journal of Production Research*, *37*(15), 3445–3468. doi:10.1080/002075499190130

Süer, G. A., Santos, J., & Vazquez, R. (1999). Scheduling rotary injection molding machines. *Proceedings of the Second Asia-Pacific Conference on Industrial Engineering,* Kanazawa, Japan (pp. 319-322).

Süer, G. A., Subramanian, A., & Huang, J. (2009). Heuristic procedures and mathematical models for cell loading and scheduling in a shoe manufacturing company. *Computers & Industrial Engineering*, *56*(2), 462–475. doi:10.1016/j.cie.2008.10.008

Süer, G. A., Subramanian, A., & Huang, J. (2009). Heuristic procedures and mathematical models for cell loading and scheduling in a shoe manufacturing company. *Computers & Industrial Engineering*, *56*(2), 462–475. doi:10.1016/j.cie.2008.10.008

Tavakkoli-Moghaddam, R., Ranjbar-Bourani, M., Amin, G. R., & Siadat, A. (2012). A cell formation problem considering machine utilization and alternative process routes by scatter search. *Journal of Intelligent Manufacturing*, *23*(4), 1127–1139. doi:10.1007/s10845-010-0395-2

Wang, X., Tang, J., Gong, J., & Chen, M. (2008). Formation of Multi-Objective Dynamic Cells Using Random Weight Multi-Objective Genetic Algorithm. *Chinese Journal of Management*, *5*(4), 516–521.

Wei, J. C., & Gaither, N. (1990). An optimal model for cell formation decisions. *Decision Sciences*, *21*(2), 416–433. doi:10.1111/j.1540-5915.1990.tb01694.x

Won, Y., & Chang Lee, K. (2004). Modified p-median approach for efficient GT cell formation. *Computers & Industrial Engineering*, *46*(3), 495–510. doi:10.1016/j.cie.2004.01.010

Chapter 10
Hybrid Supply Chain Strategies in Wind Business

Jordi Castelló
Universitat de Girona, Spain

Rodolfo de Castro
Universitat de Girona, Spain

Andrea Bikfalvi
Universitat de Girona, Spain

ABSTRACT

The present chapter aims to increase the understanding of possible supply chain models and their fit and most effective configuration in a typically engineer-to-order sector, namely the wind business. Our findings argue the pertinence of a Hybrid Supply Chain model combining elements of the Lean, Agile and Leagile paradigms upstream and the Project Management paradigm downstream in the case of a Wind Turbine Manufacturer. We also find that, depending on the complexity of the market and supply diversity, multiple decoupling points emerge. Through the present Wind Energy Supply Chain case, we find that for effective management of a global supply chain characterized by a high number of references and multiple technologies, it is necessary to mix different manufacturing paradigms in order to provide a quick, agile and competitive response to the customer.

INTRODUCTION

Key enabling technologies generating smart industries and alternative sources of energy contributing to sustainable societies have become an economic priority.

The wind business (WB) is a relatively new industry and in Western Europe its origins date back to the 1970s. Despite its relative infancy, the global wind energy market has experienced a steady 25% annual growth rate over the past 10 years. Nowadays, wind energy represents a major trend as countries

DOI: 10.4018/978-1-5225-0021-6.ch010

try to establish renewable energy technologies and diversify their energy mix. Wind energy is the world's fastest growing energy source and will power industry, businesses and homes with clean, renewable electricity for many years to come.

In terms of manufacturing and operations, the wind energy business has the following characteristics: i) low-volume, ii) high-variability, iii) highly-customized, iv) high-technology, v) high-quality and vi) short life wind turbines. The supply chain model (SCM) becomes a powerful strategic weapon, but is also an order-winning criterion when contracts are awarded.

Currently, one of the principle challenges facing wind turbine manufacturers (WTM) is trying to balance the capacity of their supply chain with demand uncertainty and variability, while attempting to ensure supply capacity, lower working capital, reduce work in process, decrease costs, lessen the lead time for components and reduce delivery times to final customers.

It is in this complex panorama that the questions, W*hich manufacturing strategy best fits the manufacturer?* and *What are the critical elements in the supply chain?* arise. The aim of the present research is to reach a better understanding as to which supply chain model is best in a complex and typically engineered-to-order sector such as the WB. In consequence, the research questions (RQ) we formulate are:

RQ1: How can a supply chain co-exist with other manufacturing paradigms e.g. Lean, Agile, Leagile and Project Management?
RQ2: What are order decoupling points in the supply chain and how do they vary?
RQ3: What type of strategy needs to be adopted when the product can require variable lead-times? Does a one-size-fits-all supply strategy apply?

Literature Review

According to Russell and Taylor (2006), several authors assume that organizations are facing some challenges at present, which are interrelated and include: intense competition, global markets, global sourcing, global financing, global strategies, enhanced product variety, mass customization, service businesses, quality improvement, flexibility, advances in technology, employee involvement, and environmental and ethical issues. To cope with these challenges, an organization use strategies and required technologies to form a virtual organization, in which participants collaboratively respond in a more agile manner to any unexpected changes in customer demand.

The present contribution focuses on developing a flexible understanding of supply chains to enable mass customization of products and services. The main idea is to use concepts which, while are not new, have either not been integrated or have not been included in the same supply chain.

The literature review is divided into two main sections. We focus first on supply chain management to explore paradigms related to the supply of components and raw materials, and second on project management because project management has become the main generator of supply orders in this complex industry. The final integration in the supply chain is monitored by project managers because of the complex business of selling power capacity rather than selling a product which generates wind energy.

Supply Chain Management

The field of supply chain management (SCM) has gained importance over the last two decades and there are many indications that it will continue to do so. A supply chain is a network of facilities that produce raw materials, transform them into intermediate goods and then final products, which are delivered to customers through a distribution system. However, while SCM has been around for more than 25 years, there is as yet no consensus on its definition or what it entails. Stock and Boyer (2009) provide a definition of SCM after performing a qualitative analysis of 173 individual definitions of the field. Their definition, which is the most comprehensive available, as it is based on the qualitative analysis of an extensive literature review, is as follows: "The management of a network of relationships within a firm and between interdependent organizations and business units consisting of material suppliers, purchasing, production facilities, logistics, marketing, and related systems that facilitate the forward and reverse flow of materials, services, finances and information from the original producer to final customer, with the benefits of adding value, maximizing profitability through efficiencies, and achieving customer satisfaction" (Stock, 2009:706).

Supply chain strategy requires an end-to-end focus on the integration of business processes throughout the value chain for the purpose of providing optimum value to the end-customer (Qrunfleh and Tarafdar, 2013; Green et al., 2008). However, it is possible to account for these processes using different paradigms and evolving knowledge. A number of paradigms are presented below.

Lean Supply Chain

The origins of Lean can be found on the shop-floors of Japanese manufacturers and, in particular, innovations at Toyota Motor Corporation (Womack and Jones, 1996; Monden, 1983). These innovations are summarized in some tools that include the just-in-time production system, the Kanban method of pull production, respect for employees and high levels of employee problem-solving / automated mistake-proofing (Hines et al, 2004).

Based on Lean principles, the Lean supply chain strategies are a logical extension to Lean Thinking. These strategies focus on waste reduction, helping firms to eliminate those activities that do not add value because they consume excess time, labour, equipment, space, and inventory across the supply chain. Christopher and Towill (2000) suggest that Lean principles are applicable in markets where demand is "relatively stable" and therefore "predictable", and where "variety is low" and "volume is high".

Agile Supply Chain

In a report published by the Iacocca Institute of Lehigh University and entitled "21st Century Manufacturing Enterprise Strategy", Nagel and Dove (1991), originally introduced the concept of agile manufacturing as an option for managing firms in a dynamic world. The origins of "agility" as a business concept lie partly in flexible manufacturing systems which, through automation (i.e. reduced setup times) enable rapid changeovers and, as a result, can be responsive to changes in product mix and volume (Christopher and Towill, 2000). Later, this idea of manufacturing flexibility was extended to a wider business context by Nagel and Dove (1991), and the concept of agility as a supply chain paradigm was born.

Many authors have a definition of agility. Gunasekaran (1998) defined agility as the ability to respond to market changes in a cost-efficient and profitable manner, while Christopher and Towill (2000) define it as, "a business-wide capability that embraces organizational structures, information systems, logistics processes and in particular, mind-sets." Agility involves the utilization of market knowledge and a responsive organization in order to exploit profitable opportunities in a volatile marketplace (Agarwal et al., 2006). Essentially, it is about being demand-driven rather than forecast-driven.

The focus of improvement efforts in the agile approach is the integration of information flow across the supply chain with the objective of creating a market-responsive supply chain that reacts quickly to unpredictable demands and thereby minimize lost sales, forced markdowns and obsolescent inventory (Hoek, 2000). A market-responsive supply chain emphasizes market mediation to a greater extent than the efficient physical supply of the product (de Treville et al, 2004). An important point is that agile manufacturing implementation requires the company to be a world class manufacturer using lean manufacturing methods (Bayraktar et al., 2007).

Leagile Supply Chain

In recent years, one of the most interesting debates concerning supply chain strategies has centered on the relative merits of the Lean and Agile philosophies. Lean and Agile approaches are discussed as opposing paradigms, even though they share a common objective: meeting customer demand at the least total cost (Goldsby et al., 2006; Hilletofth, 2012). Therefore agility should not be confused with leanness and the main difference between the two approaches is that lean supply is associated with level scheduling, whereas agile supply reserves capacity to cope with volatile demand.

Many researchers have suggested that the Lean and Agile approaches can be integrated in a variety of ways to create so-called Leagile strategies (Aitken et al., 2003; Childerhouse et al., 2002; Christopher and Towill, 2000; Mistry, 2005). Thus, instead of focusing on a definition of a Lean or an Agile strategy, attention is directed to the selection and integration of the most appropriate aspects of these paradigms in a specific supply chain strategy (Hilletofth, 2012).

Integration of Lean and Agile concepts in "Leagile" implies the use of postponement, a practice which has been used increasingly by industry in recent years. The main idea is to use the Lean concept until the decoupling point, working to enhance efficiency indicators, and then to use the Agile concept beyond that point, seeking mass customization. Many European industrial companies are currently implementing Leagile supply chain systems (Hoek et al., 2001).

Finally, the ability to customize – by being close to the customer – means that a higher level of variety can be offered at a lower total cost, enabling strategies of "mass-customization" to be pursued. In order for Leagile supply chain systems to succeed, a reliable supplier network that can supply parts and services is essential (Feitzinger & Lee, 1997).

Decoupling Points in a Supply Chain Structure

One problem in supply chains is the limited visibility of real demand. A linking point between production and demand is required. This is the point where "real demand" penetrates upstream in a supply chain, and market demand "pull" meets upstream supply "push". This may be called the decoupling point, because upstream from that point real demand cannot be felt. This point has also been called the "order penetration point" by several authors (Hallgren et al., 2011; Olhager, 2003; Christopher, 2000).

The decoupling point separates that part of the supply chain which responds directly to the customer from the part of the supply chain which uses forward planning and strategic stock management to respond to any variability in the demand of the supply chain. This concept has been applied in many environments: supply chain (Christopher et al.., 2006) or manufacturing strategy (Olhager, 2003). Nevertheless, the decoupling point connects directly with characteristics of the Lean and Agile paradigms (Naylor et al., 1999). The Lean paradigm can be applied to the supply chain upstream of the decoupling point if demand is smooth and standard products flow through a number of value streams. But, downstream from the decoupling point a number of products flow through just one value stream. The Agile paradigm must be applied downstream from the decoupling point as demand is variable and product variety per value stream is increased. The decoupling point is where the two paradigms meet and the challenge of SCM is to seek to develop Lean strategies up to the decoupling point, and then Agile strategies beyond that point. By using generic or modular inventory to "postpone" final commitment to specific, customized products, it should be possible to achieve volume-oriented economies of scale through product standardization. The point of flow up to the decoupling point should be forecast driven and after the decoupling point it should be demand driven (Chistopher, 2000).

However, in "real-world" supply chains there are actually two decoupling points (Mason-Jones and Towill, 1999; Mason-Jones et al., 2000). The first is the material decoupling point, where strategic inventory is held in the most generic form possible. This point should ideally lie as far downstream in the supply chain and as close to the final marketplace as possible in order to reduce delivery time. The second decoupling point is the information decoupling point. The idea here is that this should lie as far as possible upstream in the supply chain; it is, in effect, the furthest point to which information on real final demand penetrates. The penetration of the "information" decoupling point is promoted by sharing mechanisms such as design coordination, partnerships or in-depth collaborations.

The Postponement Concept

Christopher (2000: 42) states that, "Postponement, or delayed configuration, is based on the principle of seeking to design products using common platforms, components or modules, but where the final assembly or customization does not take place until the final market destination and/or customer requirement is known". In other words, postponement is moving the order decoupling point (ODP) to delay final definition of the product.

Several authors suggest that the combined effect of delayed configuration through "postponement" and shared information in a supply chain is a significant improvement in responsiveness. There is a strong relationship between product design and Agile supply chain performance. Product design must introduce information about which activities should be performed after orders are received and managed according to Agile principles (i.e. responsive, order-driven and customized activities) and which activities should be performed before orders are received and managed according to Lean principles (i.e. efficient, planned and standardized activities).

Project Management

Cicmil and Hodgson (2006) provide a clear description of the origins of this concept and practice, which emerged as a social practice in the post-World War II development of technology and infrastructure, following its development in the Manhattan Project in the 1940s. The majority of the groundwork was done in the USA defense and aeronautics industries in the 1950s, including its use in the Apollo space programs, although the US oil and chemical industry also played a major role in this period.

Bakouros and Kelessidis (2000: 2) state that, "Project Management emerged because of the growing demand for complex, sophisticated, customized goods and services and the exponential expansion of human knowledge. It is a set of principles, methods and techniques for effective planning of objective-oriented work, thereby establishing a sound basis for effective scheduling, controlling and re-planning in the management of programs and projects. In other words, it provides an organization with powerful tools that improve the organization's ability to plan, organize, implement and control its activities and the ways it uses its people and resources".

Research into Project Management remains heavily reliant on a functionalist approach and largely ignores theoretical aspects. It main interest is the accomplishment of some finite piece of work in a specified period of time, within a certain budget, and to agreed specifications (Atkinson, 1999).

Methodology and Research Design

It was only in the late 1990s that the development of the SCM paradigm accelerated, with the majority of the theoretical and empirical investigation starting in 1997 (Lambert et al., 1998). A meta-analysis of 405 articles carried out by Giunipero et al. (2008) revealed that 70 percent of the articles published on SCM were empirical in nature, and the rest were theoretical. Carter and Ellram (2003) found that approximately 75 percent of the research in journals, including surveys and case studies, is empirical in nature.

Empirical research contributes in the context of theory building as well in validating any proposed theories. From the practitioners point of view it is easier for business executives to understand empirical research, especially if they do not have advanced training in management science or operations research (De-Margerie & Jiang, 2011; Wong et al., 2005).

According to Soni and Kodali (2012) and Voss et al. (2002) empirical methods are imperative in Operations Management and are suitable for exploring the extension and refinement of theory, for a more in-depth examination and for validating empirical results. Operations Management is a very dynamic field in which new practices are continually emerging and subsequently being applied to other fields.

In this study we use action research methodology because we want to demonstrate that integrating the paradigms into the same supply chain is the best way to manage the supply chain. A comprehensive study of the WB sector is necessary in order to explore the different behaviors along the chain. This study was conducted from the perspective of a WTM, where one of the authors was involved and participated in decisions about how to manage the supply chain.

Action research is a qualitative research technique. The researcher participates in the company where the research is developed and may influence decisions made during the case study. Moreover the same researcher can try to escape the system to obtain a broader perspective. Afterwards, the researcher can thus be described as an outside agent, building theory, while simultaneously participating as a facilitator in the firm where the case was developed. The aim is to take action, at the same time as creating knowledge or theory about action. Although action research has longer roots, it was only recently that this methodology was also applied in the supply management domain (Soni & Kodali, 2012). One example of action research in Operations Management, is the study by Karlsson and Ahlström (1996), who supported, implemented and examined lean product development in a company over a period of two years.

For the present study, the researchers actively participated in component classification and establishing relationships with suppliers who were in constant interaction with the organization. In action research the researcher can collect information which would be impossible to obtain without active participation. The authors, as participants, use observational data as well as secondary data to develop the proposed model. To ensure methodological rigor, the action research cycle described by Coughlan and Coghlan (2002), which prescribes process steps for conducting quality action research, was followed. Action research consists of the deliberate implementation of a cyclical four-step process: planning, taking action, evaluating the action, which leads to further planning, taking action and so on.

The company concerned was a global WTM firm located in Spain. The action research started in October 2008 with planning. This was followed by the initial intervention in May 2009, and then the evaluation which lead to further action that concluded in November 2009. Finally the concluding phase was completed in March 2010.

The methodology used was guided by external academic sources and based on the perspectives of both practitioners and managers. The case study itself is based on more than fifteen years' experience of decision-making in the WB field, of data acquisition, of analyzing component lead times and manufacturing capabilities, dealing with supply chain bottlenecks and so on, as well as problem solving, to attain an efficient supply chain, while keeping in touch with suppliers, assembly plants and wind farm installations. The data collected related to 1) negotiating suppliers' prices and delivery times, 2) supplier capacity audits, 3) work-in-process and working capital for different pipelines, 4) company supply strategies, 5) workshops supply strategies, 6) supplier agreements, and 7) procurement delivery lead times.

THE CASE OF THE WIND BUSINESS

The wind business is a highly complex industry. Since our aim is to analyze the supply chain, we need to consider the perspective of a key agent, for which purpose we selected a WTM.

The WB is a relatively new industry and in Western Europe its origins date back to the 1970s. Despite its relative infancy, the global wind energy market has experienced a steady 25% annual growth rate over the past 10 years. Wind energy represents a major trend as countries try to establish renewable energy technologies and diversify their energy mix.

Sector. The investment required to install a wind farm is massive: the cost of an onshore 2MW turbine is estimated to be 1.3 million Euros per MW. Besides the determination of governments to rely less on unsustainable fossil fuels, a further reason as to why the WB future is bright is because the wind energy input is free and unlimited, so long as the wind farm is strategically located in a region with strong winds.

The lessons learned from more than a decade of operating wind power plants, along with continuing R&D, mean that wind-generated electricity is very close to being cost-competitive with conventional utility power generation in some locations. Wind energy is the world's fastest growing energy source and will power industry, businesses and homes with clean, renewable electricity for many years to come.

The WB has expanded exponentially in recent years. The power of wind turbines is constantly increasing (Figure 1). Furthermore, customers' technological specifications and product customization have also increased (Kaldellis & Zafirakis, 2011).

Most customers are large power generating "public utilities" (e.g. Iberdrola, EON, EDP, EDF), although there are also independent investors (IPPs - Independent Power Producers) who are wind farm owner-operators who sell the energy they generate to the public utilities.

One aspect of the complexity of the sector is that customers do not actually buy wind turbines; they buy energy (MWh). Moreover, demand is uncertain, not only in terms of concrete contracts, but also in terms of in executing wind farm contracts. This uncertainty may be exacerbated by administrative

Figure 1. Representative turbine evolution from 1980 to 2010
Source: Lantz et al 2012 - NREL

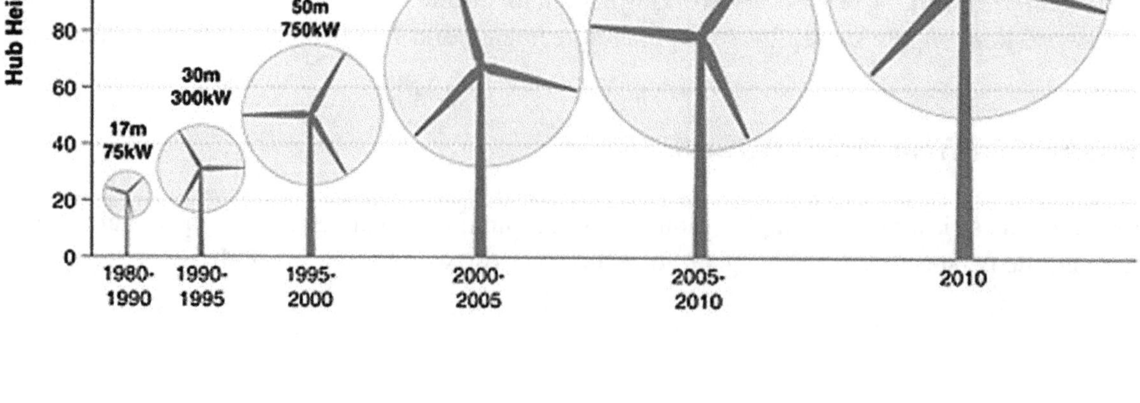

barriers concerned with the project licensing process. Such barriers to licensing arise from the public administration and regulation of each individual country, environmental impact, and so on, and sometimes stem from infrastructure or electric network connection problems. Consequently they can delay permits to construct wind farms for years, making it difficult to forecast demand, which makes it even more difficult to align WTM production capacity.

One of the main challenges to meet is the high technological requirements for wind turbines, typically availability of 97.5% to 98% of the time of the contract, although this is a median value (Wilburn et al., 2011), and a cycle life of approximately 20 years. This technological demand on the efficiency and machine cycle-life required for a wind turbine is reflected at individual component level. Developing key components can only be entrusted to a few skilled suppliers. These assumptions about risk for this technology cause a shortage of key components, which in turn creates bottlenecks in the supply chain (Aubrey, 2007; BTM Consult, 2011; MAKE Consulting, 2013).

In other words, the long lead times of these key components, along with demand uncertainty, force the WTM to slot their short-term planning work (backlogs of administrative authorizations and forecasts) in between orders. In terms of manufacturing and operations, the WB can be described as: (1) Low-volume, (2) High-variability, (3) Highly-customized, (4) High-technology, (5) High-quality and (6) Short life cycle. As a result, SCM is a powerful strategic weapon, but also an order winning criterion when contracts are awarded.

Currently, one of the principle challenges facing WTMs is trying to balance the capacity of their supply chain with demand uncertainty and variability and attempting 1) to ensure supply capacity, (2) to lower working capital, (3) to reduce work in process, (4) to decrease costs, (5) to reduce the lead time for components, and (6) to reduce delivery times to final customers.

Figure 2. Typical onshore wind turbine development timeline
Source: BTM Consultant (2013)

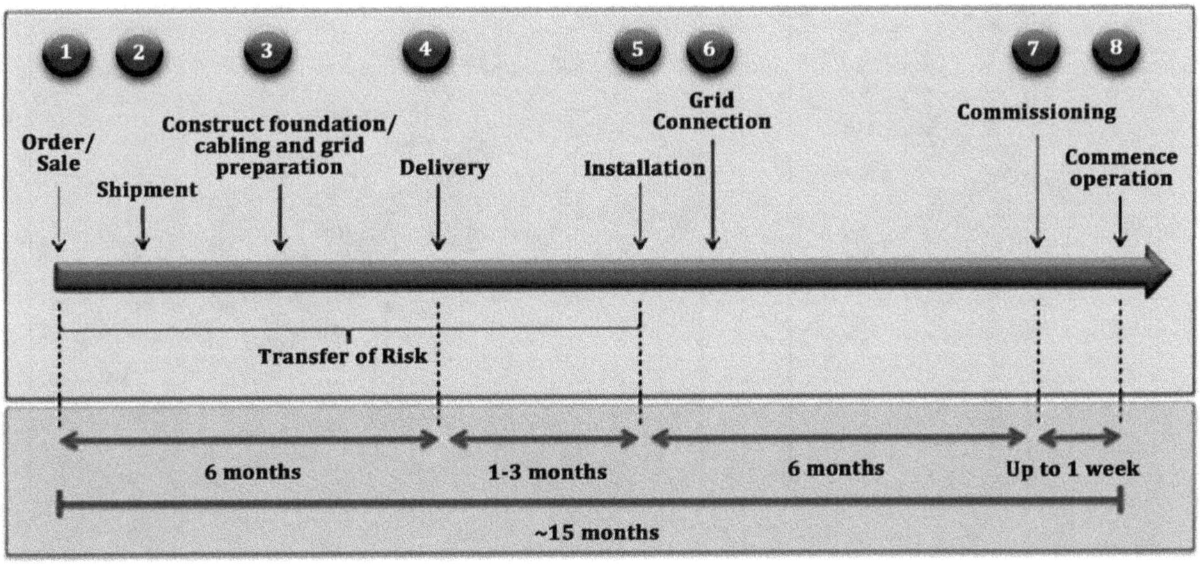

Products

IRENA (2012: 4) stated that, “Wind power technologies transform the kinetic energy of the wind into useful mechanical power. The kinetic energy of the air flow provides the motive force that turns the wind turbine blades which, via a drive shaft, provide the mechanical energy to power the generator in the wind turbine”.

Historically, the era of wind power as we know today began in 1979 with the mass production of wind turbines by the Danish manufacturers Nec Micon, Vestas, Nordex, Bonus, and some others. Those wind turbines had small capacities (10 kW to 30 kW) by today’s standards. The current average size of grid-connected wind turbines is from 1.16 MW to 3 MW (BTM Consult; 2011; IRENA, 2012).

There are many configurations but horizontal-axis wind turbine designs are more common, usually a three bladed stall or pitch-regulated machine operating at near-fixed rotational speed.

Wind turbines consist of three principal components (Wilburn et al.., 2011), (1) the nacelle, (2) the rotor, and (3) the tower (Figure 3). Willburn et al. (2011:10-11) state that, “The nacelle contains much of the equipment required for energy generation and conversion and typically accounts for 25 to 40 percent of the weight of the wind turbine. The rotor typically makes up 10 to 14 percent of the weight of the wind turbine… The rotor of a wind turbine consists of four principal components: the blades, the blade extender, the hub, and the pitch drive system… The tower provides the support system for the turbine blades and the nacelle and serves as the conduit for electrical and electronic transmission and grounding”.

A typical wind turbine contains up to 8,000 different components (Aubrey, 2007) and is reported to contain, by weight, 89.1% steel, 5.8% fiberglass, 1.6% copper, 1.3% percent concrete (primarily cement,

Figure 3. Diagrams of a typical large wind turbine and major components
Source: Wilburn (2011)

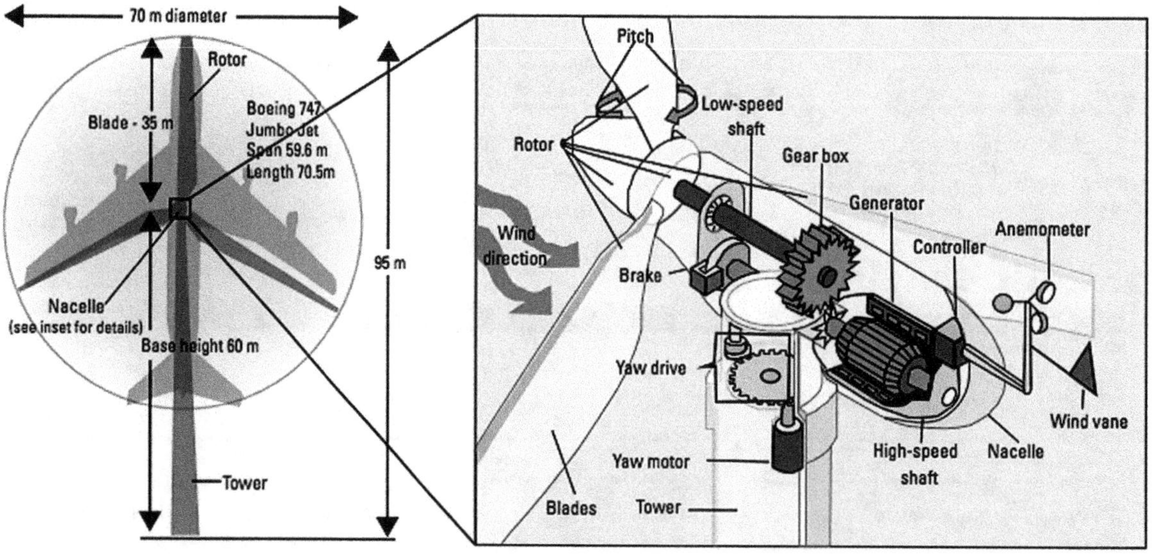

water, aggregates, and steel reinforcement), 1.1% adhesives, 0.8% aluminium, and 0.4% core materials (primarily foam, plastic, and wood) (Wilburn et al.., 2011). The cost of the turbine itself accounts for 76% of the total capital cost of a wind farm project (EWEA, 2009).

WTM as a Key Agent in Wind Business

Our focus is on the WTM, because it is the key agent in the supply chain and also where we are carrying out our action research. The WTM is interested in selling electrical production capacity rather than just a wind turbine. The concept of ODP can be used when the product is clearly defined, but, in the case of a highly complex product, postponement must deal with uncertainty in some of the parameters defined by the final project. We therefore propose to use the ODP concept twice: firstly downstream from the WTM, when the project management determines the final requirements, and secondly in the different supply pipelines depending on the component involved.

Traditionally, companies have conducted their business according to an MTS (Make to Stock) approach. This implies that they perform all the supply chain activities, including design, sourcing, manufacturing, assembly, packaging, labelling and distribution, based on forecast and speculation. However, in the case of the WTM, suppliers have to postpone committing to specific components, depending on the available degree of definition. Manufacturing postponement by suppliers aims to maintain the product in a neutral and non-committed status in the supply chain for as long as possible (Bowersox & Closs; 1996). Customization of components for a particular project is postponed to the last possible moment. This means that suppliers delay assembly to order (ATO), make to order (MTO), and source to order (STO), and may even engineer to order (ETO) until orders have been defined and received. Moving the second ODP upstream in the supply chain depends on the classification of components and is coordinated by the WTM.

Globalization is revolutionizing the scope of the supply chain. The suppliers' relative importance and their contributions in terms of technological developments, capacity, delivery time, and cost are crucial in the WB. The capacity of the WTM conditional upon the capacity of its supply base and its capacity to supply the key components required by the WTM on time (Aubrey, 2007; BTM Consult, 2011; MAKE Consulting, 2013).

In fact, the term "delivery time" is a critical factor when the final customer awards contracts. Customers may penalize delays. There are no limitations to the WTM's assembling and installing ability, and these are currently neither a bottleneck nor a constraint. The crucial limitations are in component supply management. The current lack of component supply capacity requires the WTM to make decisions whether to make or buy some key components, in order to ensure supply, control costs, and absorb demand surges, respond quickly to customers, and improve supply chain responsiveness.

While many WTMs have chosen a vertical integration strategy for some key components, not all WTMs have, and prefer instead to ensure their responsiveness by increasing their supply base for key components, following an outsourcing strategy that allows suppliers to develop their core competence in a sector where investment in R&D and equipment (CAPEX), including blades, gearboxes, converters, and generators, is much needed. Many key suppliers use the same design platforms for different WTMs customizing their components at the end of their production process, thus making them more cost competitive and more efficient through economies of scale.

Findings

The action research methodology involves various stages during the investigation. Table 1 shows these stages, which are also outlined in the following sections. The planning and taking action stage corresponds to the initial supply chain configuration and the classification of component suppliers, based on supplier attributes. This was followed by evaluation of the action, where, because of the complexity of the business and demand uncertainty, orders have to be anticipated using a mix of customer confirmed project orders and forecasting based on historical data. Finally, this led to further planning where project management was proposed to monitor wind farm construction.

Supply Chain Configuration

The WB landscape is composed of developers, manufacturers and operators. WTMs generally embrace a range of activities including site development, wind turbine design, manufacturing, wind farm construction, operation, maintenance and technical service. The present study focuses on a WTM.

The wind energy supply chain (Figure 4) is composed of five entities: (1) raw material suppliers; (2) component suppliers; (3) assembly plants – WTM; (4) wind farm construction, including civil works, installation and commissioning, and wind farm technical services; and finally (5) the customer, who is usually responsible for the operation and maintenance of the wind farm. However, utility companies might also be involved in generating, transmitting and distributing the wind energy.

In the following paragraphs we briefly describe the main activities of the supply chain illustrated in Figure 4.

- **Raw Material Suppliers:** Provide the basic materials such as steel, forged castings, plastics, fibers and copper to manufacture different components and systems for the wind turbines.

Table 1. Phases in Action Research Planning

Research Stage	Objective	Section in Paper
Planning	Problem statement and first approximation of Wind Business Supply Chain Model.	Supply Chain configuration.
Taking action	Development of WTM sourcing model based on components attributes and determination of different manufacturing paradigms and determination of different decoupling points.	Components supplier classification based on supply attributes
Evaluating the action	Perform, managing and monitoring different pipelines and use of rolling forecast and Master Planning Scheduling	Lean Assembly Plants
Leading to further planning	Introduction of Project Management paradigm downstream Wind Turbine Manufacturer. Model fully integrated	Wind Farm construction

Source: the authors

Figure 4. First approach to wind business supply chain
Source: the authors

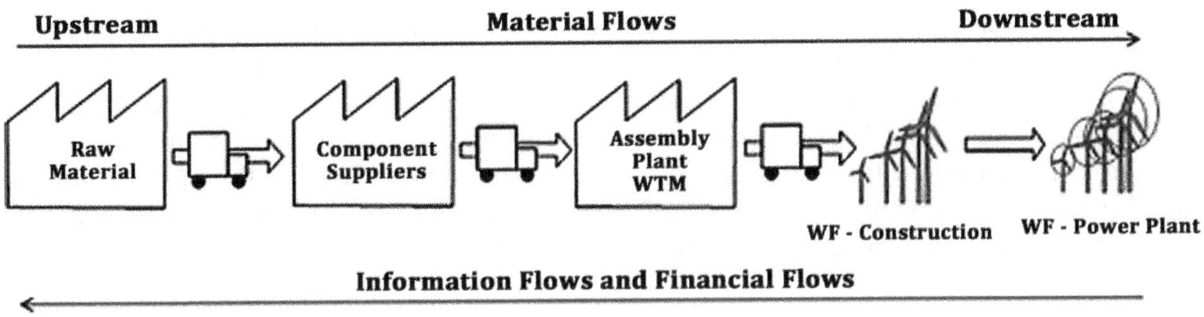

- **Component Suppliers:** Are a key strategic point in the supply chain in the WB sector, because many components are developed in collaboration with suppliers (Aubrey, 2007; BTM Consult, 2011; MAKE Consulting, 2013), since the technology of both product and process belongs, in most cases, to the suppliers. Hence, supplier management and development is extremely important.
- **Assembly Plants (Wind Turbine Assembly Plants):** Are responsible for the assembly of the different components following a master planning schedule (MPS) to produce wind turbines ready for delivery.
- **Wind Farm Construction:** Includes civil work, commissioning and erecting and installing the wind turbines on site according to the customer's contractual requirements. In this phase of the supply chain, the customer closely monitors the installation process as it occurs on the customer's premises and under their supervision.
- As the final step in the supply chain is the *customer*, who is usually a utility company or IPP, and they are responsible for the wind farm operation and maintenance, although this can vary depending on the type of customer and contract.

Component Supplier Classification based on Supply Attributes

This section focuses on the supply strategy proposed to a WTM which then leads to four pipeline strategies. As this study focuses on supply strategy, we focus on the sourcing carried out by the WTM. This activity includes the overall management of suppliers, finding new supply sources, selecting and rating suppliers, negotiating and procuring new contracts, monitoring the quality of contract purchases, as well as meeting deadlines, negotiating contracts and technical service assistance guarantees for those components that affect the availability of the wind turbines.

In the partnership between the WTM and its suppliers there are many factors that determine the different strategies adopted. The foundation of the relationship is based on determining the position of the ODP, as this establishes the point where forecasts are replaced by the assignment of a specific wind farm project. As noted above, some authors (Ben Naylor et al.., 1999; Mason-Jones et al, 2000) distinguish between Lean and Agile supply chains, where a Lean supply chain should be applied upstream of the ODP, while an Agile supply chain is more suitable for downstream operations.

According to the characteristics of the components, a different strategy should be followed in each component pipeline, as each has different needs and characteristics. While there are some models for existing supply chain design in the literature review (Christopher et al.., 2006; Fisher, 1997; Kraljic, 1983; Mason-Jones et al.., 2000) and while they are very rich in their content, they are simply not specific enough when the design is required for a particular sector. By designing a sourcing strategy in the WB, we develop a simplified model based on Kraljic (1983) and Christopher et al. (2006) to apply to this case study. The particular attributes have been listed above and highlight the fact that, in this sector the supply of components is fundamental (Aubrey, 2007; BTM Consult, 2011; MAKE Consulting, 2013).

Therefore we classify families of components on two dimensions:

- **Profit Impact:** This is the cost of the component that, in turn, influences the working capital and work in process supported by the WTM. The cost of the ex-work components is around 70% to 76% of the cost of an installed wind turbine and, as a wind turbine has around 8,000 components (Aubrey, 2007), we can reduce costs by classifying and monitoring these components carefully.
- **Supply Risk:** This is the delivery lead time of a component, which determines the time we must wait for a particular component. It is determined by the few key components suppliers, who may have such limited capacity that it could increase uncertainty. In fact, this limits the capacity for key components and causes business uncertainty.

These two attributes classified as either "low" or "high" impact, producing the component portfolio (2x2 matrix) depicted in Figure 5. This classification allows us to apply different strategies for the sup-

Figure 5. Component Segmentation Matrix for Wind Turbine Manufacturer
Source: the authors

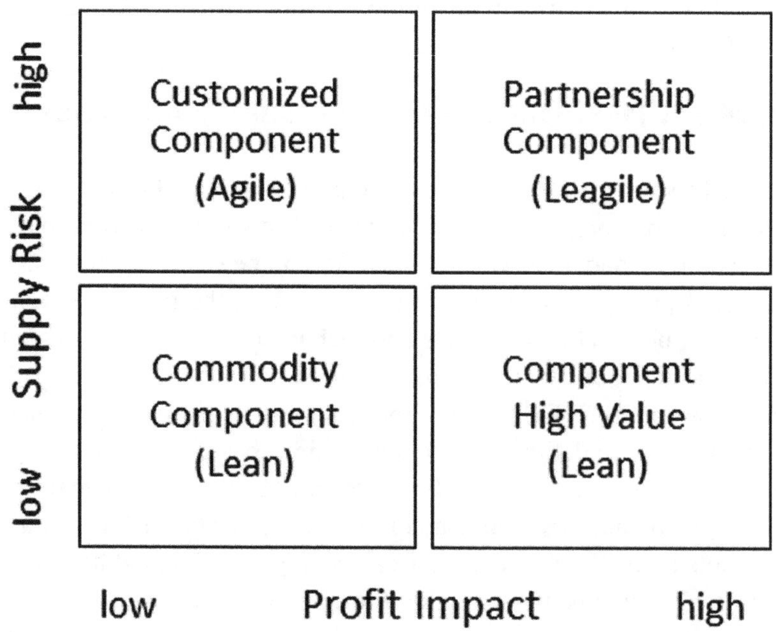

ply of each family of components (Lean, Agile, Leagile, etc.) and model the different paradigms, all the while seeking the maximum overall efficiency. The resulting four strategies are described below.

Commodities: These are commercial components that are purchased from a catalogue and have low economic value, compared to the cost of wind turbines, as well as relatively short lead times (weeks or sometimes only days). Usually there is no purchasing strategy for these components and they are purchased at the best price and the best delivery time, and customer-supplier relationships are temporary, project by project "win-lose" deals. The lean supply strategy is continuous replenishment (Christopher et al.., 2006), for example under a Kanban system. Component replenishment is the responsibility of the supplier who has to maintain a point of order according to the demand of the assembly plant.

Customized: These are bought according to specific commercial references or technical characteristics as specified in a catalogue or, generally, where the WTM can choose certain attributes to configure them to their particular requirements (e.g. low voltage electric motors, electrical equipment, high voltage cables, hydraulic power, instrumentation). These components typically have lead times of about 2 months. There is a clear purchasing strategy and the supplier relationships are "win-lose", and alternative suppliers are monitored to help source at better prices and delivery times. In this kind of agile supply, the supplier customizes the components and gives a quick response to customer-driven demand.

High Value: These components are purchased from drawings or according to particular technical specifications because, as they are usually structural components, their fabrication requires meticulous attention and high quality. They are complex in terms of the technology, such as casting, forging, machining, and welding (e.g. hubs, main frames, towers, flanges, covers). Suppliers of these components are involved from the beginning in the design phase, as the methods and processes used have a huge influence on the final quality and attributes of the product. They generally have lead times between 2 and 3 months. There is a clear purchasing strategy for these components and supplier relationships are "win- win" with long-term deals. Such components often cause bottlenecks and some may limit the ability of the whole chain to meet its deadlines. Hence some WTMs have decided to follow a vertical integration strategy (BTM Consult, 2011; MAKE Consulting, 2013). This type of supply is lean and there is an annual agreement with annual forecasts and quarterly adjustments based on demand.

Partnership: To develop this type of components, close cooperation with suppliers is required from the beginning of the design (Kraljic, 1983), since it is actually the supplier who designs the component based on the stated requirements of the WTM, so that it can be integrated with other components. The correct and accurate definition of specifications by the WTM, determines the results of the design (e.g. blades, gearbox, converter, pitch system, generator) and, to a large extent, the reliability of the final components. There is a clear sourcing strategy for these components, based on long-standing and mutual cooperation. The relationship is "win-win" as it is based on long-term co-design and partnership. These components have long lead times of 5 to 7 months, because of the supply of the raw materials and the technological complexity of the processes. The supply strategy should usually be Leagile. As there are framework capacity agreements lasting for several years, and technological developments, the supplier can plan their factories and raw material supplies accordingly. Products are customized based on real demand and quarterly planning. Suppliers may use the same product platform for different customers as a strategy to exploit economies of scale. Some WTMs have also opted for vertical integration (e.g. Gamesa, Vestas, Enercon, GEC), because some components can produce supply chain bottlenecks.

Based on these descriptions it is possible to determine four different pipelines (Figure 6) depending on the classification of the components. The place where the ODP is located will be fixed by where the pipeline components are situated in the model.

Results

Applying the model in the wind energy supply chain, we classify WTM components as in Figure 7. All the WTM components can be classified, determining a clear supply strategy for each. In summary, as shown in Figures 6 and 7, depending on the nature of the components and their attributes, four different pipelines for a WTM have been defined which in turn determine four different ways of managing supplier relationships.

Lean Assembly Plants

The introduction of different supply strategies (i.e. Lean, Agile and Leagile), depending on the nature of components, facilitates the implementation of the Lean paradigm in assembly plants, since it is possible to smooth production assembly requirements. The aim is to have components arrive just in time, according to the supply strategies defined and the required buffer flows, to smooth wind farm demand.

Due to the complexity of the business and demand uncertainty, purchasing orders are based on a mix of confirmed customer orders and forecasting based on historical data, resulting in the rolling forecast represented in Figure 8.

This mix of confirmed customer orders and forecasting orders allow us to smooth the schedule of the WTM assembly plan and also manage the assembly capacity according to confirmed wind farm project requirements. At the same time, it facilitates the implementation of lean manufacturing tools and techniques in the assembly plants.

In this supply chain stage, the WTM postponement strategy upstream is a crucial element as it makes it possible to combine confirmed orders with forecast orders until the final customization point of the

Figure 6. Postponement Strategy for Key Component Suppliers
Source: the authors

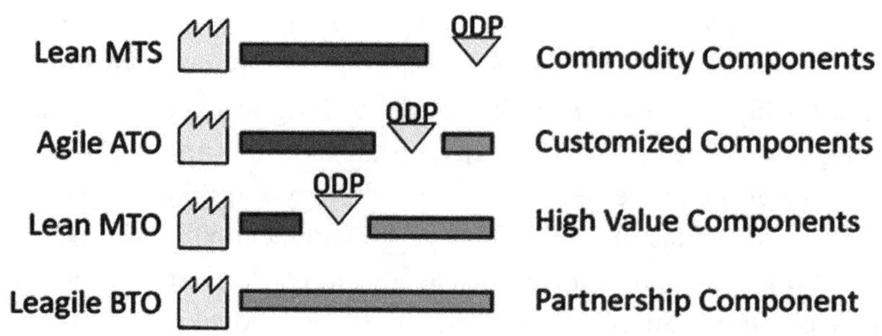

Figure 7. Matrix results from application model. Segmentation of components
Source: the authors

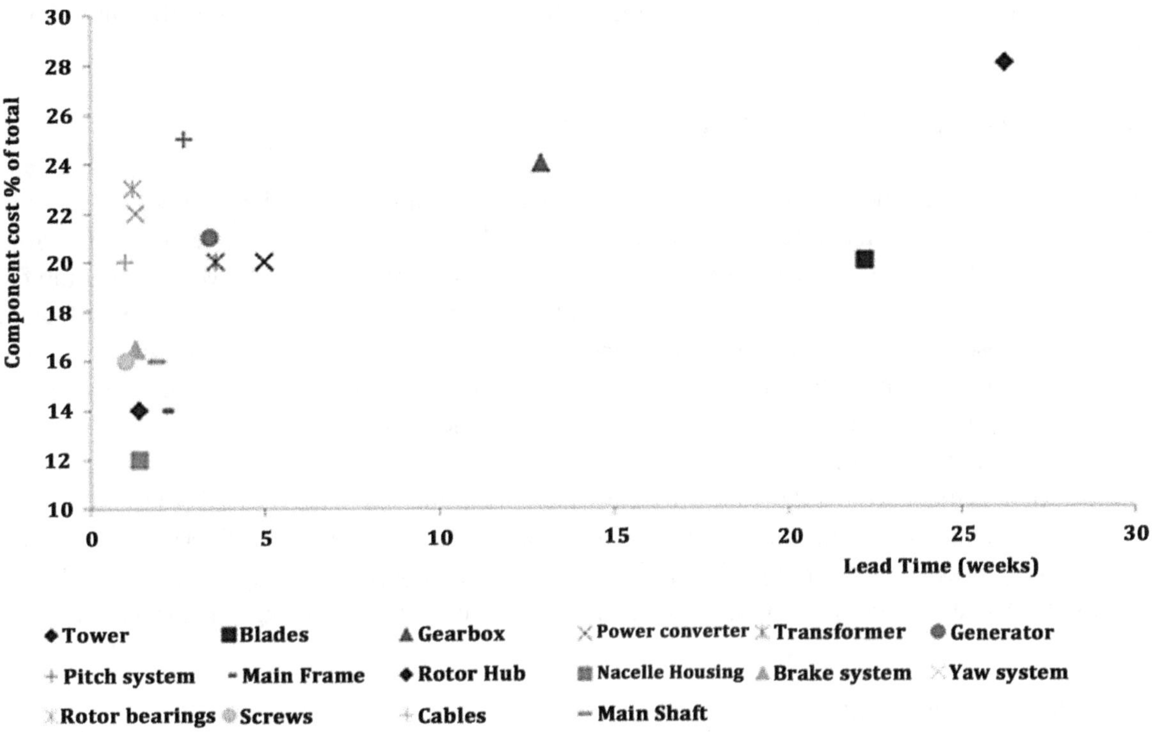

Figure 8. Rolling Master Planning, combining Forecasts and Confirmed Orders
Source: the authors

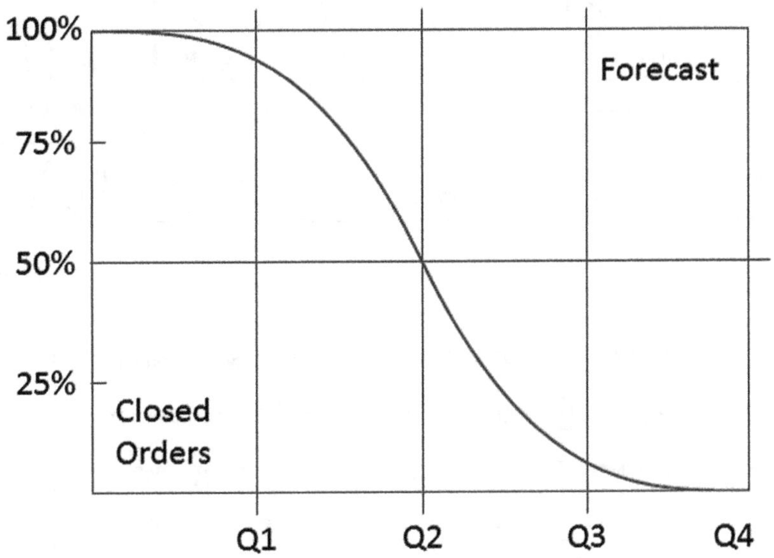

wind turbine. This provides flexibility as to when an expected order becomes a confirmed, planned order. This key concept of postponement is represented in Figure 6 and varies depending on the component supply strategies described above. However, these strategies can be integrated into the entire supply chain (Figure 9).

Wind Farm Construction (Project Management)

In this stage of the wind energy supply chain, we introduce the paradigm of Project Management, which, although not a new concept, has not been applied in an SCM context. At this supply chain stage, Project Management handles wind farm construction, installation and commissioning, delivering the wind farm to the final customer. The management at this stage of the supply chain is completely different from the supply stream upstream of the WTM since, as mentioned above, the construction of the wind farm is in a location determined by the customer and on the customer's site. The main parameters are delivery time, cost and quality in compliance with contract specifications and, in this context, interactions/interfaces with the customer are crucial.

Project Management is the last planning tool downstream of the WTM and is used to monitor the first ODP close to the customer. There are many specifications to determine in the delivery and installation stages, and they impact the sourcing of components. As appearing in figure 9, we defined ODP1 and ODP2, as two decoupling points firms need to manage the components according to Project management

Figure 9. Hybrid Supply Chain
Source: the authors

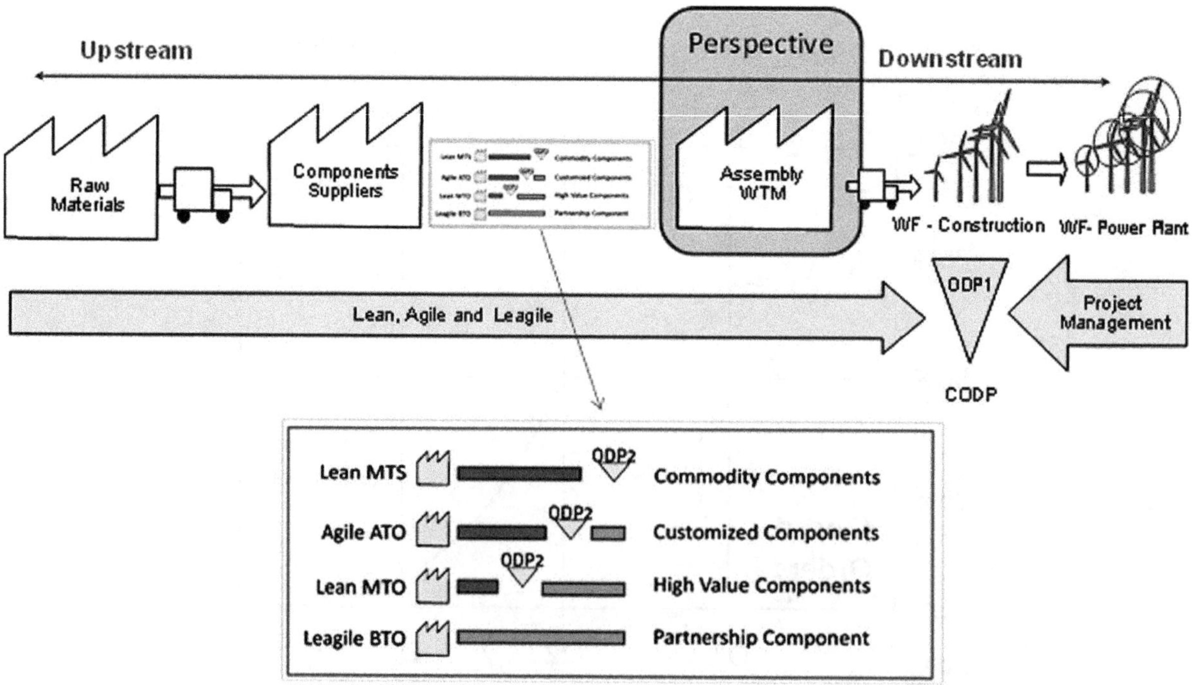

orders. Accordingly, activities defined backward from ODP1 inform the forecast of orders in the rolling plan and forward activities impact on the postponement strategy for key components related to ODP2.

CONCLUSION

According to our findings, the wind energy supply chain model consists in a mixed supply chain containing several paradigms in management, and two order decoupling points. The main feature is the separation of decisions related to Project Management downstream from the WTM from component sourcing strategy upstream of the WTM. Up to four manufacturing paradigms can co-exist in a hybrid supply chain (RQ1). The supply chain can have more than one ODP, depending on the complexity of the components (RQ2). There is a need to apply different manufacturing strategies to different supply chain pipelines for components, categorized by supply risk and profit impact.

Through this wind energy supply chain case, the effective management of a global supply chain characterized by a high number of components (approximately 8,000) and multiple technologies, different manufacturing paradigms must be combined, to give a quick, agile and competitive response to the customer. However, we believe that the implications of our findings go beyond the wind energy supply chain and could be valid in other settings with similar characteristics, such as complex projects/products with multiple and varied suppliers.

The WTM must be skilled in management and supply chain integration and needs to be able to manage the different Lean, Agile and Leagile paradigms efficiently based on key component attributes. Project Management should be a supply chain integrator so as to gain a competitive advantage over competitors, the cost of the components being 75% of the total cost of an installed wind turbine. In this sector where lead times are long and margins are narrow, SCM is a key success factor.

The needs and requirements, and in particular administrative barriers, involved in planning a wind farm installation, makes component supply a critical element. Component sourcing has to be as fast and efficient as possible following the component classification. Depending on each of the strategies, it is possible to determine a different ODP for each pipeline so that the allocation of a purchased component for a specific wind farm installation project is the most appropriate and efficient way possible to manage the working capital and work in process, for both WTM and suppliers.

This WB case opens up new opportunities for further research, to investigate broad scope of SCM and the variety of management paradigms (depending on the type of product, the complexity of markets and the nature of components) that can be found beyond those required in wind energy. Another line of research would be to test the results with a survey, as So and Sun (2010) did, even though the particular attributes of the sector make it difficult to acquire a big enough sample.

ACKNOWLEDGEMENT

The authors are grateful for the financial support from the Spanish Ministry of Economy and Competitiveness ECO2013-46954-C3-3-R and from the Autonomous Government of Catalonia 2014 SGR 868.

REFERENCES

Agarwal, A., Shankar, R., & Tiwari, M. K. (2006). Modeling the metrics of lean, agile and leagile supply chain: An ANP-based approach. *European Journal of Operational Research*, *173*(1), 211–225. doi:10.1016/j.ejor.2004.12.005

Aitken, J., Childerhouse, P., & Towill, D. (2003). The impact of product life cycle on supply chain strategy. *International Journal of Production Economics*, *85*(2), 127–140. doi:10.1016/S0925-5273(03)00105-1

Ancona, D., & McVeigh, J. (2001). *Wind turbine—Materials and manufacturing*. Princeton Energy Resources International LLC.

Body of Knowledge (BoK) Revised January 1995 (version 2). (1995). *APM - Association of Project Management*.

Atkinson, R. (1999). Project management: Cost, time and quality, two best guesses and a phenomenon, its time to accept other success criteria. *International Journal of Project Management*, *17*(6), 337–342. doi:10.1016/S0263-7863(98)00069-6

Aubrey, C. (2007). Supply Chain: The race to meet demand. *Wind Directions*, (February 2007), 27–34.

Bakouros, Y., & Kelessidis, V. (2000) Project Management. Report produced for the EC funded project INNOREGIO. Retrieved from http://www.adi.pt/docs/innoregio_pmanagement.pdf

Bayraktar, E., Jothishankar, M. C., Tatoglu, E., & Wu, T. (2007). Evolution of operations management: Past, present and future. *Management Research News*, *30*(11), 843–871. doi:10.1108/01409170710832278

Billington, P. J. (2004). A Classroom Exercise to Illustrate Lean Manufacturing Pull Concepts. *Decision Sciences Journal of Innovative Education*, *2*(1), 71–76. doi:10.1111/j.0011-7315.2004.00021.x

Bowersox, D. J., & Closs, D. J. (1996). *Logistical Management: The Integrated Supply Chain Process*. New York, NY: McGraw-Hill.

Carter, C. R., & Ellram, L. M. (2003). Thirty-five years of the journal of supply chain management: Where we have been and where we going? *Journal of Supply Chain Management*, *39*(2), 27–39. doi:10.1111/j.1745-493X.2003.tb00152.x

Childerhouse, P., Aitken, J., & Towill, D. R. (2002). Analysis and design of focused demand chains. *Journal of Operations Management*, *20*(6), 675–689. doi:10.1016/S0272-6963(02)00034-7

Christopher, M. (2000). The Agile Supply Chain: Competing in Volatile Markets. *Industrial Marketing Management*, *29*(1), 37–44. doi:10.1016/S0019-8501(99)00110-8

Christopher, M., Peck, H., & Towill, D. (2006). A taxonomy for selecting global supply chain strategies. *The International Journal of Logistics Management*, *17*(2), 277–287. doi:10.1108/09574090610689998

Christopher, M., & Towill, D. (2000). Supply Chain migration from lean and functional to agile and customised, *Supply Chain Management. International Journal (Toronto, Ont.)*, *5*(4), 206–213.

Cicmil, S., & Hodgson, D. (2006). New possibilities for project management theory : A critical engagement. *Project Management Journal, 37*(3), 111–122.

Market Report Global Wind Power Supply Chain. (2013). *MAKE Consulting.*

Coughlan, P., & Coghlan, D. (2002). Action research for operations management. *International Journal of Operations & Production Management, 22*(2), 220–240. doi:10.1108/01443570210417515

De-Margerie, V., & Jiang, B. (2011). How relevant is OM research to managerial practice? An empirical study of top executives perceptions. *International Journal of Operations & Production Management, 31*(2), 124–147. doi:10.1108/01443571111104737

De Treville, S., Shapiro, R. D., & Hameri, A.-P. (2004). From supply chain to demand chain: The role of lead time reduction in improving demand chain performance. *Journal of Operations Management, 21*(6), 613–627. doi:10.1016/j.jom.2003.10.001

Wind Energy - The facts. The economics of Wind Power. (2009). *EWEA - European Wind Energy Association.*

Feitzinger, E., & Lee, H. L. (1997). Mass Customization at Hewlett-Packard: The Power of Postponement. *Harvard Business Review, 75*(1), 116–121. PMID:10174793

Fisher, M. (1997). What is the right supply chain for your product? *Harvard Business Review*, (March), 105–116.

Gavireni, S. & Tayur, S. (1997, August). *Delayed product differentiation versus information sharing.* Carnegie Mellon University.

Giunipero, L. C., Hooker, R. E., Matthews, S. C., Yoon, T. E., & Brudvig, S. (2008). A decade of SCM literature: Past, present and future implications. *Journal of Supply Chain Management, 44*(1), 66–86. doi:10.1111/j.1745-493X.2008.00073.x

Goldsby, T. J., Griffis, S. E., & Roath, A. S. (2006). Modeling Lean, Agile, and Leagile Supply Chain Strategies. *Journal of Business Logistics, 27*(1), 57–80. doi:10.1002/j.2158-1592.2006.tb00241.x

Green, K. Jr, Whitten, W. D., & Inman, R. A. (2008). The impact of logistics performance on organizational performance in a supply chain context. *Supply Chain Management, International Journal (Toronto, Ont.), 13*(4), 317–327.

Gunasekaran, A. (1998). Agile manufacturing: Enablers and an implementation framework. *International Journal of Production Research, 36*(5), 1223–1247. doi:10.1080/002075498193291

Hallgren, M., Olhager, J., & Schroeder, R. G. (2011). A hybrid model of competitive capabilities. *International Journal of Operations & Production Management, 31*(5), 511–526. doi:10.1108/01443571111126300

Hilletofth (2012). Differentiation focused supply chain design, *Industrial Management & Data Systems,* 112(9), 1274 – 1291.

Hines, P., Holweg, M., & Rich, N. (2004). Learning to evolve: A review of contemporary lean thinking. *International Journal of Operations & Production Management*, *24*(10), 994–1011. doi:10.1108/01443570410558049

International Wind Energy Development. Supply Chain Assessment 2012-2015. (2011). *BTM Consult.*

IRENA working paper: renewable energy technologies. (2012). *Wind Power*, 5(5). Retrieved from www.irena.org/Publications

Kaldellis, J. K., & Zafirakis, D. (2011). The wind energy (r)evolution: A short review of a long history. *Renewable Energy*, *36*(7), 1887–1901. doi:10.1016/j.renene.2011.01.002

Karlsson, C., & Ahlström, P. (1996). The difficult path to lean product development. *Journal of Product Innovation Management*, *13*(4), 283–295. doi:10.1016/S0737-6782(96)00033-1

Kraljic, P. (1983). Purchasing must become supply management. *Harvard Business Review*, (Sept-Oct), 109–117.

Lambert, D. M., Cooper, M. C., & Pagh, J. D. (1998). Supply chain management: Implementation issues and research opportunities. *International Journal of Logistics Management*, *9*(2), 1–19. doi:10.1108/09574099810805807

Lantz, E., Wiser, R., & Hand, M. (2012) IEA Wind Task 26: The Past And Future Cost Of Wind Energy (Technical Report NREL/TP-6A20-53510). *National Renewable Energy Laboratory.*

Mason-Jones, R., Naylor, B., & Towill, D. R. (2000). Lean, agile or leagile? Matching your supply chain to the marketplace. *International Journal of Production Research*, *38*(17), 4061–4070. doi:10.1080/00207540050204920

Mason-Jones, R., & Towill, D. R. (1999). Total cycle time compression and the agile supply chain. *International Journal of Production Economics*, *62*(1-2), 61–73. doi:10.1016/S0925-5273(98)00221-7

McCutcheon, D. M., & Meredith, J. R. (1993). Conducting case study research in operations management. *Journal of Operations Management*, *11*(3), 239–256. doi:10.1016/0272-6963(93)90002-7

Mistry, J. J. (2005). Supply Chain Management: A Case Study of an Integrated Lean and Agile Model. *Qualitative Research in Accounting & Management*, *2*(2), 193–215. doi:10.1108/11766090510635442

Monden, Y. (1983). *Toyota production System.* Industrial Engineering and Management Press, Institute of Industrial Engineers.

Nagel, R. N. & Dove, R. (1991). 21st Century Manufacturing Enterprise Strategy: An Industry-Led View. Iaccoca Institute.

Naylor, B. J., Naim, M. M., & Berry, D. (1999). Leagility: Integrating the lean and agile manufacturing paradigms in the total supply chain. *International Journal of Production Economics*, *62*(1-2), 107–118. doi:10.1016/S0925-5273(98)00223-0

Olhager, J. (2003). Strategic positioning of the order penetration point. *International Journal of Production Economics*, *85*(3), 319–329. doi:10.1016/S0925-5273(03)00119-1

Qrunfleh, S., & Tarafdar, M. (2013). Lean and agile supply chain strategies and supply chain responsiveness: The role of strategic supplier partnership and postponement. *Supply Chain Management: An International Journal, 18*(6), 571–582. doi:10.1108/SCM-01-2013-0015

Russell, R. S., & Taylor, B. W. (2003). *Operations Management* (4th ed.). New Jersey, USA: Prentice and Hall.

So, S., & Sun, H. (2010). Supplier integration strategy for lean manufacturing adoption in electronic-enabled supply chains. *Supply Chain Management: An International Journal, 15*(6), 474–487. doi:10.1108/13598541011080455

Soni, G., & Kodali, R. (2012). A critical review of empirical research methodology in supply chain. *Journal of Manufacturing Technology Management, 23*(6), 753–779. doi:10.1108/17410381211253326

Stock, J., & Boyer, S. (2009). Developing a consensus definition of supply chain management: A qualitative study. *International Journal of Physical Distribution & Logistics Management, 39*(8), 690–711. doi:10.1108/09600030910996323

Van Hoek, R. I. (2000). The thesis of leagility revisited. *International Journal of Agile Management Systems*, 2(3), 196–201.

van Hoek, R. I., Harrison, A., & Christopher, M. (2001). Measuring agile capabilities in the supply chain. *International Journal of Operations & Production Management, 21*(1/2). doi:10.1108/01443570110358495

Voss, C., Tsikriktsis, N., & Frohlich, M. (2002). Case research in operations management. *International Journal of Operations & Production Management*, 22(2), 195–219. doi:10.1108/01443570210414329

Wilburn, D., Salazar, K., & McNutt, M. (2011). Wind energy in the United States and materials required for the land-based wind turbine industry from 2010 through 2030. *U.S. Geological Survey Scientific Investigations Report* (2011–5036).

Womack, J. P., & Jones, D. T. (1996). *Lean Thinking*. New York, NY: Simon & Schuster.

Yew Wong, C., Stentoft Arlbjørn, J., & Johansen, J. (2005). Supply chain management practices in toy supply chains. *Supply Chain Management: An International Journal, 10*(5), 367–378. doi:10.1108/13598540510624197

Chapter 11
Physical and Digital Integration Strategies of Electronic Device Supply Chains and Their Applicability to ETO Supply Chains

Claudia-Maria Wagner
Dublin Institute of Technology, Ireland

Colm Ryan
Dublin Institute of Technology, Ireland

ABSTRACT

The growth in the manufacture and distribution of electronic devices presents a source of continuing innovation. Electronic devices are products that integrate physical forms (i.e. hardware) and virtual forms (e.g. software) to deliver value to customers. These forms are very different from a product design and supply chain perspective, but nevertheless they need to work closely together in order to create value for the customers. For electronic device manufacturers, it is important that processes are in place to facilitate the seamless integration of both forms throughout the engineering, production, distribution and support stages of the product lifecycle. This chapter examines the role of physical and virtual supply chain innovation strategies in electronic device supply chains by exploring the commonalities and differences between the design, manufacturing, and distribution models of digital and physical elements. It also explores to what extent such strategies can be employed for engineer-to-order (ETO) supply chains.

DOI: 10.4018/978-1-5225-0021-6.ch011

INTRODUCTION

One of the most significant trends over the last few years has been the massive increase in the number of electronic devices in use. Examples of such products are smart phones, tablets, watches, camcorders, vehicles and many different kinds of industrial, military and healthcare devices. The growth in the manufacture and distribution of electronic devices presents a source of continuing innovation. Electronic devices are products that integrate physical forms (i.e. hardware) and virtual forms (e.g. software) to deliver value to customers. These forms are very different from a product design and supply chain perspective, but nevertheless they need to work closely together in order to create value for the customers (Wagner & Ryan, 2009). For electronic device manufacturers, it is important that processes are in place to facilitate the seamless integration of both forms throughout the engineering, production, distribution and support stages of the product lifecycle.

Although digital products potentially offer considerable advantages to innovators as they are less constrained by physical properties, they also present significant challenges. Hybrid approaches are explored, where the best of both worlds, physical and digital, can be applied. The merging of physical and digital elements into an integrated product creates a supply chain that is very different to a purely physical situation. Supply chain management principles of such a hybrid product approach can be derived from the concept of servitisation, which refers to the bundling of products and services into one harmonised product offering. Services and software are two very much overlapping concepts. Typically, in a hybrid product bundle, software as the digital element of the product is an enabler behind a service. While the software can be copied, a service is more difficult to emulate because it often incorporates non-digital elements. The authors introduce a digital SC framework by outlining some current physical and digital supply chain integration and distribution examples and by determining likely innovative supply chain scenarios on how these strategies might develop over the coming years.

Thus, this chapter examines the role of physical and virtual supply chain innovation strategies in electronic device supply chains by exploring the commonalities and differences between the design, manufacturing, and distribution models of digital and physical elements. The aim is to identify to what extent such strategies can be employed for engineer-to-order (ETO) supply chains. Gosling and Naim (2009) see ETO as a supply chain where the decoupling point is located at the design phase of a product. Integration between design and downstream supply chain activities is paramount.

BACKGROUND

Research in operations management, marketing and supply chain management has predominantly studied the relationships between upstream manufacturers and downstream retailers in supply chains for physical goods (Chellappa & Shivendu, 2003; Padmanabhan & Png, 1997). Physical goods are so defined because they obey physical laws. They possess mass and occupy volume. Over time, they can break down and degrade. It is not possible to create products without first sourcing appropriate materials. Energy (which also needs to be sourced) is expended in transformation and transportation. It could be said, therefore, that a large proportion of SCM has to do with overcoming the constraints imposed by physics on the products under consideration.

However, with the rise of information and communication technology (ICT), not all products can be longer considered physical in the traditional sense of word. An increasingly important class of products, described variously as digital products, virtual goods or information goods (Shapiro & Varian, 1999) has emerged that cannot be considered physical in the traditional sense of the word. Surprisingly, little is known about digital product supply chains, where products such as software, movies, music and newspapers are created, stored and delivered in a digital form over a network (Chellappa, 2000).

The chapter begins with giving n outline of ETO. It then goes on to investigate the properties of physical products and examines how these properties influence the supply chain. It then contrasts physical products with digital products such as software. The chapter ends by outlining some innovative integration strategies and how they can be employed in an ETO context.

ETO AND THE MOVE TO MASS CUSTOMISATION

Various definitions of Engineer-to-Order (ETO) exist in literature. ETO is a type of manufacturing process for highly customized products, which are required to be designed and engineered in detail as per the specifications in the order placed by customers (Pandit & Zhu, 2007). The challenge for ETO supply chains stems from the social, technical and economic difficulty of synergising various parties along the supply chain. Hicks et al. (2000) point out that a typical ETO manufacturing process can be defined as a supply chain. Design and project management are considered to be core competencies in most ETO organisations. Pandit and Zhu (2007) argue that some of the most common problems that ETO supply chains are facing are difficulties in estimating lead-times, delivery dates, extensive rework due to late realisation of errors, poor product quality and material wastage, as well as conflicts between project and manufacture schedules. Most of those challenges are caused due to inefficient and ineffective information sharing among supply chain partners. Elfving (2003) found that long lead times in particular has more than half of its causes associated directly or indirectly with the design phase. Some examples can be summarised under the following: tedious collection and poor reliability of design input, changes due to early commitment and lack of knowledge, changes due to design error, coupling ETO product design to other systems design, outdated practice of auxiliary design and approvals, and complexity requiring large number of specialists.

Haug, Ladeby, Edwards (2009) define engineering companies that deliver products, which are engineered to specific requirements of their customers as ETO companies. Thus, they researched the potential transition of ETO companies towards mass customisation. Wikner and Rudberg (2005) propose a framework where they decouple the engineering and the design aspect ETO supply chains and are hence able to differentiate between manufacturing approaches (see Figure 1). This so called customer order decoupling, or order penetration point defines where in the manufacturing process a product is linked to a customer order. Wikner and Rudberg (2001) identified four main order penetration points: Engineer-to-order (ETO), Make-to-order (MTO), Assemble-to-order (ATO), and Make-to-Stock (MTS). One particular trait of ETO is that the engineering work has to be done for each order, while for the other strategies, the engineering work have already been carried out, meaning the product design is already ‘in stock’ before a customer order.

Figure 1. Customer Order Decoupling Point Concept, (Rudberg and Wikner, 2004)

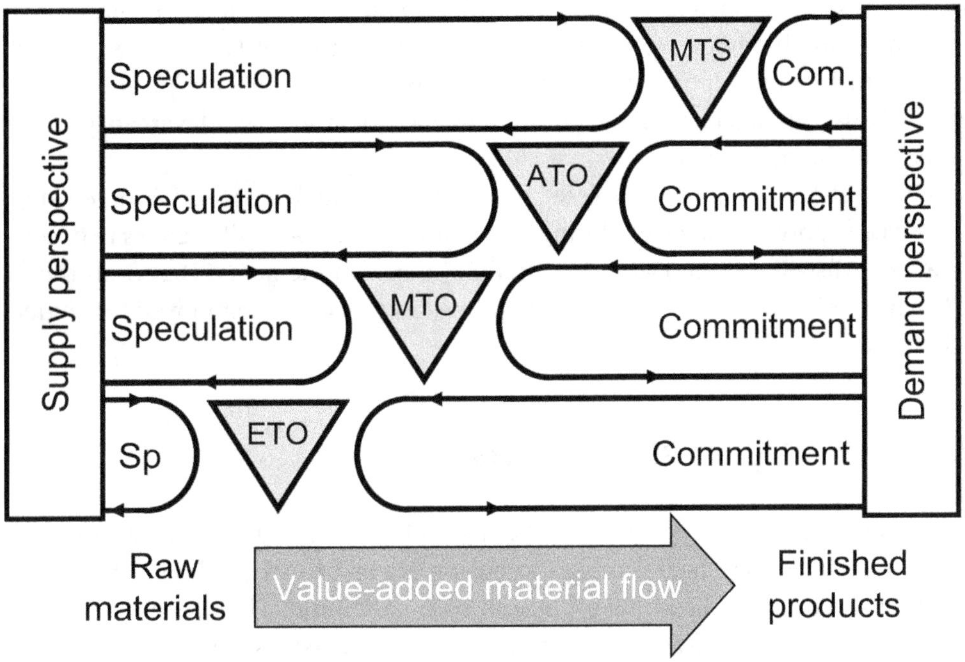

Literature discusses the move of some organisation from ETO companies to having become mass customizers. In particular, the increasing demand for more customized products sees mass customisation as a natural step from mass production for many organisations. According to Haug, Ladeby, Edwards (2009) the motivation for an ETO organisation to move towards mass customization is a decision based on many factors. They emphasize that an ETO company that moves towards mass customisation will have standardise their engineering work, which means that these companies need to reach a ETS (Engineer-to-Stock) state by being able to define parts of the solution space before receiving specific customer requirements. This implies some sort of a postponement of the customer order decoupling point. This can reduce delivery times, lead to more precise cost calculations, reduced specification and manufacturing costs, less training needed, and less errors, among other benefits. Mass customisation can be defined as "the ability to provide customized products or services through flexible processes in high volumes and at reasonably low cost" (Silveira et al., 2001). Piller (2004) goes one step further and defines mass customisation as a "customer co-design process of products and services, which meet the needs of each individual customer with regard to certain product features." To summarise, Haug, Ladeby and Edwards' (2009) research points out that some ETO companies move to mass customisation for parts of their product portfolio. This implies that such companies need to move the time of differentiation closer to the time of delivery.

In addition, in ETO, decisions as to what is to be digitized and what is to be part of the physical design need to be considered at the very beginning. For a smart phone, the casing, lenses, switches, and touchscreens will encompass the physical design, while installed apps, the operating system and firmware (e.g. the camera management system) will be components in the software design. For automotive products such as cars, digital systems such as stability control, climate control, engine management and safety systems are an increasingly important part of the value proposition. Software allows greater flexibility in how physical components will operate in the field – often beyond what can be anticipated during the design stage - and this presents new challenges as to how to software should be managed throughout its lifecycle and beyond the point of sale.

The following sections examine the role of physical and virtual supply chain innovation strategies in electronic device supply chains by exploring the commonalities and differences between the design, manufacturing, and distribution models of digital and physical elements. Based on this, the chapter will conclude with a brief analysis to what extent such strategies can be employed for engineer-to-order (ETO) supply chains.

PHYSICAL VS. DIGITAL SUPPLY CHAINS

Practitioners are familiar with physical product supply chains, in that they must overcome the basic laws of physics in order to distribute products to customers in a timely fashion. Physical products are composed of atoms, located at a particular place in space and time. They cannot typically be created virtually: they must be sourced from somewhere through processes such as farming or mining, and created through manufacturing processes. The aim of supply chain management is to move these physical entities (transportation management), re-constitute them (manufacturing), store them (warehouse management) and make the products available (distribution), all the while taking into consideration the amount of time and energy requirements necessary to achieve the desired outcomes. Overcoming such physical constraints requires extensive knowledge in disciplines ranging from forecasting, planning, procurement, cost management, manufacturing and engineering. All of this is overlaid by complex IT management systems. The limitations of physical products are furthermore constrained by the Second Law of Thermodynamics, which states that all items have a tendency to break down, tarnish or spoil over time. This requires the need for comprehensive maintenance processes and reverse logistics processes such as recycling, refurbishment, salvage and scrapping.

A number of different definitions of supply chains and their effective management exist in literature. Supply Chain Management in a traditional sense is concerned with the strategic management of a firm's activities, which together provide customers with the appropriate level of service at optimum cost (Christopher, 2005; Lambert et al., 1998; Sweeney et al., 2005). Chou and Ruchika (2006) note that in short, the aim of all entities in a supply chain is to ensure that a product moves down the chain in the most effective manner, so as to maximise profits earned by the whole chain as well as minimise wastage of raw materials, labour and time. Simchi-Levi et al. (2000, p.1) define supply chain management (SCM) as "a set of approaches utilised to effectively integrate suppliers, manufacturers, warehouses, and stores, so that merchandise is produced and distributed at the right quantities, to the right locations, and at the right time in order to minimise system-wide costs while satisfying service level requirements". It traditionally focuses on the hardware (material or goods) flow and the related information flow (Chou

& Ruchika, 2006) that can be both upstream and downstream. Within a supply chain, upstream suppliers provide input; the company then adds value to these inputs, before passing them downstream to the next actor, which can be either another company or the end user (Porter, 1985). The effective management of a supply chain transforms it into a value chain, where organisations tied in it derive a number of benefits (Baltacioglu et al., 2007). Traditionally, a value chain is the combination of nine generic value-added activities that provide (product and service) value within an organisation to be passed on to a customer. However, Ramsay (2005) notes that value is a perceived quality that is associated with the benefits that can occur at various points along a supply chain. Thus, value chains can operate in both directions.

Sweeney (2005) lays down four fundamentals of Supply Chain Management, citing problems with existing definitions of Supply Chain. These fundamentals are as follows:

1. **Objectives:** The objectives are to achieve high levels of customer service while optimising total supply chain investment and cost. Customer service, in effect, sets the specification for the supply chain and there is a strong focus on eliminating non-value-added activities in order to achieve this.
2. **Philosophy:** The success by which organisations can deliver products and services to customers is determined by the weakest link in that chain. It's not companies that compete with other companies, but supply chains competing with supply chains.
3. **Flows:** There are three key flows in any supply chain that need to be managed in a *holistic* manner. These are a) materials, b) information and c) money.
4. **Relationships:** This holistic approach requires that the participants in that supply chain look at each other differently. Confrontational approaches introduce inefficiencies and cause the entire supply chain to work sub-optimally. Relationships based on sharing and trust are required instead.

The concept of Supply Chain Management arose from the area of logistics, first of all focusing on the management of cost, then through the achievement of customer service improvements using advances in information technology, and in later years entering into management consciousness as a key strategic driver for businesses (Gattorna & Walters, 1996). Supply Chain Management has allowed for "new breeds" of companies to emerge, that use their ability to supply products efficiently to act as a competitive weapon, with companies like Dell, Amazon and 7-Eleven (Chopra & Meindl, 2001) emerging as major players on the global scale through highly innovative uses of Supply Chain Management techniques and technologies. As supply chains become ever more complex, due in part to outsourcing and today's global competitive environment, the way in which companies are structuring and managing their relationships can make the difference between profit and loss (Christopher & Towill, 2002). In particular, globalization of supply chains creates logistical problems along three dimensions, which are replenishment level, time, and distance (McCullen & Towill, 2000). In addition to that, customer choice and the nature of the product often implies that "one size does not fit all" (Shewchuck, 1998).

In contrast to physical products, digital products refer to any goods or services that can be digitised, which means converted into a binary format (Hui and Chao, 2002). They can be compressed without losing much information or quality. Thus, digital copies have a technical quality similar to the original (Peitz & Waelbroeck, 2003). Whinston et al. (1997) identifies three properties of digital goods:

- Indestructibility (the tendency of a digital product to maintain its form ad-infinitum).
- Transmutability (the ease by which a digital product can be modified).
- Reproducibility (the ease by which digital products can be reproduced, stored and transferred).

From an SCM perspective it is the property of reproducibility in particular that makes digital products so different from physical products. It has huge consequences so long as there is sufficient storage and bandwidth available to copy, move and transport the product (Eldering et al., 1999; Grochowski and Halem, 2003). Traditionally, beside the cost of material, pricing strategies of physical goods has always been based on factors such inventory cost and return policies (Chellappa & Shivendu, 2003). For digital products, no raw materials or energy need to be sourced or used up in the process and replication is instantaneous. Moreover, digital products have almost zero marginal cost (Shapiro & Varian, 1999; Whinston & Choi, 1997) because they suffer neither from storage costs nor can they be returned for a salvage price.

Digital products constitute items such as movies, music, e-books, videos, podcasts, documents, apps and computer programs, reports and packages of diagnostic information. Such products normally exist as bits inside a wider physical architecture such as servers or electronic devices such as computers and smartphones. New digital products can be created simply by duplicating bits inside such architectures, with no requirement in most cases now for the creation of new physical media to support this process. (In the past, with CD ROMs and magnetic tape, this requirement did exist to a large extent). The creation of a new movie inside a server does not require any change to the atoms in that computer, merely a change of electronic state in a small part of its circuitry. The structuring of such bits is determined by developers and product creators, but once the final version is available, duplication is trivial. Transportation of the bits to the customer is relatively trivial also. Bits do not expend significant amounts energy in moving from source servers to their destinations. Neither do they require significant amounts of time, as the transfer processes are relatively instantaneous, or at least largely governed by the bandwidth speeds from source to delivery.

Managing the supply chain in traditional markets mainly provides challenges related to cost-effectiveness and physical barriers. By contrast, in a digital marketplace supply chain management encounters barriers that are not physical but rather strategic in nature (Nath, Saha & Salehi-Sangari, 2007). Chou and Ruchika (2006) argue that a digital supply chain is a supply chain where the digital product constitutes a significant part of the total value of the product and where goods are not physically flowing through the supply chain. In a software supply chain, for example, a significant proportion of the supply chain is comprised by only one entity, which is usually one single company handling the design, development and production of a software product. It is usually at the distribution stage that a software company might engage the services of another company to market its product. Moreover, channel coordination policies such as returns are not applicable in the digital world. They do not need to be sent back for repair or rework like physical products do. So long as the physical device that enables the product runs correctly, so too will the digital asset.

Table 1 summarises the differences between digital and physical products by their characteristics.

Table 1. Difference of Physical and Digital Products

	Physical	Digital
Composition	Atoms	Bits
Environment	Physical Resources	Computer Architectures
Availability	Limited by physical supply	Limited by bandwidth
Ease of manufacture	Difficult	Trivial
Speed to Market	Days, Weeks, Months	Seconds, Minutes
Energy Requirements in Manufacturing	Significant	Insignificant
Energy Requirements in Transportation	Significant	Insignificant
Human Involvement in Manufacture / Transportation	Significant	Insignificant
Costs	Significant % of product value	Almost zero

DIGITAL PRODUCT SUPPLY CHAIN FRAMEWORK

Managing the supply chain in traditional markets mainly provides challenges related to cost-effectiveness and physical barriers. By contrast, in a digital marketplace supply chain management encounters barriers that are not physical but rather strategic in nature (Nath, Saha & Salehi-Sangari, 2007). Chou and Ruchika (2006) argue that a digital software-focused supply chain is a supply chain where the digital product (software) constitutes a significant part of the total value of the product and where goods are not physically flowing through the supply chain. In a software supply chain, for example, a significant proportion of the supply chain is comprised by only one entity, which is usually one single company handling the design, development and production of a digital product. It is usually at the distribution stage that such a company might engage the services of another company to market its product. Moreover, channel coordination policies such as returns are not applicable in the digital world.

In order to outline the differences of a traditional and digital supply chain, and to describe a distinctive distribution framework, certain characteristics that determine a supply chain need to be contrasted. Hardaker and Graham (2001) have identified four interrelated dimensions that are involved in designing a supply chain, which are: (1) activities within a SC; (2) actors within a SC; (3) SC governance and (4) SC co-ordination. Thus, the authors held a number of in-depth, one-on-one interviews with practitioners in 7 companies with a significant software and electronic device business and that predominantly supply to enterprise customers on a global scale. The nature of our research was to look at some contemporary hybrid supply chains to understand key characteristics of the distribution approach. The companies examined are outlined in Table 2.

All companies are involved in the manufacture and distribution of electronic equipment that has both a software and a hardware component. Interviews were conducted with the above companies to understand how their supply chains were structured. The companies were a mix of consumer market suppliers and suppliers to the enterprise market, but it can be seen that the software distribution approach was quite

Table 2. Company Description

Company	Description
A	Supplier of network computing solutions, including servers, storage, software, and services
B	Organisation engaged in the development and manufacture of advanced information technologies, including computer systems, software, storage systems, and microelectronics.
C	Organisation engaged in the development, manufacture, licensing, and support of software products for various computing devices.
D	Organisation engaged in the development, delivery, and support of information infrastructure technologies and solutions.
E	Supplier of information, and care management products and services for the healthcare industry.
F	Organisation engaged in the design, manufacture, and marketing of computing systems and devices and related software, services, peripherals, and networking solutions.
G	Supplier of high-quality tool, diagnostic and equipment solutions for professional tool and equipment users.

similar irrespective of the nature of the customer base. Almost all used a mixed distribution approach. Companies bundled software with the original hardware sale, but when new software or software updates are being distributed, there is a tendency to distribute over the Internet. Table 3 summarises our findings.

These four dimensions identified by Hardaker and Graham (2001) will be used in the following to establish idiosyncratic features of combined hardware and software supply chains and describe unique digital distribution patterns apparent in the software industry based on the interviews.

Dimension 1: Supply Chain Activities and Software Distribution

The structure of activities is related to the natural sequence of processes inherent in a supply chain. In many industries, activities can be serially interdependent (Hardaker & Graham, 2001). The traditional software product supplied by physical means, such as a CD represents such a structure. Chou and Ruchika (2006) have identified several stages of a software planning and production process. A cycle time up to three years could be required. The first three stages of the process involve design and development and testing of the software. The next three stages involve preparing the support infrastructure that accompanies the main software product. This might include a user-guide or documentation, media production and other add-on features, as well as the final product rollout. Finally, marketing and distribution, as well as retailing efforts to sell the software ensue (see figure 2).

In the increasingly digitised software industry supply chain, the flow of material is declining in importance, due to the availability of the Internet as a free distribution medium. Our research has elicited that a digital product can be characterised as a movement of value instead. Thus, key flows represent information, ideas and innovation, from requirements into development, from release through to distribution and finally to the end-consumer. As such, information is shared mainly through electronic means within and outside of an organisation. Moreover, network technologies have the potential to transform the supply chains of many industries because of the abolition of the trade-off between richness and reach of

Table 3. Summary of Company Characteristics

	A	B	C	D	E	F	G
Hardware	Y	Y	Y	Y	Y	Y	Y
Software	Y	Y	Y	Y	Y	Y	Y
Customer	Consumer & Enterprise	Enterprise	Consumer & Enterprise	Enterprise	Enterprise	Consumer	Enterprise
Separate software distribution from hardware	Y	Y	Y	Y	Y	Y	N
Internet Distribution	Y	Y	Y	Y	Y	Y	N
Initial distribution method	Physical with H/W	Physical with H/W	Boxed S/W and Internet	Physical with H/W	Physical with H/W	Physical with H/W	Physical with H/W
Internet Distribution of New SW to Install Base	Y	Y	N	N	Y	Y	N
Distribution of Updates over Internet	Y	Y	Y	Y	Y	Y	N
Distribution of Updates by physical means	Y	N	Y	Y	Y	N	Y
Nature of Internet distribution	DL	DL	DL / Streaming	DL	DL	DL/Streaming	CD-ROM
Multiple platforms	Yes – Proprietary & nonprop	No - Proprietary	Yes – Proprietary & nonprop	Yes – Proprietary & nonprop	Yes – All non-propietary	Yes – proprietary & non-proprietary	No – Proprietary
Software Revenue Process	Subscription based	Subscription based	Pay per new s/w purchase	Pay per new s/w purchase	Subscription based	Free	Pay per new s/w purchase
Secure Distribution	Temporary weblinks	Temporary weblinks	Licence Key	License Key	None	None	CD with License Key

Figure 2. Software Supply Chain Activities and Processes

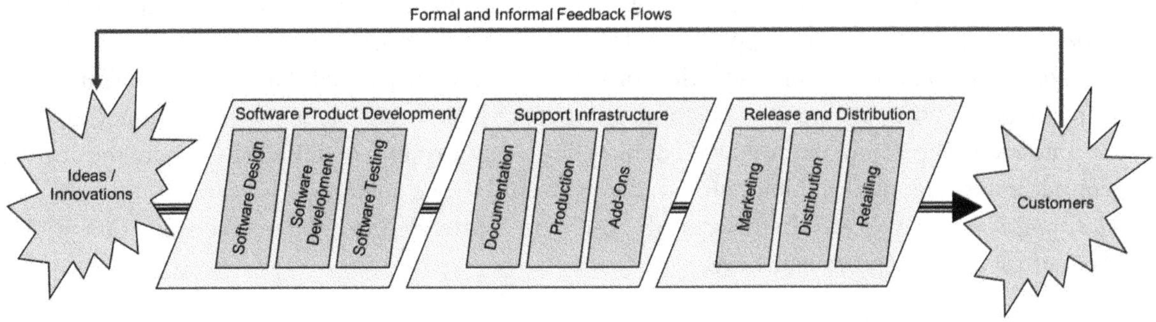

information, which means that everybody can communicate with everybody at almost zero cost, without constraints on the richness of information. (Graham et al., 2004 cite Evans & Wurster, 1997). Richness of information includes characteristics such as bandwidth, customisation, and interactivity. Reach is defined as the connectivity, and is the number of agents involved in exchanging information. Before the Internet, to reach large numbers of people with rich information was a costly and time-consuming process. Traditionally, information could only be transferred in the form of a physical carrier such as a document, book, CD, etc.; it could not travel by itself. Therefore, costs and physical constraints limited the size of the audience to which the information could be sent and the nature of the information itself (Graham et al., 2004).

The Internet has an important impact on how software is distributed. Through the interviews three major distribution characteristics could be identified: (1) Physical and pseudo-physical distribution; and (2) free distribution.

1. Pseudo-Physical Distribution

3 out of the 7 interviewed companies have chosen to pursue the first approach: born out of supply chains, legacy infrastructures and a competitive landscape that was designed to deal with physical product manufacture. Companies pursuing this strategy act to limit supply through technological copy protection measures backed up by legal enforcement. The customer's use of the software is impaired from the outset by highly restrictive licenses that need to be agreed to before the product can be used at all. Then, copy protection measures such as activation keys are included in the software to prevent duplicate copies being made. Complex supply chains need to be developed to support many of these initiatives. For instance, an infrastructure may be required to distribute activation keys in a secure fashion to the correct customers. These processes are complicated where third parties (distributors or resellers) need to be involved. Billing mechanisms may need to be integrated with the distribution mechanisms. Such processes require significant customer support overhead.

2. Free Distribution

In the second approach, few or no controls are imposed on the distribution of software. Some or all versions of the software are distributed to users with few control mechanisms in place. Although 3 of respondents offered free distribution of some sort, a number of different approaches were used:

1. *The total-service based model.* In this model, the software is free to distribute and install. However, associated services such as consultancy, bug fixing, hardware, training, technical support and customer support (all limited in supply) are available to the customer for a fee.
2. *The customised model.* In this case, the software is freely distributed. However, the nature of the software is that without extensive customisation, it has little value except to the customer for which it is intended. The companies involved deliver large-scale system installations, where the software requires extensive modification before it can be useful to the customer.
3. *The complementary model.* In this case the software is provided to enhance the value of a physical product such as computer hardware.

Dimension 2: Supply Chain Actors

This dimension means the level of dynamism in the choice of actors in a supply chain (Hardaker and Graham, 2001). The chain is seen as very static when actors are relatively established. On the other hand it is very dynamic, when actors vary from one market opportunity to another. There are various levels of flexibility in the choice of partners in-between. (Graham, Burnes, Lewis, & Langer, 2004). Traditionally, the software industry has been dominated by a few well-established major players and thus has been very static in character. Chou and Ruchika (2006) argue that most stages of the traditional software supply chain framework is taking place in-house and decisions to outsource are only made when production and distribution is infeasible in-house due to investment constraints. Thus, a small supply chain is less complex than having a variety of different actors. Therefore, it can be seen that little effort is being made in liaising with a wide array of partners with regards to inventory management throughout the supply chain. This also ensures that profits stay in-house rather than being shared with additional actors within the chain.

As outlined above, the Internet is eliminating the need for physical distribution and the retail of software products and has let to the invention of new forms of distribution for software products. All interviewees highlighted that it facilitates communications between the different actors. Moreover, new online portals have entered the online market providing download capabilities for a wide range of software products. As such, the traditional software supply chain is developing into a more dynamic marketplace, allowing new start-ups to market and distribute their products at a low cost point. However, our interviews revealed that poor management of internal and external relationships within a software network could have a detrimental effect on supply chain performance. Relationship management becomes ever important, especially in the software distribution process. Many software companies collaborate together to produce customer deliverables and a much stronger customer interaction is advocated, than it has been with physical supply chains.

Dimension 3: Supply Chain Governance

This relates to the ownership and the control of various actors in the supply chain. Dyer (2000) argues that in line with transaction cost theory (Williamson, 1979), three alternatives for governing a transaction or supply chain relationship exist: (1) vertical integration; (2) arm's length relationship; and (3) partnerships. In a vertically-integrated supply chain the organisation that commissions other organisations to help create the final software product is the main force in the supply chain and controls as such key channels and value steps; at the extreme, a company would produce all inputs itself. This is opposed to independent parties that operate at arm's length relationships (Graham, Burnes, Lewis, & Langer, 2004), which mean that a company buys inputs from outside through short-term suppliers for the best conditions each time an input is required. The concept of partnership, by contrast, means that a company sources from a small number of suppliers by building co-operative and long-term relationships (Dyer, 2000). If transaction costs are high, it is more favourable to own suppliers and distributors and as such produce in-house. On the other hand, if transaction costs are low, it is more favourable to outsource and procure materials and services.

Traditionally, the software supply chain has been characterised by a small number of very powerful companies that control all aspects of the software supply chain, from creation to test to distribution and installation. Companies such as Microsoft and SAP have made use of supply side economies of scale (availability of trained developers and IP), plus demand side economies of scale (customer lock-in and network effects) to achieve disproportional market shares, effectively pushing out the competition. To a large extent, such effects are still in play, although our interviews elicited that incumbents have to increasingly deal with new development models and revenue models that are resulting in a shift in the power dynamics of the industry. The rise of Open Source development (Hars & Ou, 2002) has enabled communities of programmers, motivated by non-financial incentives, to challenge commercial incumbents (e.g. Mozilla Firefox, Linux, Apache Tomcat). The movement of revenue from pay-per-unit business models to advertising business models has seen companies such as Google begin to challenge powerful companies such as Microsoft, particularly on the consumer side of the business.

Dimension 4: Supply Chain Coordination

Sweeney (2007) depicts that managing information flows in the supply chain is one of the most crucial activities in supply chain management as the flow of materials and money is usually initiated by information movements. Hardaker and Graham (2001) reinforce this by outlining that co-ordination in a supply chain occurs through the communication of orders, stock level and demand feedback. Poor information flows essentially leads to the so-called bullwhip-effect that requires holding excessive levels of inventory. As such, high demand visibility plays a strategic role in reducing inventory levels (Sweeney, 2007). Thus, network-based communication structures in a supply chain might offset these effects.

The new co-ordination scenario surrounding Internet distribution is much different. Our interviews highlighted that in some respects co-ordination is less of an issue as the latest product can always be picked up by just clicking on a web-link, and because there is an abundance of copies to pick up, there are no critical inventory or purchasing issues to contend with. Companies that wish to retain control over their software through restrictions to copying have to contend with more complex co-ordination challenges, e.g. the supply of activation keys and the monitoring of software assets remotely to ensure that copyright is not being breached. Many companies we interviewed are not interested in the extra overhead associated with such oversight.

To summarise the above, the method of distributing new software to the existing hardware base tended to correlate with the software revenue process. If new software was free of charge or if a support contract was in place via subscription, there was a tendency to use Internet update mechanisms as a primary means of distribution. If there was an up-front license charge for new software, physical distribution mechanisms were more commonplace. For software updates, almost all respondents tended to use an Internet distribution framework. Only in one case, Company G, distribution of updates was purely physical. This is because an Internet distribution infrastructure had not yet been built for the company at this time. Updates and bug-fixes were difficult to distribute and therefore every possible effort was made to ensure the products were tested thoroughly before release. What was interesting was that for many companies, physical distribution mechanisms were in place for software, even for updates. Many companies cited customer preference and concerns over Internet security. However, many of the companies were moving customers over to Internet distribution for updates because of the cost-efficiencies involved. Security around software distribution tended to be strongest where there was a pay-per-item software contract in place. In these cases, customers needed to activate a license key, which required a completely separate supply chain to distribute these keys to customers. In the case of subscription customers and free distribution, no such supply chain was required. Customers were given access to the new software via temporary web pages or password protected customer web-accounts. It is clear from our interviews that there was a clear advantage to companies in moving away from pay-per-item digital purchase mechanism and instead using subscription based mechanisms. Pay-per-item required complex licence management infrastructures, high security, Digital Rights Management (DRM) and payment mechanisms that we tied to the software sale. There was a greater tendency to rely on physical distribution compared to subscription. For subscription based or free distribution, the management systems were less complex and updates tended to be more regular. Subscription based companies also tended to be more comfortable with centralising software distribution - creating a specialised skill-base within the company for software distribution. This lead to standardisation and the implementation of strong processes with distributed hardware engineering teams. Pay-per-item companies tended to be less mature in terms of experience with software distribution.

For pure physical goods manufacturers or purely digital product producers, totally different strategies can be adopted to deliver products to end-customers. Traditional supply chains will continue to work for physical goods, while Internet based strategies can be exploited for digital products. There is a growing class of products that will require both physical and digital distribution strategies. These product contain hardware and software elements and they currently include computers, smart phones, smart TVs and various industrial equipment. As the "Internet of Things" continues to emerge we are seeing watches, lightbulbs, cars, glasses and cameras acquire internet capability. More and more products will feature these dual modes in the future. The demand for new electronic devices has not abated. The demand for devices has been spurred by new websites such as IndieGogo, that link the customer directly to new and innovative electronic product designs. This is a kind of ETO approach where informal investors take a financial stake in the product design.

To conclude, the differences between physical and digital supply chains outlined above are very large. This affects how we need to approach managing the supply chains for these types of products. Table 4 summarises the differences between digital and physical products by their supply chains.

The role of engineering is important for both product types. Even though the nature of the engineering may be somewhat different, good product design is a paramount consideration in all cases. Beyond that – in terms of distribution operations – there is a strong divergence. Sourcing of digital products is primarily a mental activity, driven by ideas. While ideas are important for physical products, raw materials must be purchased or extracted in order to commence the process. Beyond this, the logistics for digital products is Internet and server driven. The next section outlines the role of innovation in electronic device supply chains to illustrate the linkage of our findings with the theory of innovation.

Table 4. SC Properties of digital and physical products

	Physical	Digital
Design	Materials Engineering, Electronic Engineering, Mechanical Engineering, Chemical Engineering, Food Engineering	Software Engineering, Authoring, Development, Game Development, Composing
Sourcing	Mining, Farming, Forestry, Ideas	Ideas, Requirements
Inbound Logistics	Trucking, Shipping, Air Freight	N/A
Manufacturing	Assembly, Process, Incubation, Germination, Extraction	Duplication
Outbound Logistics	Trucking, Shipping, Air Freight	Digital Networks
Warehousing	Distribution Centres, Silos, Wholesalers, 3PLs.	Distributed Servers
Distribution	Retail shops, Malls, Auctions, Online	Online, Streaming, Retail shops, Movie Theatres
Reverse Logistics	Returns, Repair, Refurbishment, Re-purposing,	Version Management

INNOVATIVE INTEGRATION STRATEGIES IN ELECTRONIC DEVICE SUPPLY CHAINS

Innovation is generally considered to be one of the key drivers of corporate success (Cardozo et al., 1993). Modern definitions of supply chain management expand beyond the activities that are related to working capital as outlined in the traditional supply chain management definitions (Lummus et al., 2001). Innovative creation and promotion of value and the maximisation of this value over the longer term is becoming more and more critical elements in supply chain management (Wagner and Ryan, 2009). Supplier involvement and alliances are routes to innovation generation. The generation of an innovation typically occurs in the development phase of a new product. This includes the generation of initial ideas, their critical evaluation, the testing of derived concepts, the product development, and final launch (Crawford & Di Benedetto, 2000). Roy, Sivakumar and Wilkinson (2004) argue that nevertheless, fundamental changes in the configuration of an existing product may as well be considered as radical for the affected supply chain members. If the innovations involving less radical or incremental changes entail for example quicker delivery periods, or reduction of material thickness and cost. Innovation generation in a supply chain context involves changes in product, process, or service that either reduce cost or improve efficiency.

In Roger`s (1995) seminal research the process is detailed how innovations move from one population to another and the role of key factors in the individual decision to adopt is discussed. In this context, Rogers (1995) defines adoption as *"(...) the decision to make full use of an innovation as the best course of action available"*, while diffusion refers to the spread over time of an innovation within the unit or to other units. The explicit motivators for the organisational adoption and innovation diffusion are described by the following set of variables:

- **Relative Advantage:** The extent to which the innovation is perceived to have significant advantages over current alternatives.

- **Compatibility:** The degree to which the innovation is seen as being consistent with past practices, current values and existing needs.
- **Complexity:** The extent to which the innovation can readily be understood and easily implemented.
- **Trialability:** New ideas that can be tried out at low costs before wholesale adoption are more likely to be taken up.
- **Observability:** The degree to which the use and benefits of the innovation are visible to others, and therefore act as a further stimulus to uptake by others.

In this context it is worthwhile to mention the theory of S-curve in innovation adoption (Rogers, 1995). It suggests that technologies are a conceptualisation of S-curve sequences that drive various new product introductions. The S-curve emerges because a new technology or product is rather limited to a small group of early adopters and small niche markets that derive benefits from this early introduction (Chandy & Tellis, 2000). With its diffusion to a wider user group, the product or technology gains ascendancy and new capabilities are introduced and refined. During this middle phase a dominant allegiance of the product or technology in the market place emerges, which is effecting standardization from design to manufacturing. This in turn allows heightened competition as new entrants realize opportunities for further innovation based on cost and scale as well as product performance. This is the period of rapid and greatest growth. During the final phase the product reaches market saturation and hits a plateau. Successful companies begin to consolidate their market share, seek further economies of scale, but also shift focus by targeting other growth opportunities (Essa, 2005). Harrison and Waite (2005) therefore argue that the extent to which adoption of an innovation can yield a competitive advantage is relatively short-lived. As the innovation becomes more widespread, competitive advantage diminishes and innovation becomes a necessary competitive requirement. This means that new product development, its maintenance and updates need to be effectively and flexibly managed from the initial idea through to final delivery to the customer in a closed loop fashion. The management of closed innovation cycles is seen as a key component in sustaining competitive advantage within supply chains (Utterback, 1994; Tidd et al., 2005). Few companies have achieved mastery over their innovation cycles. Highly innovative companies are characterised by regular updates of new products and the ability to introduce new products that alter the basis of competition (Anthony et al., 2008). Fast evolving companies are for example electronic device manufacturers in the infotainment industry. The Internet in particular has accelerated this sectors' evolutionary speed and consumers of digital infotainment devices are experiencing a true revolution in the supply and distribution chains for many of such products. The reason for this rapid change can be attributed to groundbreaking developments in conductor technologies, such as increased transmission capacities in broadband, together with the evolution of so called boxes, including DVD players, digital recorders, digital TV decoders, music and video digital players, PDAs and digitally enabled mobile phones. This gave rise to develop hybrid devices that combine physical and digital product innovations to create a market-leading product idea, which is able to dictate the alignment of conductor and content providers to its own standards (Graeml, Engelbert & Weiler, 2008). Graeml, Engelbert and Weiler (2008) describe Apple's iPod (box) and the iTunes (conductor) store as one such example.

Wagner and Ryan (2009) point out that digital product manufacturers must work with customers and developers in a tightly managed cyclical fashion. While original ideas may be generated internally within the developer network, once those ideas are transformed into products the customer takes a central role in the future direction of the product and the further changes that ensue from there. Within this extended concept, the role of partnership remains paramount. Product developers operate within complex webs and networks of players – engineers, product managers, developers, industrial designers etc. – each of whom need to work together to turn around new products quickly and efficiently. Information management is required to connect producers to customers, thus ensuring that the products meet changing customer requirements. A kind of reverse logistics flow needs to be factored into digital supply chain – the movement of customer feedback into the hands of the manufacturers in order to enable products to be updated and improved. Customer responses will inevitably form part of the development of subsequent versions of the product, so mechanisms that gather feedback from the customer, be it formal (for example, surveys or automated bug-reporting) or informal (for example, phone calls, emails, web postings), need to be in place. In essence the key to competition in the digital realm is not really a single sale, but the long-term relationship with the customer base and the securing of a profitable income stream over time.

Innovation cycles need to be incorporated when considering supply chains for products that require physical and digital elements. Changes required by customers from a software perspective may force modifications to the hardware. More commonly, changes in the hardware may require further software development and careful consideration of software roadmaps in the light of rapid advances in physical technology. Software and hardware developments need to be closely aligned for the lifecycle of the product because the customer may not be as concerned by the distinction between the two as the developers might be.

INTEGRATION STRATEGIES FOR ETO COMPANIES

Digital product manufacturers must work with customers and developers in a tightly managed cyclical fashion (Wagner & Ryan, 2009). The ETO approach aligns with this perspective for many electronic products, if not at initial product design, then often during the development of new versions of the same product. While original ideas may be generated internally within the developer network, once those ideas are transformed into products the customer takes a central role in the future direction of the product and the further changes that ensue from there. A high degree of partnership is required, and processes that facilitate this partnership are essential. Information management is required to connect producers to customers, thus ensuring that the products meet changing customer requirements. A kind of reverse logistics flow needs to be factored into the digital supply chain – the movement of customer feedback into the hands of the manufacturers in order to enable products to be updated and improved.

New websites such as IndieGogo link the customer directly to new and innovative electronic product designs, which need to be transformed into usable and highly functional products in a short space of time. These products need to deliver not only high quality functionality, but also with well thought out software distribution and update mechanisms. ETO companies using bundled electronics will require software and hardware distribution mechanisms. Our analysis suggests that two separate strategies are important – one for physical distribution, the other for software distribution. While there needs to be a touchpoint at initial product sale, i.e. with the hardware getting the latest software prior to product shipment, from then the distribution strategies can diverge.

Relevant companies in the ETO space must consider the longer term aspects of supporting their install base, and in such cases they should be able to regularly provide new software and software updates to their customers. In circumstances where a software distribution strategy is not included, even if only for firmware bug-fixes, the risks to the company could be sizable. Our analysis sees clear benefits to companies in using Internet based distribution strategies for software. Our analysis also suggests that there are advantages in utilising a subscription mechanism for bundled software or distributing the software to customers for free. In either case the software distribution infrastructure is simpler and the customer experience is better. Where a pay-per-item licensing mechanism is required, costs will be significant and copyright protection / security will reduce customer satisfaction.

Companies using ETO are encouraged to use password-based web accounts with customers. This will assist with communications and sharing of information and specifications, etc. Through such accounts, new software and new software updates can be provided securely and safely to the customer install base and for a minimum of hassle. Within ETO companies, a centre of expertise for software distribution is recommended. Such a centre will concentrate skills and create standards around which hardware engineers can work effectively. It will facilitate a common approach to software distribution and help with Internet strategy development for the company as it expands into new products.

FUTURE RESEARCH DIRECTION

While this analysis mainly concentrated on software distribution within the context of electronics, future areas of research could include an examination of other types of digital product, e.g. movies, music and podcasts etc., to see if the recommendations still hold for such goods. A number of in-depth case-studies of mature ETO suppliers of electronic goods and electronic devices would provide additional insights into particular issues faced by these companies. The interfaces between hardware engineering and software engineering for mature ETO suppliers, particularly around the management of the launch and update process would be an interesting focus of further research.

CONCLUSION

The chapter has discussed the role of the Internet and how it has facilitated new modes of distribution and support for electronic products. Supply chains with a digital content have distinctive ways of being managed. Integration and innovation are important concepts in digital and hybrid supply chains. Digital product manufacturers must work with customers and developers in a tightly managed cyclical fashion. The Internet and the virtualisation of functionality means that electronically based ETO products need to accommodate a much greater degree of complexity that was traditionally the case. Software development and distribution, while separate from physical supply chains, need to integrate seamlessly at all stages of the process up to the eventual replacement of the product.

REFERENCES

Anthony, S. D., Johnson, M. W., Sinfield, J. V., & Altman, E. J. (2008). *The Innovator's Guide to Growth: Putting Disruptive Innovation to Work*. Boston, MA: Harvard Business Press.

Baltacioglu, E. A., Kaplan, M. D., Yurt And, O., & Cem Kaplan, Y. (2007). A new framework for service supply chains. *Service Industries Journal*, *27*(2), 105–124. doi:10.1080/02642060601122629

Cardozo, R., McLaughlin, K., Harmon, B., Reynolds, P., & Miller, B. (1993). Product- market choices and growth of new businesses. *Journal of Product Innovation Management*, *10*(4), 331–340. doi:10.1016/0737-6782(93)90075-2

Chandy, R. K., & Tellis, G. J. (2000). The incumbent's curse? Incumbency, size, and radical product innovation. *Journal of Marketing*, *6*(3), 1–17. doi:10.1509/jmkg.64.3.1.18033

Chellappa, R. K. (2000). Digital Products and E-Business. Marshall Magazine, Spring, 48-49.

Chellappa, R. K., & Shivendu, S. (2003). Pay Now or Pay Later? *Managing Digital Product Supply Chains. ACM International Conference Proceeding Series* (Vol. 50, pp. 230-234).

Chopra, S., & Meindl, P. (2001). *Supply Chain Management: Strategy, planning and operation*. Prentice Hall.

Chou, M., & Ruchika, A. (2006). An in-depth study of the software supply chains. *Proceedings of the IEEE International Conference on Industrial Informatics*. doi:10.1109/INDIN.2006.275656

Christopher, M., & Towill, D. R. (2002). Developing market specific supply chain strategies. *International Journal of Logistics Management*, *13*(1), 1–4. doi:10.1108/09574090210806324

Crawford, C. M., & Di Benedetto, C. A. (2000). *New Product Management*. Boston: McGraw-Hill.

Da Silveira, G., Borenstein, D., & Fogliatto, F. S. (2001). Mass customisation: Literature review and research directions. *International Journal of Production Economics*, *72*(1), 1–13. doi:10.1016/S0925-5273(00)00079-7

Dyer, J. H. (2000). *Collaborative Advantage: Winning Through Extended Enterprise Supplier Networks*. Oxford: Oxford University Press.

Eldering, C. A., Sylla, M. L., & Eisenach, J. A. (1999). Is there a Moore's law for bandwidth? *IEEE Communications Magazine*, *37*(10), 117–121. doi:10.1109/35.795601

Elfving, J. (2003). Exploration of opportunities to reduce lead times for engineering-to-order products [Ph.D. dissertation]. University of California, Berkley.

Essa, A. (2005). *Innovation Life Cycle and the S-Curve*. Retrieved from http://tatler.typepad.com/nose/2005/11/innovation_life.html

Evans, P. B., & Wurster, T. S. (1997). Strategy and the new economics of information. *Harvard Business Review*, 5, 70–82. PMID:10170332

Gattorna, J. L., & Walters, D. W. (1996). *Managing the Supply Chain: A Strategic Perspective.* Macmillan Business.

Gosling, J., & Naim, M. M. (2009). Engineer-to-order supply chain management: A literature review and research agenda. *International Journal of Production Economics*, *122*(2), 741–754. doi:10.1016/j.ijpe.2009.07.002

Graeml, A. R., Ebgelbert, R., & Weiler, A. L. G. (2008, June 15-18). Understanding value chain strategies for the iPod through the perspective of Fine's evolutionary model. *Proceedings of the 15th Annual International Conference of the European Operations Management Association (EurOMA)*, Gronigen, Netherlands.

Graham, G., Burnes, B., Lewis, G., & Langer, J. (2004). The transformation of the music industry supply chain: A major label perspective. *International Journal of Operations & Production Management*, *24*(11), 1087–1103. doi:10.1108/01443570410563241

Grochowski, W., & Halem, R. D. (2003). Technological Impact of magnetic hard disk drives on storage systems. *IBM Systems Journal*, *42*(2), 338–346. doi:10.1147/sj.422.0338

Hardaker, G., & Graham, G. (2001). *Wired marketing: energizing business for e-Commerce. Chicester*: John Wiley & Sons.

Harrison, T., & Waite, K. (2005). Critical factors affecting intermediary web site adoption: Understanding how to extend e-participation. *Journal of Business and Industrial Marketing*, *20*(4/5), 187–199. doi:10.1108/08858620510603864

Hars, A., & Ou, S. (2002). Working for Free? - Motivations for Participating in Open Source Projects. *Proceedings International Journal of Electronic Commerce*, *6*(3), 25–39.

Haug, A., Ladeby, K., & Edwards, K. (2009). From engineer-to-order to mass customisation. *Management Research News*, *32*(7), 633–644. doi:10.1108/01409170910965233

Hicks, C., McGovern, T., & Earl, C. F. (2000). Supply chain management: A strategic issue in engineer to order manufacturing. *International Journal of Production Economics*, *65*(2), 179–190. doi:10.1016/S0925-5273(99)00026-2

Hui, K. L., & Chau, P. Y. K. (2002). Classifying Digital Products. *Communications of the ACM*, *45*(6), 73–79. doi:10.1145/508448.508451

Lummus, R. R., Krumwiede, D. W., & Vokurka, R. J. (2001). The relationship of logistics to supply chain management: Developing a common industry definition. *Industrial Management & Data Systems*, *101*(8), 426–432. doi:10.1108/02635570110406730

McCullen, P., & Towill, D. R. (2000). Practical ways of reducing bullwhip: The case of the Glosuch global supply chain. *International Journal of Operations Management*, *26*(10), 24–30.

Merrifield, D. B. (2000). Changing nature of competitive advantage. *Research Technology Management*, *41*(1), 41–45.

Nath, A. K., Saha, P., & Salehi-Sangari, E. (2007). Transforming Supply Chains in Digital Content Delivery: A Case Study in Apple. In L. Xu, A. Tjoa, S. Chaudhry (Eds.), Research and Practical Issues of Enterprise Information Systems II (Vol. 2, pp. 1079-1089). Boston: Springer.

Padmanabhan, V., & Png, I. P. L. (1997). Manufacturer's Returns Policies and Retail Competition. *Marketing Science*, *16*(1), 81–94. doi:10.1287/mksc.16.1.81

Pandit, A., & Zhu, Y. (2007). An ontology-based approach to support decision-making for the design of ETO (engineer-to-order) products. *Automation in Construction*, *16*(6), 759–770. doi:10.1016/j.autcon.2007.02.003

Peitz, M., & Waelbroeck, P. (2003). Piracy of Digital Products: A critical Review of the Economics Literature. *CESifo Working Paper, No. 1071.*

Piller, F. (2004). Mass customisation: Reflections on the state of the concept. *International Journal of Flexible Manufacturing Systems*, *16*(4), 313–334. doi:10.1007/s10696-005-5170-x

Porter, M. E. (1985). *Competitive Advantage*. New York: The Free Press.

Porter, M. E., & Stern, S. (2001). Innovation: Location matters. *MIT Sloan Management Review*, *42*(4), 28–36.

Ramsay, J. (2005). The real meaning of value in trading relationships. *International Journal of Operations & Production Management*, *25*(6), 549–565. doi:10.1108/01443570510599719

Rogers, E. M. (1995). Diffusion of innovations (4 ed.). New York: The Free Press.

Roy, S., Sivakumar, K., & Wilkinson, I.F. (2004). Innovation generation in supply chain relationships: a conceptual model and research propositions. *Journal of the Academy of Marketing Supply chainience*, *32*(1), 61-79.

Shapiro, C., & Varian, H. (1999). *Information Rules*. Boston: Harvard Business School Press.

Shewchuck, J. P. (1998). Agile manufacturing: one size does not fit all. *Proc. Int. Conference Manufacturing Value Chains*, Troon (pp. 143-150). doi:10.1007/978-0-387-35321-0_16

Simchi-Levi, D., Kaminsky, P., & Simchi-Levi, E. (2001). *Designing and managing the supply chain: Concepts, strategies, and cases*. New York: McGraw-Hill.

Sweeney, E. (2007). Introduction. In Perspectives on Supply Chain Management and Logistics – Creating Competitive Organisations in The 21st Century. Dublin: Blackhall Publishers.

Tidd, J., Bessant, J., & Pavitt, K. (2005). *Managing innovation: integrating technological, market and organizational change*. Chichester: Wiley.

Utterback, J. M. (1994). *Mastering the dynamics of innovation: how companies can seize opportunities in the face of technological change*. Boston, MA: Harvard Business Press.

Wagner, C., & Ryan, C. (2009). Innovation Cycle Management of Digital Supply Chains. In E. Sweeney (Ed.), *Supply Chain Management and Logistics in a Volatile Global Environment* (pp. 169–182). Dublin: Blackhall Publishers.

Whinston, A. O. D., & Choi, S. (1997). *The Economics of Electronic Commerce*. New York: Macmillan Publishing Company.

Wikner, J., & Rudberg, M. (2001). On the customer order decoupling point (Working Paper No. WP-284). *Linkoeping Institute of Technology*, Sweden.

Wikner, J., & Rudberg, M. (2005). Integrating production and engineering perspectives on the customer order decoupling point. *International Journal of Operations & Production Management*, *25*(7), 623–641. doi:10.1108/01443570510605072

Williamson, O. E. (1979). Transaction cost economics: The governance of contractual relations. *The Journal of Law & Economics*, *22*(2), 233–261. doi:10.1086/466942

Compilation of References

About the PPP Unit. (n. d.). *Public Private Partnership*. Retrieved from http://www.ppp.gov.za/Pages/About.aspx

Addo-Tenkorang, R. (2014). *Conceptual Framework for Large-Scale Complex Engineering-Design & Delivery Processes: A Case of Enterprise SCM Network Activities and Analysis* [Doctoral Dissertation].

Addo-Tenkorang, R., & Eyob, E. (2012). Engineer-to-Order: A Maturity Con-current Engineering Best Practice in Improving Supply Chains. In E. Eyob, & E.G. Tetteh (Eds.), Customer-Oriented Global Supply Chains: Concepts for Effective Management (Ch 7, pp. 112-122). doi:10.4018/978-1-4666-0246-5.ch007

Addo-Tenkorang, R. (2011). Concurrent Engineering (CE): A Review Literature Report.*Proceedings of the World Congress on Engineering and Computer Science Vol IIWCECS 2011*, San Francisco, USA.

Agarwal, A., Shankar, R., & Tiwari, M. K. (2006). Modeling the metrics of lean, agile and leagile supply chain: An ANP-based approach. *European Journal of Operational Research*, *173*(1), 211–225. doi:10.1016/j.ejor.2004.12.005

Agarwal, M. (2010). *Business Communication*. India.

Agouridas, V., Allen, M., McKay, A., Pennington, A., & Holland, S. (2001). Using Information Structures to Gain Competitive Advantage.*Proceedings of the 2001 PICMET - Portland International Conference on Management of Engineering and Technology, Portland, OR* (Vol. 2, pp. 681–692). doi:10.1109/PICMET.2001.952416

Ahuja, G. (2000). The duality of collaboration: Inducements and opportunities in the formation of interfirm linkages. *Strategic Management Journal*, *21*(3), 317–343. doi:10.1002/(SICI)1097-0266(200003)21:3<317::AID-SMJ90>3.0.CO;2-B

Aitken, J., Childerhouse, P., & Towill, D. (2003). The impact of product life cycle on supply chain strategy. *International Journal of Production Economics*, *85*(2), 127–140. doi:10.1016/S0925-5273(03)00105-1

Akturk, M. S. (2011). Joint cell loading and scheduling approach to cellular manufacturing systems. *International Journal of Production Research*, *49*(21), 6321–6341. doi:10.1080/00207543.2010.532165

Akturk, M. S., & Turkcan, A. (2000). Cellular manufacturing system design using a holonistic approach. *International Journal of Production Research*, *38*(10), 21. doi:10.1080/00207540050028124

Albadawi, Z., Bashir, H. A., & Chen, M. (2005). A mathematical approach for the formation of manufacturing cells. *Computers & Industrial Engineering*, *48*(1), 3–21. doi:10.1016/j.cie.2004.06.008

Alkemade, F., Kleinschmidt, C., & Hekkert, M. P. (2007). Analysing emerging innovation systems: A functions approach to foresight. *International Journal of Foresight and Innovation Policy*, *3*(2), 139–168. doi:10.1504/IJFIP.2007.011622

Altiok, T. (1996). *Performance Analysis of Manufacturing Systems*. Springer.

Amiri, A. (2006). Designing a distribution network in a supply chain system: Formulation and efficient solution procedure. *European Journal of Operational Research*, *171*(2), 567–576. doi:10.1016/j.ejor.2004.09.018

Ancona, D., & McVeigh, J. (2001). *Wind turbine—Materials and manufacturing*. Princeton Energy Resources International LLC.

Anthony, S. D., Johnson, M. W., Sinfield, J. V., & Altman, E. J. (2008). *The Innovator's Guide to Growth: Putting Disruptive Innovation to Work*. Boston, MA: Harvard Business Press.

Anvuur, A. M., & Kumaraswamy, M. M. (2007). Conceptual Model of Partnering and Alliancing. *Journal of Construction Engineering and Management*, *133*(3), 225–234. doi:10.1061/(ASCE)0733-9364(2007)133:3(225)

Armstrong, M. (2006). *A Handbook of Human Resources Management Practice*. USA: Kogan Page.

Arthus, M. B., & Rousseau, D. M. (1996). *The Boundaryless Career: A New Employment Principle for a New Organizational Era*. NY, USA: Oxford University Press.

Askin, R. G. (2013). Contributions to the design and analysis of cellular manufacturing systems. *International Journal of Production Research*, *51*(23-24), 6778–6787. doi:10.1080/00207543.2013.825745

Askin, R. G., & Estrada, S. (1999). An Investigation of Cellular Manufacturing Practices. In S. Irani (Ed.), *Handbook of Cellular Manufacturing* (pp. 25–34). New York: Wiley. doi:10.1002/9780470172476.ch1

Askin, R. M., & Standridge, C. R. (1993). *Modeling and analysis of manufacturing systems* (1st ed.). New York: Wiley.

Ates, O. K. (2013). Global Supply Chain and Competitive Business Strategies: A Case Study of Blood Sugar Monitoring Industry [Doctoral dissertation]. Ohio University.

Atkinson, R. (1999). Project management: Cost, time and quality, two best guesses and a phenomenon, its time to accept other success criteria. *International Journal of Project Management*, *17*(6), 337–342. doi:10.1016/S0263-7863(98)00069-6

Aubrey, C. (2007). Supply Chain: The race to meet demand. *Wind Directions*, (February 2007), 27–34.

Bach, S. (2005). *Managing Human Resources: Personnel Management in Transition*. USA: Blackwell Pub.

Bakouros, Y., & Kelessidis, V. (2000) Project Management. Report produced for the EC funded project INNOREGIO. Retrieved from http://www.adi.pt/docs/innoregio_pmanagement.pdf

Bakshi, U. A., & Godse, A. P. (2009). *Communication Engineering*. India: Technical Pub. Pune.

Balsamo, S., Harrison, P. G., & Marin, A. (2012). Methodological construction of product-form stochastic Petri nets for performance evaluation. *Journal of Systems and Software*, *85*(7), 1520–1539. doi:10.1016/j.jss.2011.11.1042

Baltacioglu, E. A., Kaplan, M. D., Yurt And, O., & Cem Kaplan, Y. (2007). A new framework for service supply chains. *Service Industries Journal*, *27*(2), 105–124. doi:10.1080/02642060601122629

Bankvall, L., Bygballe, L. E., Dubois, A., & Jahre, M. (2010). Interdependence in supply chains and projects in construction. *Supply Chain Management: An International Journal*, *15*(5), 385–393. doi:10.1108/13598541011068314

Bayliss, R., Cheung, S., Suen, H. C. H., & Wong, S. (2004). Effective partnering tools in construction: A case study on MTRC TKE contract 604 in Hong Kong. *International Journal of Project Management*, *22*(3), 253–263. doi:10.1016/S0263-7863(03)00069-3

Bayraktar, E., Jothishankar, M. C., Tatoglu, E., & Wu, T. (2007). Evolution of operations management: Past, present and future. *Management Research News*, *30*(11), 843–871. doi:10.1108/01409170710832278

Becattini, G. (1979). Dal Settore Industriale al Distretto Industriale. Alcune considerazioni sull'unitá di indagine in economia industrial. *Revista di Economia e Politica Industriale*, *1*, 7–14.

Becattini, G. (1990). The Marshallian industrial district as a socio-economic notion. In F. Pyke, G. Becattini, & W. Sengenberger (Eds.), *Industrial districts & inter-firm co-operation in Italy* (pp. 37–51). Genève: International Institute for Labour Studies.

Becattini, G. (2002). Industrial sectors & industrial districts: Tools for industrial analysis. *European Planning Studies*, *10*(4), 483–493. doi:10.1080/09654310220130194

Beck, C. H., & Friedman, T. L. (2005). *The World Is Flat. A Brief History of the 21st Century*. NY: Farra, Straus and Giroux.

Bellandi, M. (1996). Innovation & change in the Marshallian Industrial District. *European Planning Studies*, *4*(3), 357–369. doi:10.1080/09654319608720351

Bellandi, M. (2002). Italian industrial districts: An industrial economics interpretation'. *European Planning Studies*, *10*(4), 425–437. doi:10.1080/09654310220130158

Bellandi, M. (2006). A perspective on clusters, localities, & specific public goods. In C. Pitelis, R. Sugden, & J. R. Wilson (Eds.), *Clusters & Globalisation: the Development of Urban & Regional Economies* (pp. 96–113). Cheltenham: Edward Elgar. doi:10.4337/9781847200136.00012

Benchemam, F., & Galindo, G. (2009). *Gestion des resources humaines*. Paris: Gualino.

Bergek, A., Jacobsson, S., Carlsson, B., Lindmark, S., & Rickne, A. (2008). Analyzing the functional dynamics of technological innovation systems: A scheme of analysis. *Research Policy*, *37*(3), 407–429. doi:10.1016/j.respol.2007.12.003

Bertrand, J. W. M., & Munslag, D. R. (1993). Production Control in Engineer-to-Order firms. *International Journal of Production Economics*, *30/31*, 3–22. doi:10.1016/0925-5273(93)90077-X

Bhattacharya, C. (2009). *Services marketing*. New Delhi: Excel Book.

Bidgoli, H. (2004). *The Internet Encyclopedia* (Vol. 1). USA: John Wiley&Sons. doi:10.1002/047148296X

Billington, P. J. (2004). A Classroom Exercise to Illustrate Lean Manufacturing Pull Concepts. *Decision Sciences Journal of Innovative Education*, *2*(1), 71–76. doi:10.1111/j.0011-7315.2004.00021.x

Bindzár, P., Kačmáry, P., & Lavrin, A. (2009). Projekt KASSETTS ako pomoc malým a stredným podnikom k riešeniu dopravného problému. *Transport & Logistics*, *6*, 28–31.

Body of Knowledge (BoK) Revised January 1995 (version 2). (1995). *APM - Association of Project Management*.

Bondarouk, T. (2009). *Handbook of Research on E-Transformation and Human Resources Management*. Netherlands: Idea Group Inc.

Borgatti, S. P., Everett, M. G., & Freeman, L. C. (2002). UCINET for Windows: Software for Social Network Analysis. Harvard: Analytic Technologies. Cambridge, MA.

Borgatti, S. P., & Halgin, D. S. (2011). *On Network Theory. OrganizationScience, INFORMS.*

Bournois, F., Point, S., Rojot, J., & Scaringella, J. L. (2007). *RH - les meilleures pratiques. Eyrolles.* Paris: Editions d'Organisation.

Bowersox, D. J., & Closs, D. J. (1996). *Logistical Management: The Integrated Supply Chain Process.* New York, NY: McGraw-Hill.

Bradley, P. (1996). *A Performance Measurement Approach to the Re-engineering of Manufacturing Enterprises* [PhD Thesis]. CIMRU, NUI Galway, Ireland.

Brass, D. J. (2002). *Social networks in organizations: Antecedents and consequences* (Unpublished). University of Kentucky, Lexington.

Brentani, U., Kleinschmidt, E. J., & Salomo, S. (2010). Success in Global-New Product Development: Impact of Strategy and the Behavioral Environment of the Firm. *Journal of Product Innovation Management, 27*(2), 143–160. doi:10.1111/j.1540-5885.2010.00707.x

Breschi, S., & Malerba, F. (1997). Sectoral systems of innovation: Technological regimes, Schumpeterian dynamics & spatial boundaries. In C. Edquist (Ed.), *Systems of Innovation* (pp. 130–156). London: Frances Pinter.

Bresnen, M. (1996). An Organizational Perspective on Changing Buyer-Supplier Relation: A Critical Review of the evidence. *Organization, 3*(1), 121–146. doi:10.1177/135050849631007

Briscoe, G., & Dainty, A. (2005). Construction supply chain integration: An elusive goal? *Supply Chain Management: An International Journal, 10*(4), 319–326. doi:10.1108/13598540510612794

Brusco, S. (1990). The idea of the industrial district: its genesis. In *Pyke, F., Becattini, G. & Sengenberger, W. Industrial districts & inter-firm co-operation in Italy* (pp. 10–19). Genève: International Institute for Labour Studies.

Brusco, S., & Pezzini, M. (1990). *Small-scale enterprise in the ideology of the Italian left' n Pyke. In F., Becattini, G. & Sengenberger, W. (Eds.), Industrial districts & inter-firm co-operation in Italy* (pp. 142–159). Genève: International Institute for Labour Studies.

Burri, M., & Cottier, T. (2012). *Trade governance in the digital age.* NY, USA: Cambridge University Press. doi:10.1017/CBO9781139136716

Bygballe, L. E., Jahre, M., & Swärd, A. (2010). Partnering relationships in construction: A literature review. *Journal of Purchasing and Supply Management, 16*(4), 239–253. doi:10.1016/j.pursup.2010.08.002

Cadin, L., Guerin, F., & Pigeyre, F. (2009). Gestion des ressources humaines. Paris: Ed.

Callon, M. (1992). The dynamics of techno-economic networks. In R. Coombs & V. Walsh (Eds.), *Technical change & company strategies: Economic & sociological perspectives* (pp. 72–102). San Diego: Harcourt Brace Jovanovich Publishers.

Camarinha-Matos, L. M. (2009). Collaborative networked organizations: Status and trends in manufacturing'. *Annual Reviews in Control, 33*(2), 199–208. doi:10.1016/j.arcontrol.2009.05.006

Camarinha-Matos, L. M. (2014). Collaborative networks: a mechanism for enterprise agility and resilience. In K. Mertins, F. Benaben, R. Poler, & J.-P. Bourriers (Eds.), *Enterprise Interoperability VI: Interoperability for Agility, Resilience and Plasticity of Collaborations.* Switzerland: Springer International Publishing. doi:10.1007/978-3-319-04948-9_1

Cardozo, R., McLaughlin, K., Harmon, B., Reynolds, P., & Miller, B. (1993). Product- market choices and growth of new businesses. *Journal of Product Innovation Management, 10*(4), 331–340. doi:10.1016/0737-6782(93)90075-2

Carlsson, B. (Ed.). (1997). *Technological systems & industrial dynamics (No. 10)*. Springer Science & Business Media. doi:10.1007/978-1-4615-6133-0

Carlsson, B., & Stankiewicz, R. (1991). On the nature, function & composition of technological systems. *Journal of Evolutionary Economics*, *1*(2), 93–118. doi:10.1007/BF01224915

Carlsson, B., & Stankiewicz, R. (1995). On the Nature, Function & Composition of Technological Systems. In B. Carlsson (Ed.), *Technological systems & economic performance: the case of factory automation*. Boston, Dordrecht, London: Kluwer Academic Publishers. doi:10.1007/978-94-011-0145-5_2

Caron, F., & Fiore, A. (1995). 'Engineer to Order' companies: How to integrate manufacturing and innovative processes. *International Journal of Project Management*, *13*(5), 313–319. doi:10.1016/0263-7863(95)00023-J

Carter, C. R., & Ellram, L. M. (2003). Thirty-five years of the journal of supply chain management: Where we have been and where we going? *Journal of Supply Chain Management*, *39*(2), 27–39. doi:10.1111/j.1745-493X.2003.tb00152.x

Chandra, C., & Grabis, J. (2007). *Supply chain configuration*. New York: Springer.

Chandra, P., & Fisher, M. L. (1994). Coordination of Production and Distribution Planning. *European Journal of Operational Research*, *72*(3), 503–517. doi:10.1016/0377-2217(94)90419-7

Chandy, R. K., & Tellis, G. J. (2000). The incumbent's curse? Incumbency, size, and radical product innovation. *Journal of Marketing*, *6*(3), 1–17. doi:10.1509/jmkg.64.3.1.18033

Chan, F.T.S., Lau, K.W., Chan, L.Y., & Lo, V.H.Y. (2008). Cell formation problem with consideration of both intracellular and intercellular movements. *International Journal of Production Research*, *46*(10), 2589–2620. doi:10.1080/00207540500478843

Cheese, P., Thomas, R., & Craig, C. (2008). *The Talent Powered Organizations*. London: Kogan Page.

Chellappa, R. K. (2000). Digital Products and E-Business. Marshall Magazine, Spring, 48-49.

Chellappa, R. K., & Shivendu, S. (2003). Pay Now or Pay Later?*Managing Digital Product Supply Chains. ACM International Conference Proceeding Series* (Vol. 50, pp. 230-234).

Cheng, E. W. L., & Li, H. (2002). Construction Partnering Process and Associated Critical Success Factors: Quantitative Investigation. *Journal of Management Engineering*, *18*(4), 194–202. doi:10.1061/(ASCE)0742-597X(2002)18:4(194)

Chen, M., & Cao, D. (2004). Coordinating production planning in cellular manufacturing environment using Tabu search. *Computers & Industrial Engineering*, *46*(3), 571–588. doi:10.1016/j.cie.2004.02.002

Chen, M., & Wang, W. (1997). A Linear Programming Model for Integrated Steel Production and Distribution Planning. *International Journal of Operations & Production Management*, *17*(6), 592–610. doi:10.1108/01443579710167276

Chen, Y.-J., & Chen, Y.-M. (2009). An XML-based Modular System Analysis and Design for Supply Chain Simulation. *Robotics and Computer-integrated Manufacturing*, *25*(2), 289–302. doi:10.1016/j.rcim.2007.12.005

Chen, Y.-J., Chen, Y.-M., & Chu, H.-C. (2009). Development of a mechanism for ontology-based product lifecycle knowledge integration. *Expert Systems with Applications*, *36*(2), 2759–2779. doi:10.1016/j.eswa.2008.01.049

Cheung, S., Ng, T. S. T., Wong, S., & Suen, H. C. H. (2003). Behavioral aspects in construction partnering. *International Journal of Project Management*, *21*(5), 333–343. doi:10.1016/S0263-7863(02)00052-2

Childerhouse, P., Aitken, J., & Towill, D. R. (2002). Analysis and design of focused demand chains. *Journal of Operations Management, 20*(6), 675–689. doi:10.1016/S0272-6963(02)00034-7

Cho, K., & Lim, J. (1999). Mixed centralized/decentralized supervisory control of discrete events dynamic systems. *Automatica, 35*(1), 121–128. doi:10.1016/S0005-1098(98)00137-X

Choobineh, F. (1988). A framework for the design of cellular manufacturing systems. *International Journal of Production Research, 26*(7), 1161–1172. doi:10.1080/00207548808947932

Chopra, S. & Sodhi, M. (2004). Managing risks to avoid supply-chain breakdown. *MITSloan Management Review*, 2004(Fall).

Chopra, S., & Meindl, P. (2001). *Supply Chain Management: Strategy, planning and operation.* Prentice Hall.

Chopra, S., & Meindl, P. (2007). *Supply chain management. Strategy, planning & operation.* New Jersey, Pearson Prentice Hall.

Chou, M., & Ruchika, A. (2006). An in-depth study of the software supply chains.*Proceedings of the IEEE International Conference on Industrial Informatics.* doi:10.1109/INDIN.2006.275656

Christopher, M. (1998). *Logistics and supply chain management: Strategies for reducing costs and improving services.* Financial Times Publishing, London.

Christopher, M. (2004). Creating resilient supply chains. *Logistics Europe*, 2004, 14-21.

Christopher, M. (2000). The Agile Supply Chain: Competing in Volatile Markets. *Industrial Marketing Management, 29*(1), 37–44. doi:10.1016/S0019-8501(99)00110-8

Christopher, M., & Lee, H. (2004). Mitigating supply chain risk through improved confidence. *International Journal of Physical Distribution & Logistics Management, 34*(5), 388–396. doi:10.1108/09600030410545436

Christopher, M., Mena, C., Khan, O., & Yurt, O. (2011). Approaches to managing global sourcing risk. *Supply Chain Management: An International Journal, 16*(2), 67–81. doi:10.1108/13598541111115338

Christopher, M., & Peck, H. (2004). Building the resilient supply chain. *The International Journal of Logistics Management, 15*(2), 1–14. doi:10.1108/09574090410700275

Christopher, M., Peck, H., & Towill, D. (2006). A taxonomy for selecting global supply chain strategies. *The International Journal of Logistics Management, 17*(2), 277–287. doi:10.1108/09574090610689998

Christopher, M., & Towill, D. (2000). Supply Chain migration from lean and functional to agile and customised, *Supply Chain Management. International Journal (Toronto, Ont.), 5*(4), 206–213.

Christopher, M., & Towill, D. R. (2002). Developing market specific supply chain strategies. *International Journal of Logistics Management, 13*(1), 1–4. doi:10.1108/09574090210806324

Chui, M., Manyika, J., Bughin, J., Dobbs, R., Roxburgh, C., Sarrazin, H., & Westergren, M. et al. (2012). *The Social Economy: Unlocking Value and Productivity through Social Technologies.* McKinsey Global Institute.

Cicmil, S., & Hodgson, D. (2006). New possibilities for project management theory : A critical engagement. *Project Management Journal, 37*(3), 111–122.

Clempner, J. (2014). Verifying soundness of business processes: A decision process Petri nets approach. *Expert Systems with Applications, 41*(11), 5030–5040. doi:10.1016/j.eswa.2014.03.005

Cohen, M.A. and H.L. Lee, (1989). Resource Deployment Analysis of Global Manufacturing and Distribution Networks. *Journal of Manufacturing and Operations Management*, 1989, 81-104.

Construction Industry Institute (CII). (1991). In search of partnering excellence. Special Publication No. 17-1, Rep., Partnering Task Force of CII, Austin, Tex.

Construction of the National Museum of Australia and the Australian Institute of Aboriginal and Torres Strait Islander Studies. Audit Report (No.34) . (2000Auditor-General of the Australian National Audit Office (AGANAO).Canberra, Australia: Australian National Audit Office.

Cooke, P., Uranga, M. G., & Etxebarria, G. (1997). Regional innovation systems: Institutional & organisational dimensions. *Research Policy, 26*(4), 475–491. doi:10.1016/S0048-7333(97)00025-5

Cooper, A. H. (1957). Integration of computers with factory processes. *Journal of the British Institution of Radio Engineers, 17*(8), 431–441. doi:10.1049/jbire.1957.0044

Cornelissen, J. (2011). *Corporate Communication: A Guide to Theory and Practice*. CA, USA: Sage Pub. Ltd.

Cornelissen, J. P., & Lock, A. R. (2000). Theoretical Concept or Management Fashion? Examining the Significance of IMC. *Journal of Advertising Research, 40*(5), 7–15. doi:10.2501/JAR-40-5-7-15

Cornet, A., & Warlaud, Ph. (2008). *Gestion de ressources humaines et gestion de la differentité*. Paris: Dunod.

Coronas, T. T. (2008). *Encyclopedia of HR Information Systems: Challenges in E-HRM*. Hershey, PA, USA: IGI Global.

Costa, A., Barbosa, P., Gomes, L., Romalho, F., Figueiredo, J., & Junior, A. (2010). Properties preservation in distributed execution of Petri nets models. *Emerging Trends in Technological Innovation, 314*, 241–250. doi:10.1007/978-3-642-11628-5_26

Coughlan, P., & Coghlan, D. (2002). Action research for operations management. *International Journal of Operations & Production Management, 22*(2), 220–240. doi:10.1108/01443570210417515

Crawford, C. M., & Di Benedetto, C. A. (2000). *New Product Management*. Boston: McGraw-Hill.

Cullen, J., & Parboteeah, P. (2013). *Multinational management*. Cengage Learning.

Da Silveira, G., Borenstein, D., & Fogliatto, F. S. (2001). Mass customisation: Literature review and research directions. *International Journal of Production Economics, 72*(1), 1–13. doi:10.1016/S0925-5273(00)00079-7

Darbyshire, P. (2014). *Hampton D., Hedge fund modelling and analysis using MATLAB*. John Wiley & Sons.

Das, T. K., & Teng, B.-S. (1999). Managing risks in strategic alliances. *The Academy of Management Perspectives, 13*(4), 50–62. doi:10.5465/AME.1999.2570554

David, R., & Alla, H. (2005). *Discrete, Continuous and Hybrid Petri Nets*. Berlin: Springer.

Davis, P. & Love, P. (2011). Alliance contracting: Adding value through relationship development. *Engineering, Construction, and Architectural Management, 15*(11), 444–461.

de Souza, D. M., Lafontaine, M., Charron-Doucet, F., Bengoa, X., Chappert, B., Duarte, F., & Lima, L. (2015). Comparative Life Cycle Assessment of ceramic versus concrete roof tiles in the Brazilian context. *Journal of Cleaner Production, 89*, 165–173.

De Treville, S., Shapiro, R. D., & Hameri, A.-P. (2004). From supply chain to demand chain: The role of lead time reduction in improving demand chain performance. *Journal of Operations Management, 21*(6), 613–627. doi:10.1016/j.jom.2003.10.001

Dean, P.R., Tu, Y.L., Xue, D. (2008). A Framework for Generating Product Production Information for Mass Customization. *International Journal of Advanced Manufacturing Technology*, 38(11-12), 1244 - 1259.

De-Margerie, V., & Jiang, B. (2011). How relevant is OM research to managerial practice? An empirical study of top executives perceptions. *International Journal of Operations & Production Management, 31*(2), 124–147. doi:10.1108/01443571111104737

Dhaenens-Flipo, C., & Finke, G. (2001). An Integrated Model for an Industrial Production- Distribution Problem. *IIE Transactions, 33*(9), 705–715. doi:10.1080/07408170108936867

Dixon, K. R. (2011). *Modelling and simulation in Ecotoxicology with applications in MATLAB and Simulink.* CRC Press. doi:10.1201/b11089

Dong, M., & Chen, F. F. (2001). Process modeling and analysis of manufacturing supply chain networks using object-oriented Petri nets. *Robotics and Computer-integrated Manufacturing, 17*(1-2), 121–129. doi:10.1016/S0736-5845(00)00045-4

Doran, A. (2001). *E-Work Arhitect: How HR Leads the Way Using the Internet.* Austin, Texas, USA: IHRIM Press Book.

Dotoli, M., Fanti, M. P., Meloni, C., & Zhou, M. C. (2005). A multi-level approach for network design of integrated supply chains. *International Journal of Production Research, 43*(20), 4267–4287. doi:10.1080/00207540500142316

Drucker, P. (2003). Management's new paradigms.

Duffee, N. A., & Prabhu, V. V. (1996). Heterarchical control of highly distributed manufacturing systems. *International Journal of Computer Integrated Manufacturing, 9*(4), 270–281. doi:10.1080/095119296131562

Dunod, C. M. C. (2005). *Introduction to retailing.* Cape Town, South Africa: Juta&Co. Ltd.

Dyer, J. H. (2000). *Collaborative Advantage: Winning Through Extended Enterprise Supplier Networks.* Oxford: Oxford University Press.

Dyer, J. H., & Singh, H. (1998). The relational view: Cooperative strategy and sources of Inter-organisational competitive advantage. *Academy of Management Review, 23*(4), 660–679.

Easton, G., & Araujo, L. (1997). Interfirm responses to heterogeneity of demand over time. In M. Ebers (Ed.), *The Formation of Inter-organisational Networks* (pp. 66–94). Oxford: Oxford University Press.

Ebers, M. (1999). The dynamics of inter-organizational relationships. In *Research in the Sociology of Organisations* (Vol. 16, pp. 31–56). Stamford: JAI Press.

Edquist, C. (2004). Systems of Innovation: Perspectives & Challenges. In J. Fagerberg, D. Mowery, & R. R. Nelson (Eds.), *Oxford Handbook of Innovation* (pp. 181–208). Oxford: Oxford University Press.

Edquist, C. (Ed.), (1997). *Systems of innovation; Technologies, Institutions, & Organizations*. London: Pinter.

Egilmez, G., & Süer, G. (2013). The impact of risk on the integrated cellular design and control. *International Journal of Production Research*.

Egilmez, G., Süer, G. A., & Huang, J. (2012). Stochastic cellular manufacturing system design subject to maximum acceptable risk level. *Computers & Industrial Engineering*, *63*(4), 842–854. doi:10.1016/j.cie.2012.05.006

Eldering, C. A., Sylla, M. L., & Eisenach, J. A. (1999). Is there a Moore's law for bandwidth? *IEEE Communications Magazine*, *37*(10), 117–121. doi:10.1109/35.795601

Elfving, J. (2003). Exploration of opportunities to reduce lead times for engineering-to-order products [Ph.D. dissertation]. University of California, Berkley.

Ermilova, E., & Afsarmanesh, H. (2006). *Competency and profiling management in virtual organization breeding environments. Network-centric collaboration and supporting frameworks* (pp. 131–142). Boston: Springer. doi:10.1007/978-0-387-38269-2_14

Essa, A. (2005). *Innovation Life Cycle and the S-Curve*. Retrieved from http://tatler.typepad.com/nose/2005/11/innovation_life.html

Evans, P. B., & Wurster, T. S. (1997). Strategy and the new economics of information. *Harvard Business Review*, 5, 70–82. PMID:10170332

Fagerberg, J. (2003). Innovation: A Guide to the Literature. *Working Papers on Innovation Studies 20031012, Centre for Technology, Innovation and Culture*, University of Oslo.

Feitzinger, E., & Lee, H. L. (1997). Mass Customization at Hewlett-Packard: The Power of Postponement. *Harvard Business Review*, *75*(1), 116–121. PMID:10174793

Fernández de Lucio, I., & Gabaldón-Estevan, D. (2007). Nous reptes per a la indústria ceràmica. At Observatori Estadístic de l'EURAM; Institut d'Economia i Empresa Ignasi Villalonga (Eds.) Recerca i innovació a l'EURAM. València: Tres i Quatre, pp. 37-50.

Fernández, I. Gabaldón, D. & Gómez, C. (2005). *La innovación en el sector de pavimentos y revestimientos cerámicos de la Comunidad Valenciana*. Valencia: Alto Consejo Consultivo en I+D de la Presidencia de la Generalitat Valenciana.

Fischer, K. (1999). Agent based design of holonic manufacturing systems. *Robotics and Autonomous Systems*, *27*(1-2), 3–13. doi:10.1016/S0921-8890(98)00079-7

Fisher, M. (1997). What is the right supply chain for your product? *Harvard Business Review*, (March), 105–116.

Florea, N. V. (2014). Training, Coaching, Mentoring. Methods and Models. Bucuresti: Ed.

Flyvbjerg, B., Skamris Holm, M. K., & Buhl, S. L. (2003). How common and how large are cost overruns in transport infrastructure projects. *Transport Reviews*, *23*(1), 71–88. doi:10.1080/01441640309904

Folan, P., & Browne, J. (2005). A Review of Performance Measurement: Towards Performance Management. *Computers in Industry*, *56*(7), 663–680. doi:10.1016/j.compind.2005.03.001

Freeman, C. (1982). *The Economics of Industrial Innovation*. London: Pinter.

Freeman, C. (1995). The 'National System of Innovation' in historical perspective. *Cambridge Journal of Economics, 19*(1), 5–24.

Freeman, C., & Soete, L. (Eds.). (1987). *Technical Change & Full Employment*. Oxford: Basil Blackwell.

Fullana i Palmer, P., Puig, R., Bala, A., Baquero, G., Riba, J., & Raugei, M. (2011). From Life Cycle Assessment to Life Cycle Management. *Journal of Industrial Ecology, 15*(3), 458–475. doi:10.1111/j.1530-9290.2011.00338.x

Furrer, O., Thomas, H., & Goussevskaia, A. (2008). The structure and evolution of the strategic management field: A content analysis of 26 years of strategic management research. *International Journal of Management Reviews, 10*(1), 1–23. doi:10.1111/j.1468-2370.2007.00217.x

Gabaldón-Estevan, D., Fernández de Lucio, I., & Tortajada-Esparza, E. (2007, June 18-20) How Does Structure Affect Industrial Districts Innovation? *Danish Research Unit for Industrial Dynamics Summer Conference*. Copenhagen, (Denmark). Retrieved from http://digital.csic.es/handle/10261/10172

Gabaldón-Estevan, D., Mezquita, A., Ferrer, S., & Monfort, E. (2014b) Is European Union Environmental Policy Efficient at Promoting a Post-carbon Industry? The Case of Energy in the European Ceramic Tile Sector. *Proceedings of the 11th International Conference on Innovation & Management* (pp. 104-113). Wuhan University of Technology Press.

Gabaldón-Estevan, D., Molina Morales, F. X., & Fernández de Lucio, I. (2009). Research contracts and patents in the Spanish ceramic tile district innovation system. *INGENIO Working Papers Series*. 2009-1

Gabaldón-Estevan, D. (2011). *El sistema distritual de innovación cerámico de Castellón*. Universitat de València. Servei de Publicacions.

Gabaldón-Estevan, D., Criado, E., & Monfort, E. (2014a). The Green Factor in European Manufacturing: A case study of the Spanish ceramic tile industry'. *Journal of Cleaner Production, 70*, 242–250. doi:10.1016/j.jclepro.2014.02.018

Gabaldón-Estevan, D., Fernández-de-Lucio, I., & Molina-Morales, F. X. (2012). Sistemas distrituales de innovación. *Arbor, 188*(753), 63–73. doi:10.3989/arbor.2012.753n1005

Gabaldón-Estevan, D., & Hekkert, M. P. (2013). How does the innovation system in the Spanish ceramic tile sector function?' *Boletín de la Sociedad Española de Cerámica y Vidrio, 52*(3), 151–158. doi:10.3989/cyv.202013

Gabaldón-Estevan, D., Tortajada-Esparza, E., & Fernández de Lucio, I. (2008). *El papel de las fritas, esmaltes y colores en la evolución del sector cerámico de la Comunidad Valenciana*. Generalitat Valenciana.

Gadde, L.-E., Huemer, L., & Hakansson, H. (2003). Strategizing in industrial networks. *Industrial Marketing Management, 32*(5), 357–364. doi:10.1016/S0019-8501(03)00009-9

Galanaki, E. (2002). The Decision to Recruit Online. *Career Development International, 7*(4), 243–251. doi:10.1108/13620430210431325

Gattorna, J. L., & Walters, D. W. (1996). *Managing the Supply Chain: A Strategic Perspective*. Macmillan Business.

Gavireni, S. & Tayur, S. (1997, August). *Delayed product differentiation versus information sharing*. Carnegie Mellon University.

Geunes, J., Pardalos, P., & Romeijn, H. E. (2002). *Supply chain management: models, applications, and research directions*. Dordrecht: Kluwer academic publishers.

Ghezavati, V., & Saidi-Mehrabad, M. (2010). Designing integrated cellular manufacturing systems with scheduling considering stochastic processing time. *International Journal of Advanced Manufacturing Technology*, *48*(5-8), 701–717. doi:10.1007/s00170-009-2322-2

Giunipero, L. C., Hooker, R. E., Matthews, S. C., Yoon, T. E., & Brudvig, S. (2008). A decade of SCM literature: Past, present and future implications. *Journal of Supply Chain Management*, *44*(1), 66–86. doi:10.1111/j.1745-493X.2008.00073.x

Goldsby, T. J., Griffis, S. E., & Roath, A. S. (2006). Modeling Lean, Agile, and Leagile Supply Chain Strategies. *Journal of Business Logistics*, *27*(1), 57–80. doi:10.1002/j.2158-1592.2006.tb00241.x

Gosling, J., & Naim, M. M. (2009). Engineer-to-order Supply Chain Management: A literature review and research agenda. *International Journal of Production Economics*, *122*(2), 741–754. doi:10.1016/j.ijpe.2009.07.002

Grabenstetter, D., & Usher, J. M. (2013). Defining Job Complexity for the Engineer-to-Order Environment. In A Krishnamurthy, & W.K.V Chan (Eds.), *Proceedings of the 2013 Annual Industrial and Systems Engineering Research Conference* (pp. 2404 – 2413).

Graeml, A. R., Ebgelbert, R., & Weiler, A. L. G. (2008, June 15-18). Understanding value chain strategies for the iPod through the perspective of Fine's evolutionary model.*Proceedings of the 15th Annual International Conference of the European Operations Management Association (EurOMA)*, Gronigen, Netherlands.

Graham, G., Burnes, B., Lewis, G., & Langer, J. (2004). The transformation of the music industry supply chain: A major label perspective. *International Journal of Operations & Production Management*, *24*(11), 1087–1103. doi:10.1108/01443570410563241

Grant, R. (1991). The resource-based theory of competitive advantage: Implications for strategy formulation. *California Management Review*, *33*(3), 114–135. doi:10.2307/41166664

Gravel, J.-P., & Kavanagh, T.F. (1957). Computers -- the Key to Modern Manufacturing Scheduling. *IRE Transactions on Industrial Electronics*, 4, 90-93. doi: .10.1109/IRE-IE.1957.5007790

Green, K. Jr, Whitten, W. D., & Inman, R. A. (2008). The impact of logistics performance on organizational performance in a supply chain context. *Supply Chain Management, International Journal (Toronto, Ont.)*, *13*(4), 317–327.

Greiffenstern, S. (2010). *The Influence of Computers, the Internet and Computer-Mediated Communication on Everyday English*. Berlin, Germany: Bibliografische Inf.

Grochowski, W., & Halem, R. D. (2003). Technological Impact of magnetic hard disk drives on storage systems. *IBM Systems Journal*, *42*(2), 338–346. doi:10.1147/sj.422.0338

Guental, H. G., & Stone, D. L. (2005). The Brave New World of eHR. HRM in the Digital Age. SF, USA: Jossey-Bass a Wiley Imprint.

Gulati, R. (1998). Alliances and networks. *Strategic Management Journal*, *19*(4), 293–317. doi:10.1002/(SICI)1097-0266(199804)19:4<293::AID-SMJ982>3.0.CO;2-M

Gulati, R., Nohria, N., & Zaheer, A. (2000). Strategic networks. *Strategic Management Journal*, *21*(3), 203–215. doi:10.1002/(SICI)1097-0266(200003)21:3<203::AID-SMJ102>3.0.CO;2-K

Gunasekaran, A. (1998). Agile manufacturing: Enablers and an implementation framework. *International Journal of Production Research*, *36*(5), 1223–1247. doi:10.1080/002075498193291

Gunasekaran, A., Patel, C., & Tirtiroglu, E. (2001). Performance Measures and Metrics in a Supply Chain Environment. *International Journal of Production and Operations Management*, *21*(1-2), 71–87. doi:10.1108/01443570110358468

Haas, M. R., & Hansen, M. T. (2005). When using knowledge can hurt performance: The value of organizational capabilities in a management consulting company. *Strategic Management Journal*, *26*(1), 1–24. doi:10.1002/smj.429

Hallgren, M., Olhager, J., & Schroeder, R. G. (2011). A hybrid model of competitive capabilities. *International Journal of Operations & Production Management*, *31*(5), 511–526. doi:10.1108/01443571111126300

Hamel, G. (2007). *The Future of Management, Harvard Business Press, Hanna, N.K., (2010). Enabling Enterprise Transformation*. USA: Springer.

Harcourt, B. (2011). *The Illusion of Free Markets*. Harvard University Press.

Hardaker, G., & Graham, G. (2001). *Wired marketing: energizing business for e-Commerce. Chicester*: John Wiley & Sons.

Harrison, T., & Waite, K. (2005). Critical factors affecting intermediary web site adoption: Understanding how to extend e-participation. *Journal of Business and Industrial Marketing*, *20*(4/5), 187–199. doi:10.1108/08858620510603864

Hars, A., & Ou, S. (2002). Working for Free? - Motivations for Participating in Open Source Projects. *Proceedings International Journal of Electronic Commerce*, *6*(3), 25–39.

Haug, A., Ladeby, K., & Edwards, K. (2009). From engineer-to-order to mass customisation. *Management Research News*, *32*(7), 633–644. doi:10.1108/01409170910965233

Headlam-Wells, J., Gosland J. & Craig, J. (2006). Beyond the organization: the design and management of e-mentoring systems. *International Journal of Information Management*, 26, 272-285.

He, F., Dong, M., & Shao, X. (2011). Modeling and analysis of material flows in re-entrant supply chain networks using modified partial differential equations. *Journal of Applied Mathematics*.

Hekkert, M. P., Suurs, R. A. A., Negro, S. O., Kuhlmann, S., & Smits, R. E. H. M. (2007). Functions of Innovation Systems: A new approach for analyzing technological change. *Technological Forecasting and Social Change*, *74*(4), 413–432. doi:10.1016/j.techfore.2006.03.002

Hertel, F. (2003). *Motivation in the Workplace*. GRIN Verlag.

Hicks, C., & Braiden, P. M. (2000). Computer-aided Production Management Issues in the Engineer-to-Order Production of Complex Capital Goods explored using a simulation approach. *International Journal of Production Research*, *38*(18), 4783–4710. doi:10.1080/00207540010001019

Hicks, C., & McGovern, T. (2009). Product Life Cycle Management in Engineer to Order Industries. *International Journal of Technology Management*, *48*(2), 153–167. doi:10.1504/IJTM.2009.024913

Hicks, C., McGovern, T., & Earl, C. F. (2000). Supply Chain Management: A Strategic Issues in Engineer to Order Manufacturing. *International Journal of Production Economics*, *65*(2), 179–190. doi:10.1016/S0925-5273(99)00026-2

Hicks, C., McGovern, T., & Earl, C. F. (2001). A Typology of UK Engineer-to-Order Companies. *International Journal of Logistics: Research and Applications*, *4*(1), 43–56. doi:10.1080/13675560110038068

Hilletofth (2012). Differentiation focused supply chain design, *Industrial Management & Data Systems*, 112(9), 1274 – 1291.

Hines, P., Holweg, M., & Rich, N. (2004). Learning to evolve: A review of contemporary lean thinking. *International Journal of Operations & Production Management, 24*(10), 994–1011. doi:10.1108/01443570410558049

Hlahla, M. (1999). The Municipal Infrastructure Investment Unit: The government's PPP-enabling strategy. *Development Southern Africa, 16*(4), 565–583. doi:10.1080/03768359908440102

Holbeche, L. (2001). Aligning Human Resources and Business Strategy. *Business Economics (Cleveland, Ohio).*

Holweg, M. (2005). The three dimensions of responsiveness. *International Journal of Operations & Production Management, 25*(7), 603–622. doi:10.1108/01443570510605063

Horton, W. K. (2001). *Leading e-learning. USA: ASTD*

Hoskisson, R. E., Hitt, M. A., Wan, W. P., & Yiu, D. (1999). Theory and research in strategic management: Swings of a pendulum. *Journal of Management, 25*(3), 417–456. doi:10.1177/014920639902500307

Hsu, C., & Wallace, W. A. (2007). An industrial network flow information integration model for supply chain management and intelligent transportation. *Enterprise Information Systems, 1*(3), 327–351. doi:10.1080/17517570701504633

Huang, J., & Süer, G. A. (2012). Long-Term Sourcing Model for Multi-Objective Supplier Selection Under Uncertainty. *Proceedings of 42nd International Conference on Computers & Industrial Engineering*, Cape Town, South Africa.

Hughes, T. (1984). The evolution of large technological systems. In W. Bijker, T. Hughes, & A. Johnson (Eds.), (2001). *Functions in innovation systems approaches. In electronic paper at the proceeding of the Nelson and Winter conference*, Aalborg. Retrieved from http://www.druid.dk/conferences/nw/paper1/a_johnson.pdf

Hui, K. L., & Chau, P. Y. K. (2002). Classifying Digital Products. *Communications of the ACM, 45*(6), 73–79. doi:10.1145/508448.508451

Husejnagic, D., & Sluga, A. (2015). A Conceptual Framework for a Ubiquitous Autonomous Work System in the Engineer-to-Order Environment. *International Journal of Advanced Manufacturing Technology, 78*(9-12), 1971–1988. doi:10.1007/s00170-015-6798-7

Hyer, N., & Vemmerlov, U. (2002). *Reorganizing the Factory - Competing Through Cellular Manufacturing. Usa.* Productivity Press.

Improved Data Visibility Drives Manufacturing Performance. (2013). *Plex White Paper*. Retrieved from http://www.plex.com/wordpress/wp-content/uploads/2013/06/Manufacturing_Visibility_WP.pdf

International Wind Energy Development. Supply Chain Assessment 2012-2015. (2011). *BTM Consult.*

IRENA working paper: renewable energy technologies. (2012). *Wind Power*, 5(5). Retrieved from www.irena.org/Publications

ISO. (2015a). ISO 14000 - Environmental Management. Retrieved from http://www.iso.org/iso/home/standards/management-standards/iso14000.htm

ISO. (2015b). ISO 14040:2006 - Environmental Management: Life Cycle Assessment - principles and framework. Retrieved from http://www.iso.org/iso/catalogue_detail?csnumber=37456

Jain , A.K. , (2011). Professional Communication Skills. ND, India: S.Chand&Co.Ltd.

Janoshka, A. (2004). *Web Advertising: New forms of communication on the Internet*. USA: John Benjamin Pub. doi:10.1075/pbns.131

Jayaswal, S., & Adil, G. K. (2004). Efficient algorithm for cell formation with sequence data, machine replications and alternative process routing. *International Journal of Production Research*, *42*(12), 2419–2433. doi:10.1080/00207540310001652914

Jeon, G., Leep, H. R., & Parsaei, H. R. (1998). A cellular manufacturing system based on new similarity coefficient which considers alternative routes during machine failure. *Computers & Industrial Engineering*, *34*(1), 21–36. doi:10.1016/S0360-8352(97)00148-4

Jiang, H., Liu, C., & Cui, Z. (2009). Research on Knowledge Management System in Enterprise.*Proceedings of the 2009 International Conference on Computational Intelligence and Software Engineering* (pp. 1-4). doi:10.1109/CISE.2009.5363935

Johns, D. T., & Harding, H. A. (1989). *Operations management – A personal skills handbook*. Southampton: Gower.

Johnson, J. C., & Wood, D. F. (1996). *Contemporary logistics*. Upper Saddle River, New Jersey: Prentice Hall.

Johnston, P. (1995). Supply chain management: The past, the present and the future. *Manufacturing Engineer*, *74*(5), 213–217. doi:10.1049/me:19950510

Jules, G. D., Saadat, M., & Saeidlou, S. (2015). Holonic ontology and interaction protocol for manufacturing network organization. *IEEE Transactions on Systems, Man and Cybernetics. Systems*, *43*(5), 1240–1247.

Kadushin, C. (2004). Introduction to Social Network Theory. In *Some Basic Network Concepts and Propositions* (Ch. 2). http://melander335.wdfiles.com/local--files/reading-history/kadushin.pdf

Kaldellis, J. K., & Zafirakis, D. (2011). The wind energy (r)evolution: A short review of a long history. *Renewable Energy*, *36*(7), 1887–1901. doi:10.1016/j.renene.2011.01.002

Kamath, , N., B., & Roy, R. (2007). Capacity augmentation of a supply chain for a short lifecycle product: A system dynamics framework. *European Journal of Operational Research*, *179*(2), 334–351. doi:10.1016/j.ejor.2006.03.045

Kaplan, R., & Norton, D. (1992). The Balanced Scorecard - measures that drive performance. *Harvard Business Review*, *70*(1), 71–79. PMID:10119714

Kaplan, R., & Norton, D. (1996). Using the Balanced Scorecard as a Strategic Management System. *Harvard Business Review*, *74*(1), 75–85.

Karlsson, C., & Ahlström, P. (1996). The difficult path to lean product development. *Journal of Product Innovation Management*, *13*(4), 283–295. doi:10.1016/S0737-6782(96)00033-1

Karpowitz, D. J., Cox, J. J., Humpherys, J. C., & Warnick, S. C. (2008). A Dynamic Workflow framework for Mass Customization using Web Service and Autonomous Agent Techniques. *Journal of Intelligent Manufacturing*, *19*(5), 537–552. doi:10.1007/s10845-008-0136-y

Kaufman, B. E., Beaumont, R. A., & Helfgott, R. B. (2003). *Industrial Relations to Human Resources and Beyond: The Evolving Process of Employee Relations Management*. M.E. Sharpe.

Kawai, T., Sakaguchi, J., & Shimizu, N. (2013). Transition of buyer-supplier relationships in Japan: Empirical Evidence from manufacturing companies. *Journal of Accounting and Organizational Change*, *9*(4), 427–447. doi:10.1108/JAOC-07-2011-0033

Khouri, S., & Kačmáry, P. (2009). Optimalizácia tokov informácií v modelovej firme prostredníctvom zavedenia ERP systému. *Transport & Logistics*, *6*, 125–127.

Kiron, D., Palmer, D., Phillips, A. N., & Kruschwitz, N. (2012). *Social Business: What Are Companies Really Doing?" MIT Sloan Management Review*. Social Business Global Executive Study and Research Project.

Kleijnen, J. P. C., & Smits, M. T. (2003). Performance Metrics in Supply Chain Management. *The Journal of the Operational Research Society*, *54*(5), 507–514. doi:10.1057/palgrave.jors.2601539

Kleindorfer, P. R., & Saad, G. H. (2005). Managing disruption risks in supply chains. *Production and Operations Management*, *14*(1), 53–65. doi:10.1111/j.1937-5956.2005.tb00009.x

Komoda, N. (2006). Service oriented architecture (SOA) in industrial systems.*Proceedings of the 2006 IEEE International Conference on Industrial Informatics* (pp. 1-5). doi:10.1109/INDIN.2006.275708

Kopczak, L. R., & Johnson, M. E. (2003). The supply-chain management effect. *MIT Sloan Management Review*, *44*(3), 27–34.

Kraljic, P. (1983). Purchasing must become supply management. *Harvard Business Review*, (Sept-Oct), 109–117.

Kulp, S. C., Lee, H. L., & Ofek, E. (2004). Manufacturer benefits from information integration with retail customers. *Management Science*, *50*(4), 431–444. doi:10.1287/mnsc.1030.0182

Kumar, R. (2010). *Basic Business Communication*. ND, India: Excel Books.

Kusiak, A. (1987). The generalized group technology concept. *International Journal of Production Research*, *25*(4), 561–569. doi:10.1080/00207548708919861

Kusluvan, S. (2003). *Managing Employee Attitudes and Behaviors in the Tourism and Hospitality Industry*. NY, USA: Nova Science Pub.

La Londe, B. (1998). *Supply Chain Management: An Opportunity for Competitive Advantage*. Cincinnati, OH: Department of Transport and Logistics, The Ohio State University.

Lakhdar, S. (2001). *Gestion de ressources humaines*. Canada: Université de Boeck.

Lamb, C. W. (2012). *Marketing*. Cengage Learning.

Lambert, D.M. (2008). *Supply Chain Management: Processes, Partnerships, Performance*. Supply Chain Management Inst.

Lambert, D. M., Cooper, M. C., & Pagh, J. D. (1998). Supply chain management: Implementation issues and research opportunities. *International Journal of Logistics Management*, *9*(2), 1–19. doi:10.1108/09574099810805807

Lantz, E., Wiser, R., & Hand, M. (2012) IEA Wind Task 26: The Past And Future Cost Of Wind Energy (Technical Report NREL/TP-6A20-53510). *National Renewable Energy Laboratory*.

Law, H. (2007). *The psychology of coaching, mentoring and learning*. England: J. Wiley & Sons.

Lee, C., Lee, K., & Pennings, J. M. (2001). Internal capabilities, external networks, and performance: A study on technology-based ventures. *Strategic Management Journal*, *22*(6/7), 615–640. doi:10.1002/smj.181

Lenort, R., Staš, D., & Samolejová, A. (2012). Heuristic Algorithm for Planning and Scheduling of Forged Pieces Heat Treatment. *Metalurgija*, *51*(2), 225–228.

Levin, R. I., Rubin, D. S., Stinson, J. P., & Gardner, E. S. r. (1989). *Quantitative approaches to Management* (7th ed.). New York: McGraw Hill Inc.

Lianfeng, Z., Xiao, Y., Jianxin, J., & Petri, H. (2009). Supply chain configuration with coordinated product, process and logistics: An approach based on Petri nets. *International Journal of Production Research*, *47*(23), 6681–6706. doi:10.1080/00207540802213427

Ling, F. Y. Y., Rahman, M. M., & Ng, T. L. (2006). Incorporating Contractual Incentive to Facilitate Relational Contracting. *Journal of Professional Issues in Engineering Education and Practice*, *132*(1), 57–66. doi:10.1061/(ASCE)1052-3928(2006)132:1(57)

Li, S. H., Rao, S. S., Nathan, R. T., & Nathan, B. R. (2006). The impact of supply chain management practices on competitive advantage and organizational performance. *Omega*, *34*(2), 107–124. doi:10.1016/j.omega.2004.08.002

Little, D., Rollins, R., Peck, M., & Porter, J. K. (2000). Integrated Planning and Scheduling in the engineer-to-order sector. *International Journal of Computer Integrated Manufacturing*, *13*(6), 545–554. doi:10.1080/09511920050195977

Lo Nigro, G., & Abbate, L. (2011). Risk assessment and profit sharing in business networks. *International Journal of Production Economics*, *131*(1), 234–241. doi:10.1016/j.ijpe.2009.08.014

Long, C., & Vickers-Koch, M. (1995). Using core capabilities to create competitive advantages. *Organizational Dynamics*, *24*(1), 7–22. doi:10.1016/0090-2616(95)90032-2

Love, P. E. D., Davis, P. R., Chevis, R., & Edwards, D. J. (2011). Risk/reward compensation model for civil engineering infrastructure alliance projects. *Journal of Construction Engineering and 616 Management*, 137(2), 127-136.

Love, P. E. D., Irani, Z., & Edwards, D. J. (2004). A seamless supply chain management model for construction. *Supply Chain Management: An International Journal*, *9*(1), 43–56. doi:10.1108/13598540410517575

Lowell, B., & Joyce, C. (2007). *Mobilizing Minds*. McGraw Hill.

Lummus, R. R., Krumwiede, D. W., & Vokurka, R. J. (2001). The relationship of logistics to supply chain management: Developing a common industry definition. *Industrial Management & Data Systems*, *101*(8), 426–432. doi:10.1108/02635570110406730

Lundvall, B. A. (1992). User-producer relationships, national systems of innovation & internationalisation. *National systems of innovation: Towards a theory of innovation & interactive learning* (pp. 45-67).

Maier, A. M., Kreimeyer, M., Hepperle, C., Eckert, C. M., Lindemann, U., & Clarkson, P. J. (2008). Exploration of Correlations between Factors Influencing Communication in Complex Product Development. *Concurrent Engineering: Research and Applications*, *16*(1), 37–59. doi:10.1177/1063293X07084638

Malerba, F. (2002). Sectoral systems of innovation & production'. *Research Policy*, *31*(2), 247–264. doi:10.1016/S0048-7333(01)00139-1

Malerba, F. (2004). Sectoral systems of innovation: basic concepts. In F. Malerba (Ed.), *Sectoral systems of innovation: Concepts, Issues & Analyses of Six Major Sectors in Europe* (pp. 9–41). Cambridge: Cambridge University Press. doi:10.1017/CBO9780511493270.002

Malindžák, D. (1983). Determination of the optimum sequence of heuristics. In production Management tasks. *Transaction TU Košice.*

Malindžák, D., & Takala, J. (2005). Projektovanie logistických systémov: Teória a prax. Express Publicit s.r.o., Košice.

Malindžák, D., ... (1993). Modely pre automatizáciu a koordináciu operatívnych plánov TŠP 1700 a narážacích pecí, Záverečná správa projektu, Košice.

Malindžák, D., Kačmáry,P., Bednárová,L., & Vrlíková, J. (2013). *The Production Capacity Planning Model in Term of Supply Chain Management.* Metal, Tanger, Brno.

Malindžák, D., Lenort, R., & Mervart, J. (2011). The logistics principles for fast flexible strategy design of the company in Crisis Time. *Managing global transition*, 9(2).

Malindžák, D. (1983). *Application of heuristic methods in manufacturing operative planning in Považské Strojárne, Považská Bystrica.* TU Košice.

Malindžák, D. (1997). *Production Logistics I.* Košice: Štroffek.

Malindžák, D. et al.. (2011). *Návrh systemu výrobnej logistiky a layout firmy Rosenberg Slovakia spol. s r.o.: analýza a návrh koncepcie. Project realisation.* TU Košice.

Malindžák, D., & Kačmáry, P. et al.. (2013). *Technical project: Systémová analýza a návrh algoritmov výrobného rozvrhovania výroby základný chfólií Chemosvit fólie a.s., TU Košice.* F-BERG.

Malindžák, D., Kačmáry, P., & Moussttfa, A. A. (2012). Models for capacity planning in the condition for SMEs connected to supply chain management.*Proceedings of Carpathian Logistic Congress, Tanger, Jeseník.*

Malindžáková, M. (2011). Significance evaluation of environmental aspects. *Communications*, *13*(3), 1335–4205.

Manley, Karen & Chen, Le. (2015). Collaborative learning model of infrastructure construction: A capability perspective. *Construction Innovation: Information, Process, Management*, *15*(3), 355–377.

Marchal, E., & Riencau, G. (2010). *Le recrutement.* Paris: Edition La Decouverte.

Market Report Global Wind Power Supply Chain. (2013). *MAKEConsulting.*

Martin, M., & Jackson, T. (2008). *Practica de personal.* Bucuresti: Editura CODECS.

Mason-Jones, R., Naylor, B., & Towill, D. R. (2000). Lean, agile or leagile? Matching your supply chain to the marketplace. *International Journal of Production Research*, *38*(17), 4061–4070. doi:10.1080/00207540050204920

Mason-Jones, R., & Towill, D. R. (1999). Total cycle time compression and the agile supply chain. *International Journal of Production Economics*, *62*(1-2), 61–73. doi:10.1016/S0925-5273(98)00221-7

Masterman, J. W. E. (2002). *Introduction to Building Procurement Systems.* London: Spon Press.

Matthews, J. R. (2009). *The customer-focused library.* ABC-CLIO.

Mayrhofer, W., Brewster, C., & Morley, M. (2007). *HRM in Europe*. Taylor&Francis.

McCleskey, S. (2010). *When Free Markets Fail: Saving the Market When It Can't Save Itself.* John Wiley&Sons.

McCullen, P., & Towill, D. R. (2000). Practical ways of reducing bullwhip: The case of the Glosuch global supply chain. *International Journal of Operations Management*, *26*(10), 24–30.

McCutcheon, D. M., & Meredith, J. R. (1993). Conducting case study research in operations management. *Journal of Operations Management*, *11*(3), 239–256. doi:10.1016/0272-6963(93)90002-7

McGovern, T., Hicks, C., and Earl, C.F. (1999). Modelling Supply Chain Management Process in Engineer-to-Order Companies. *International Journal of Logistics Research and Applications*, 2(2), 147-159.

McNeil, A. J., Frey, R., & Embrechts, P. (2015). *Quantitative Risk Management: Concepts, Techniques and Tools*. New Jersey, USA: Princeton University Press.

Mellenbergh, G. J., & Adèr, H. J. (2008). Tests and Questionnaires: Construction and administration. In H. J. Adèr & G. J. Mellenbergh (Eds.), *Advising on Research Methods: A consultant's companion* (pp. 211–236). Huizen, The Netherlands: Johannes van Kessel Publishing.

Menne, P. (2008). *Potential of Geo-Marketing-Tools for the development of advance online-marketing business models* [Master thesis].

Merrifield, D. B. (2000). Changing nature of competitive advantage. *Research Technology Management*, *41*(1), 41–45.

Meyer-Stamer, J., Maggi, C., & Seibel, S. (2004) Upgrading in the tile industry of Italy, Spain, and Brazil: insights from cluster and value chain analysis. In H. Schmitz (Ed.), Local enterprises in the global economy: issues of governance and upgrading (pp. 174-199). Cheltenham (Reino Unido): Edward Elgar.

Mezquita, A.; Monfort, E.;Vaquer, E.; Ferrer, S.; Pitarch, J.M.; Arnal, M.A.; Cobo, F. (2014) Reduction of CO2-emissions in ceramic tiles manufacture by combining energy-saving measures. *Cfi Ber. DKG 85*, 91(5), E37-E42

Michlowicz, E. (2002). *Podslawy logistyki przemyslowej UWND*. AGH Krakow.

Minguillón, M. C., Monfort, E., Escrig, A., Celades, I., Guerra, L., Busani, G., & Querol, X. et al. (2013). Air quality comparison between two European ceramic tile clusters. *Atmospheric Environment*, *74*, 311–319. doi:10.1016/j.atmosenv.2013.04.010

Mistry, J. J. (2005). Supply Chain Management: A Case Study of an Integrated Lean and Agile Model. *Qualitative Research in Accounting & Management*, *2*(2), 193–215. doi:10.1108/11766090510635442

Molina, A., Aca, J., & Wright, P. (2005). Global collaborative engineering environment for integrated product development. *International Journal of Computer Integrated Manufacturing*, *18*(8), 635–651. doi:10.1080/09511920500324472

Monden, Y. (1983). *Toyota production System*. Industrial Engineering and Management Press, Institute of Industrial Engineers.

Monfort, E., Mezquita, A., Vaquer, E., Mallol, G., & Gabaldón-Estevan, D. (2014). La evolución energética del sector español de baldosas cerámicas. *Boletín de la Sociedad Española de Cerámica y Vidrio*, *53*(3), 111–120. doi:10.3989/cyv.152014

Moreira, M., & Basilio, J. (2014). Bridging the gap between design and implementation of discrete-event controllers. *IEEE Transactions on Automation Science and Engineering*, *11*(1), 48–65. doi:10.1109/TASE.2013.2281733

Morwood, R., Scott, D., & Pitcher, I. (2008). *Alliancing A Participant's Guide: Real Life Experiences for Constructors, Designers, Facilitators and Clients*. AECOM.

Mott, W. H. (2004). *Globalization. People, Perspectives, and Progress*. USA: Greenwood Pub.Group.

Mullenders, A. (2009). *E-DRH. Outil de géstion innovant*. Bruxelles, Belgique: Edition de Boeck Universite.

Murata, T., Komoda, N., Matsumoto, K., & Haruna, K. (1986). A Petri net based controller for flexible and maintainable sequence control and its application in factory automation. *IEEE Transactions on Industrial Electronics*, *33*(1), 1–8. doi:10.1109/TIE.1986.351700

Murman, E., Rebentisch, E., & Walton, M. (2000). Challenges in the better, faster, cheaper era of aeronautical design, engineering and manufacturing. Lean Aerospace Initiative. USA: MIT. Retrieved from web.mit.edu/lean

Musa, A., Gunasekaran, A., & Yusuf, Y. (2014). Supply Chain Product Visibility: Methods, Systems and Impacts. *Expert Systems with Applications – An International Journal,* 41(1), 176–194.

Nagel , R. N. & Dove , R. (1991). 21st Century Manufacturing Enterprise Strategy: An Industry-Led View. Iaccoca Institute.

Nambiar, A. N. (2009a, July 1-3). Mass Customization - Where do we go from here? *Proceedings of the 2009 World Congress on Engineering*, London, U.K (*Vol I*).

Nambiar, A. N. (2009b, July 6-9). Agile Manufacturing: A Taxonomic Framework for Research. *Proceedings of the 2009 International Conferences on Computers and Industrial Engineering* (pp. 684-689). doi:10.1109/ICCIE.2009.5223676

Nambiar, A. N. (2010a, March 17-19). Modern Manufacturing Paradigms - A Comparison. *Proceedings of the 2010 International Multiconference of Engineers and Computer Scientists*, Hong Kong(Vol. 3).

Nambiar, A. N. (2010b, January 9-10). Challenges in Sustainable Manufacturing. *Proceedings of the 2010 International Conference on Industrial Engineering and Operations Management*, Dhaka, Bangladesh.

Nambiar, A. N. (2013). A Current Perspective of Knowledge Management in a Global Economy. *International Journal of Soft Computing and Software Engineering*, *3*(3), 236–239.

Nambiar, A. N., Judd, R. P., & Greenblatt, J. (2009c). Data Exchange among different modules in a manufacturing or service framework. *International Journal of Product Development*, *8*(2), 122–140. doi:10.1504/IJPD.2009.024184

Naoum, S. (2003). An overview into the concept of partnering. *International Journal of Project Management*, *21*(1), 71–76. doi:10.1016/S0263-7863(01)00059-X

Nath, A. K., Saha, P., & Salehi-Sangari, E. (2007). Transforming Supply Chains in Digital Content Delivery: A Case Study in Apple. In L. Xu, A. Tjoa, S. Chaudhry (Eds.), Research and Practical Issues of Enterprise Information Systems II (Vol. 2, pp. 1079-1089). Boston: Springer.

Naylor, B. J., Naim, M. M., & Berry, D. (1999). Leagility: Integrating the lean and agile manufacturing paradigms in the total supply chain. *International Journal of Production Economics*, *62*(1-2), 107–118. doi:10.1016/S0925-5273(98)00223-0

Negro, S. O., & Hekkert, M. P. (2006). Benchmarking the Functions of Innovation Systems - The Success Story of the German Biomass Digestion Innovation System. *SPRU 40th Anniversary*. Retrieved from http://www2.unil.ch/easst2006/Papers/N/Negro%20and%20Hekkert%20Paper%20EASST%202006.pdf

Negro, S. O. (2007). *Dynamics of Technological Innovation Systems - The case of biomass energy*. Innovation Studies.

Negro, S. O., & Hekkert, M. P. (2008). Explaining the success of emerging technologies by innovation system functioning: The case of biomass digestion in Germany. *Technology Analysis and Strategic Management*, *20*(4), 465–482. doi:10.1080/09537320802141437

Negro, S. O., Hekkert, M. P., & Smits, R. E. (2007). Explaining the failure of the Dutch innovation system for biomass digestion--A functional analysis. *Energy Policy*, *35*(2), 925–938. doi:10.1016/j.enpol.2006.01.027

Negro, S. O., Hekkert, M. P., & Smits, R. E. H. M. (2008a). Stimulating renewable energy technologies by innovation policy. *Science & Public Policy*, *35*(6), 403–416. doi:10.3152/030234208X323334

Negro, S. O., Suurs, R. A., & Hekkert, M. P. (2008b). The bumpy road of biomass gasification in the Netherlands: Explaining the rise & fall of an emerging innovation system. *Technological Forecasting and Social Change*, *75*(1), 57–77. doi:10.1016/j.techfore.2006.08.006

Nelson, R. R. (Ed.). (1993). *National Innovation Systems*. New York: Oxford University Press.

Nelson, R. R., & Winter, S. G. (1982). *An evolutionary theory of economic change*. Cambridge: Belknap Press.

Newman, M. E. J. (2003). The structure and function of complex networks. *SIAM Review*, *45*(2), 167–256. Retrieved from[REMOVED HYPERLINK FIELD] http://arxiv.org/abs/cond-mat/0303516v1 doi:10.1137/S003614450342480

Ni, G., Wang, W., Wang, J., Zong, Z., & Xie, M. (2010). Research on the Knowledge Management System of the Vicarious Management Corporation. *Proceedings of the 2010 International Conference on Information Science and Management Engineering* (pp. 62–67). doi:10.1109/ISME.2010.128

Nykiel, R. A. (2007). *Handbook of Marketing Research Methodologies for Hospitality and Tourism*. Routledge.

OECD. (1998). *Open Markets Matter. The Benefits of Trade and Investment Liberalisation*. OECD Pub.

Olhager, J. (2003). Strategic positioning of the order penetration point. *International Journal of Production Economics*, *85*(3), 319–329. doi:10.1016/S0925-5273(03)00119-1

Oshri, I., Kotlarsky, J., & Willcocks, L. P. (2015). *The Handbook of Global Outsourcing and Offshoring: The Definitive Guide to Strategy and Operations* (3rd ed.). UK: Palgrave-Macmillan. doi:10.1057/9781137437440

Otte, E., & Roussaeu, R. (2002). Social network analysis: A powerful strategy, also for the information sciences. *Journal of Information Science*, *28*(6), 441–453. doi:10.1177/016555150202800601

Padmanabhan, V., & Png, I. P. L. (1997). Manufacturer's Returns Policies and Retail Competition. *Marketing Science*, *16*(1), 81–94. doi:10.1287/mksc.16.1.81

Pala, M., Edum-Fotwe, F., Ruikar, K., Doughty, N., & Peters, C. (2014). Contractor practices for managing extended supply chain tiers. *Supply Chain Management: An International Journal*, *19*(1), 31–45.

Pandit, A., & Zhu, Y. (2009). An Ontology-based Approach to support Decision-making for the design of ETO (Engineer-to-order) products. *Automation in Construction*, *16*(6), 759–770. doi:10.1016/j.autcon.2007.02.003

Pan, F., & Nagi, R. (2013). Multi-echelon supply chain network design in agile manufacturing. *Omega*, *41*(6), 969–983. doi:10.1016/j.omega.2012.12.004

Parise, S., & Casher, A. (2003). Alliance portfolios: Designing and managing your network of business-partner relationships. *The Academy of Management Perspectives*, *17*(4), 25–39.

Pattanayak, B. (2005). *Human Resources Management*. New Delhi, India: Prentice Hall.

Paulson, U. (2005). Developing a supply chain flow risk model.*Proceedings of the NOFOMA 2005 Conference*, Copenhagen, Denmark (pp. 9-10).

Payne, A., & Frow, P. (2013). *Strategic Customer Management. Integrating Relationship Marketing and CRM*. UK: Cambridge University Press. doi:10.1017/CBO9781139057417

Peck, H. (2006). Reconciling supply chain vulnerability, risk and supply chain management. *International Journal of Logistics Research and Applications*, *9*(2), 127–142. doi:10.1080/13675560600673578

Peitz, M., & Waelbroeck, P. (2003). Piracy of Digital Products: A critical Review of the Economics Literature. *CESifo Working Paper, No. 1071.*

Penteado, C. S. G., de Carvalho, E. V., & Lintz, R. C. C. (2015). Reusing ceramic tile polishing waste in paving block manufacturing. *Journal of Cleaner Production*. doi:10.1016/j.jclepro.2015.06.142

Peteraf, M. A. (1993). The cornerstones of competitive advantage: A resource-based view. *Strategic Management Journal*, *14*(3), 179–192. doi:10.1002/smj.4250140303

Peterson, J. (1981). *Petri Net Theory and the Modeling of Systems*. Englewood Cliffs, NJ: Prentice-Hall International.

Petri, C. A. (1962). *Kommunikation mit Automaten [Communication with automata]*. Schriften des Rheinisch-Westfalischen Institutes fur Instrumentelle Mathematik an der Universitat Bonn, Nr.2, (PhD dissertation, Darmstadt University of Technology, 1961.7).

Pham, D. T., Pham, P. T. N., & Thomas, A. (2008). Integrated production machines and systems - beyond lean manufacturing. *Journal of Manufacturing Technology Management*, *19*(6), 695–711. doi:10.1108/17410380810888094

Phillips, J. J., & Edward, S. L. (2008). *Managing Talent Retention. An ROI Approach*. JohnWiley&Sons.

Piller, F. (2004). Mass customisation: Reflections on the state of the concept. *International Journal of Flexible Manufacturing Systems*, *16*(4), 313–334. doi:10.1007/s10696-005-5170-x

Pirkul, H., & Jayaraman, V. (1998). A multi-commodity, multi-plant, capacitated facility location problem: Formulation and efficient heuristic solution. *Computers & Operations Research*, *25*(10), 869–878. doi:10.1016/S0305-0548(97)00096-8

Porter, K., Little, D., Peck, M., & Rollins, R. (1999). Manufacturing Classifications: Relationships with production control systems. *Integrated Manufacturing Systems*, *10*(4), 189–199. doi:10.1108/09576069910280431

Porter, M. E. (1985). *Competitive Advantage*. New York: The Free Press.

Porter, M. E. (1996). What is Strategy? *Harvard Business Review*, *74*(6), 61–78. PMID:10158474

Porter, M. E. (1998). Clusters & the new economics of competition. *Harvard Business Review*, *76*(6), 77–90. PMID:10187248

Porter, M. E., & Stern, S. (2001). Innovation: Location matters. *MIT Sloan Management Review*, *42*(4), 28–36.

Powell, W. W. (2001). Competitive advantage: Logical and philosophical considerations. *Strategic Management Journal*, *22*(9), 875–888. doi:10.1002/smj.173

Puttonen, J., Lobov, A., & Martinez Lastra, J. L. (2013). Semantics-based composition of factory automation processes encapsulated by web services. *IEEE Transactions on Industrial Informatics*, *9*(4), 2349–2359. doi:10.1109/TII.2012.2220554

Puvanasvaran, P., Megat, H., Hong, T. S., & Razali, M. M. (2009). The roles of communication process for an effective lean manufacturing implementation. *Journal of Industrial Engineering and Management*, *2*(1), 128–152. doi:10.3926/jiem.2009.v2n1.p128-152

Pyke, F., & Sengenberger, W. (Eds.), (1992). Industrial districts & local economic regeneration. International Labour Organisation.

Qrunfleh, S., & Tarafdar, M. (2013). Lean and agile supply chain strategies and supply chain responsiveness: The role of strategic supplier partnership and postponement. *Supply Chain Management: An International Journal*, *18*(6), 571–582. doi:10.1108/SCM-01-2013-0015

Ramsay, J. (2005). The real meaning of value in trading relationships. *International Journal of Operations & Production Management*, *25*(6), 549–565. doi:10.1108/01443570510599719

Rankin, N. (2009). *IRS best practice in HR handbook*. Routledge.

Rayward-Smith, V. J., Osman, I. H., Reeves, C. R., & Smith, G. D. (1996). *Modern heuristic search methods*. John Wiley &Sons Ltd.

Report on the World Commission on Environment and Development: Our Common Future. (1987). *UN*. Retrieved from http://www.un-documents.net/our-common-future.pdf

Reyes, P., & Raisinghani, M. S. (2002). Integrating Information Technologies and Knowledge-based Systems: A Theoretical Approach in Action for Enhancements in Production and Inventory Control. *Knowledge and Process Management*, *9*(4), 256–263. doi:10.1002/kpm.153

Roach, G. M. (2003). *The Product Design Generator - A Next Generation Approach to Detailed Design*. Provo, UT: Brigham Young University.

Roberts, T. L., Cheney, P. H., & Sweeney, P. D. (2002). Project Characteristics and Group Communication: An Investigation. *IEEE Transactions on Professional Communication*, *45*(2), 84–98. doi:10.1109/TPC.2002.1003690

Robson, C. (2011). *Real world research: a resource for users of social research methods in applied settings*. Chichester: Wiley.

Rogers, E. M. (1995). Diffusion of innovations (4 ed.). New York: The Free Press.

Rogers, E. M. (2010). *Diffusion of innovations*. Simon and Schuster.

Rohrbeck, R., Konnertz, L., & Knab, S. (2013). Collaborative business modelling for systemic and sustainability innovations. *International Journal of Technology Management*, *63*(1/2), 4–23. doi:10.1504/IJTM.2013.055577

Rosenberg, N. (1982). *Inside the Black Box: Technology & Economics*. Cambridge: Cambridge University Press.

Rosová, A. (2010). Sústava ukazovateľov distribučnej logistiky, logistiky dopravy a materiálové hotoku ako jeden z nástrojov controllingu v logistike podniku. *Acta Montanistica Slovaca, 15*(1), 67–72.

Ross, D. F. (2002). *Introduction to e-Supply Chain Management: Engaging Technology to Build Market-Winning Business Partnerships.* NY: CRC Press. doi:10.1201/9781420025415

Rowlinson, S., & Cheung, F. (2002). *Review of the concepts and definitions of the various forms of relational contracting (2002-022-A).* Brisbane, Australia: CRC for Construction Innovation.

Roy, S., Sivakumar, K., & Wilkinson, I.F. (2004). Innovation generation in supply chain relationships: a conceptual model and research propositions. *Journal of the Academy of Marketing Supply chainience, 32*(1), 61-79.

Rusinko, C. A. (2008). Towards more sustainable management systems: Through life cycle management and integration. *Journal of Cleaner Production, 16*(10), 1071–1080. doi:10.1016/j.jclepro.2007.06.006

Russell, R. S., & Taylor, B. W. (2003). *Operations Management* (4th ed.). New Jersey, USA: Prentice and Hall.

Russell, R., Chiang, W.-C., & Zepeda, D. (2008). Integrating multi-product production and distribution in newspaper logistics. *Computers & Operations Research, 35*(5), 1576–1588. doi:10.1016/j.cor.2006.09.002

Russo, M. (1996). *Cambiamento tecnico e relazioni tra imprese. Il distretto cerámico di Sassuolo.* Turin: Rosenberg y Sellier.

Sabri, E., & Beamon, B. M. (2000). A Multi-Objective Approach to Simultaneous Strategic and Operational Planning in Supply Chain Design. *Omega, 28*(5), 581–598. doi:10.1016/S0305-0483(99)00080-8

Saemann, B. (2009). *Motivation in Second Language Acquisition.* Auflage.

Satyendra, R., & Akay, V. (2006). *Business Is the People and People Are the Business. Humanizing the Corporation.* Pine Lake Road, USA: iUniverse.

Schachtebeck, T. (2005). *Internet chat communication. A tightrope walk between oral communication and written communication. Seminar paper.* Berlin: Grin.

Schmenner, R. W. (1990). *Production / Operations Management: Concepts and Situations.* New York: Macmillan Publishing Company.

Schmitz, H. (2004). Globalized localities: introduction. In H. Schmitz (Ed.), *Local enterprises in the global economy: issues of governance and upgrading* (pp. 1–19). Cheltenham, United Kingdom: Edward Elgar. doi:10.4337/9781843769743.00008

Schonsleben, P. (2000). With agility and adequate partnership strategies towards effective logistics networks. *Computers in Industry, 42*(1), 33–42. doi:10.1016/S0166-3615(99)00059-7

Schroder, R. (1989). *Operational management.* McGraw Hill, New York.

Schultz, D. E., & Schultz, H. (2004). *The Next Generation. Five Steps for Delivering Value and Measuring Returns Using Marketing Communication* (pp. 49–66). McGraw Hill.

Scott, J. (2000). *Social Network Analysis: A Handbook* (2nd ed.). London: Sage Publications.

Sears, W. H. (2007). *The Front Line Guide to Communicating with Employees.* USA: HRD Press.

Seidmann, A., & Sundararajan (1997). Building and sustaining inter-organizational information sharing relationships: the competitive impact of interfacing supply chain operations with marketing strategy. *Proceedings of the 18th International Conference on Information Systems (ICIS-97)*, Atlanta (pp. 205-222).

Seifoddini, H. (1989). A note on the similarity coefficient method and the problem of improper machine assignment in group technology applications. *International Journal of Production Research*, *27*(7), 1161–1165. doi:10.1080/00207548908942614

Seifoddini, H. (1989). Duplication process in machine cells formation in group technology. *IIE Transactions*, *21*(4), 382–388. doi:10.1080/07408178908966245

Seifoddini, H. K. (1989). Single linkage versus average linkage clustering in machine cells formation applications. *Computers & Industrial Engineering*, *16*(3), 419–426. doi:10.1016/0360-8352(89)90160-5

Selim, H. M., Askin, R. G., & Vakharia, A. J. (1998). Cell formation in group technology: Review, evaluation and directions for future research. *Computers & Industrial Engineering*, *34*(1), 3–20. doi:10.1016/S0360-8352(97)00147-2

Serral, E., Smedt, J. D., Snoeck, M., & Vanthienen, J. (2015). Context adaptive Petri nets: Supporting adaptation for the execution context. *Expert Systems with Applications*, *42*(23), 9307–9317. doi:10.1016/j.eswa.2015.08.004

Shamsuzzoha, A., Kankaanpaa, T., Carneiro, L. M., Almeida, R., Chiodi, A., & Fornasiero, R. (2013). Dynamic and collaborative business networks in the fashion industry. *International Journal of Computer Integrated Manufacturing*, *26*(1-2), 125–139. doi:10.1080/0951192X.2012.681916

Shapiro, C., & Varian, H. (1999). *Information Rules*. Boston: Harvard Business School Press.

Sharma, C. (2011). Determinants of PPP in infrastructure in developing economies. *Transforming Government: People,Process and Policy*, *6*(2), 149–166.

Shaughnessy, J., Zechmeister, E., & Jeanne, Z. (2011). *Research methods in psychology* (9th ed.). New York, NY: Mc-Graw Hill.

Shewchuck, J. P. (1998). Agile manufacturing: one size does not fit all.*Proc. Int. Conference Manufacturing Value Chains*, Troon (pp. 143-150). doi:10.1007/978-0-387-35321-0_16

Shin, Y., An, S.-H., Cho, H.-H., Kim, G.-H., & Kang, K.-I. (2008). Application of Information Technology for Mass Customization in the housing construction industry in Korea. *Automation in Construction*, *17*(7), 831–838. doi:10.1016/j.autcon.2008.02.010

Shtubt, A. (1989). Modelling group technology cell formation as a generalized assignment problem. *International Journal of Production Research*, *27*(5), 775–782. doi:10.1080/00207548908942586

Silva, M., Teruel, E., & Colom, J. M. (1996). *Linear algebraic and linear programming techniques for the analysis of net systems. Lectures on Petri Nets,* LNCS (Vol. 1491, pp. 309–373). Springer.

Simatupang, T. M., & Sridharan, R. (2004). Benchmarking supply chain collaboration: An empirical study. *Benchmarking, International Journal (Toronto, Ont.)*, *11*, 484–503.

Simchi-Levi, D. (2009). *Designing and managing the supply chain concepts strategies and case studies*. Tata McGraw-Hill Education.

Simchi-Levi, D., Kaminsky, P., & Simchi-Levi, E. (2001). *Designing and managing the supply chain: Concepts, strategies, and cases*. New York: McGraw-Hill.

Simpson, L., Robinson, P., Fletcher, M., & Wilson, R. (2005). *E-Communication Skills*. Oxon, UK: Radcliffe Pub.Ltd.

Sims, R. R. (2002). *Organizational Success through Efective Human Resources Managemen*. USA: Greenwood Pub. Inc.Group.

Skorin-Kapov, D., & Skorin-Kapov, J. (1994). On tabu search for the location of interacting hub facilities. *European Journal of Operational Research*, *73*(3), 502–509. doi:10.1016/0377-2217(94)90245-3

Smith, P., & Goddard, M. (2002). Performance Management and Operational Research: A Marriage made in Heaven? *The Journal of the Operational Research Society*, *53*(3), 247–255. doi:10.1057/palgrave.jors.2601279

Solimanpur, M., Vrat, P., & Shankar, R. (2004). Ant colony optimization algorithm to the inter-cell layout problem in cellular manufacturing. *European Journal of Operational Research*, *157*(3), 592–606. doi:10.1016/S0377-2217(03)00248-0

Soni, G., & Kodali, R. (2012). A critical review of empirical research methodology in supply chain. *Journal of Manufacturing Technology Management*, *23*(6), 753–779. doi:10.1108/17410381211253326

So, S., & Sun, H. (2010). Supplier integration strategy for lean manufacturing adoption in electronic-enabled supply chains. *Supply Chain Management: An International Journal*, *15*(6), 474–487. doi:10.1108/13598541011080455

Sparrow, P., Brewster, C., & Harris, H. (2004). *Globalizing HRM*. London: Routledge.

Stadler, H. (2015). Supply chain management: an overview. In H. Stadler, C. Kilger, & H. Meyr (Ed.), Supply Chain Management and Advanced Planning (pp. 3-28). Springer Berlin Heidelberg. doi:10.1007/978-3-642-55309-7_1

Stegaroiu, I. & Florea, N.V., (2013). *Tehnici şi instrumente de recrutare şi selecţie*. Bucuresti: C.H.Beck.

Stock, J., & Boyer, S. (2009). Developing a consensus definition of supply chain management: A qualitative study. *International Journal of Physical Distribution & Logistics Management*, *39*(8), 690–711. doi:10.1108/09600030910996323

Subramani, M. (2004). How do suppliers benefit from information technology use in supply chain relationships? *Management Information Systems Quarterly*, *28*(1), 45–73.

Suda, H. (1989). Future factory system formulated in Japan. *Techno Japan*, *22*(10), 15–25.

Süer, G. A., & Mese, E. M. (2011). Cell Loading and Family Scheduling for Jobs with Individual Dates. In V. Modrak, & R. Sudhakara Pandian (Eds.), Operations Management Research and Cellular Manufacturing, Innovative Methods and Approaches. Hershey, PA, USA: IGI Global.

Süer, G. A. (1996). Optimal Operator Assignment and Cell Loading in Labor Intensive Manufacturing Cells. *Computers & Industrial Engineering*, *31*(1/2), 155–158. doi:10.1016/0360-8352(96)00101-5

Süer, G. A. (1999). Cell Loading in Connected Cells.*Proceedings of the 26th international conference on computers and industrial engineering,*New Orleans (pp. 438-441).

Süer, G. A., Arikan, F., & Babayiğit, C. (2008). Bi-objective cell loading problem with non-zero setup times with fuzzy aspiration levels in labour intensive manufacturing cells. *International Journal of Production Research*, *46*(2), 371–404. doi:10.1080/00207540601138460

Süer, G. A., Arikan, F., & Babayigit, C. (2009). Effects of different fuzzy operators on fuzzy bi-objective cell loading problem in labor-intensive manufacturing cells. *Computers & Industrial Engineering*, *56*(2), 476–488.

Süer, G. A., Ates, O. K., & Mese, E. M. (2014). Cell loading and family scheduling for jobs with individual due dates to minimise maximum tardiness. *International Journal of Production Research.*

Süer, G. A., Cosner, J., & Patten, A. (2009). Models for cell loading and product sequencing in labor-intensive cells. [Elsevier Ltd.]. *Computers & Industrial Engineering*, *56*(1), 97–105. doi:10.1016/j.cie.2008.04.002

Süer, G. A., Huang, J., & Maddisetty, S. (2010). Design of dedicated, shared and remainder cells in a probabilistic demand environment. *International Journal of Production Research*, *48*(19), 5613–5646. doi:10.1080/00207540903117865

Süer, G. A., & Ortega, M. (1996). *Flexibility Considerations in Designing Manufacturing Cells: A Case Study. Group Technology and Cellular Manufacturing - Methodologies and Applications* (pp. 97–127). Amsterdam, The Netherlands: Gordon and Branch Science.

Süer, G. A., Pena, Y., & Vazquez, R. (2003). Cellular design to minimise investment costs by using evolutionary programming. *International Journal of Manufacturing Technology and Management*, *5*(1/2), 118–132. doi:10.1504/IJMTM.2003.002536

Süer, G. A., & Saiz, M. (1993). Cell loading in cellular manufacturing systems. *Computers & Industrial Engineering*, *25*(1-4), 247–250. doi:10.1016/0360-8352(93)90267-2

Süer, G. A., Saiz, M., Dagli, C., & Gonzalez, W. (1995). Manufacturing cell loading rules and algorithms for connected cells. *Manufacturing Research and Technology*, *24*, 97–127. doi:10.1016/S1572-4417(06)80038-0

Süer, G. A., Saiz, M., Dagli, C., & Gonzalez, W. (1995). *Planning, Design, and Analysis of Cellular Manufacturing Systems* (Vol. 24). Elsevier.

Süer, G. A., Santos, J., & Vazquez, R. (1999). Scheduling rotary injection molding machines.*Proceedings of the Second Asia-Pacific Conference on Industrial Engineering,*Kanazawa, Japan (pp. 319-322).

Süer, G. A., Subramanian, A., & Huang, J. (2009). Heuristic procedures and mathematical models for cell loading and scheduling in a shoe manufacturing company. *Computers & Industrial Engineering*, *56*(2), 462–475. doi:10.1016/j.cie.2008.10.008

Süer, G., & Bera, I. S. (1998). Multi-period cell loading and cell size determination. *Computers & Industrial Engineering*, *35*(1-2), 85–88. doi:10.1016/S0360-8352(98)00026-6

Süer, G., & Dagli, C. (2005). Intra-cell manpower transfers and cell loading in labor-intensive manufacturing cells. *Computers & Industrial Engineering*, *48*(3), 643–655. doi:10.1016/j.cie.2003.03.006

Süer, G., Pico, F., & Santiago, A. (1997). Identical machine scheduling to minimize the number of tardy jobs when lot-splitting is allowed. *Computers & Industrial Engineering*, *33*(1-2), 277–280. doi:10.1016/S0360-8352(97)00092-2

Süer, G., Saiz, M., & Gonzalez, W. (1999a). Evaluation of manufacturing cell loading rules for independent cells. *International Journal of Production Research*, *37*(15), 3445–3468. doi:10.1080/002075499190130

Süer, G., Vazquez, R., & Cortes, M. (2005). A hybrid approach of genetic algorithms and local optimizers in cell loading. *Computers & Industrial Engineering*, *48*(3), 625–641. doi:10.1016/j.cie.2003.03.005

Sutton, S. G., Khazanchi, D., Hampton, C., & Arnold, V. (2008). Risk analysis in extended enterprise environments: Identification of critical risk factors in B2B e-commerce relationships. *Journal of the Association for Information Systems, 9*(3-4), 151–174.

Suurs, R. A. A., & Hekkert, M. P. (2007). *Cumulative Causation in the Formation of a Technological Innovation System: Biofuels Development in the Netherlands*. Working Paper /Innovation Studies Utrecht (ISU) /working paper series.

Suurs, R. A. A. (2009). *Motors of sustainable innovation - Towards a theory on the dynamics of technological innovation sytems*. Department of Innovation & Environmental Sciences.

Svetlik, I., & Ilic, B. (2006). *HRM's Contribution to Hard Work*. Bern: European Acd.Pub.

Swafford, P., Gosh, S., & Murthy, N. (2006). A framework for assessing value chain agility. *International Journal of Operations & Production Management, 26*(2), 118–140. doi:10.1108/01443570610641639

Sweeney, E. (2007). Introduction. In Perspectives on Supply Chain Management and Logistics – Creating Competitive Organisations in The 21st Century. Dublin: Blackhall Publishers.

Takala, J., Malindžák, D., & Straka, M. et al.. (2007). *Manufacturing Strategy – Applying the Logistics Models* (p. 206). Vaasan Yliopisto.

Tanasescu, D. (2008). *Relaţii publice şi de protocol în organizaţii*. Târgovişte: Editura Bibliotheca.

Tanasescu, D., Lupu, M.-M., & Petrescu, M. (2014). *Comunicarea integrata*. Targoviste: Editura Bibliotheca.

Tang, A., & Lim, S. (2010). *Retail operations. How to run your own store*. Jurong, Singapore: Prentice Hall.

Tang, O., & Nurmaya Musa, S. (2011). Identifying risk issues and research advancements in supply chain risk management. *International Journal of Production Economics, 133*(1), 25–34. doi:10.1016/j.ijpe.2010.06.013

Tao, F., Laili, Y. J., Xu, L. D., & Zhang, L. (2013). FC-PACO-RM: A parallel method for service composition optimal-selection in cloud manufacturing system. *IEEE Transactions on Industrial Informatics, 9*(4), 2023–2033. doi:10.1109/TII.2012.2232936

Tavakkoli-Moghaddam, R., Ranjbar-Bourani, M., Amin, G. R., & Siadat, A. (2012). A cell formation problem considering machine utilization and alternative process routes by scatter search. *Journal of Intelligent Manufacturing, 23*(4), 1127–1139. doi:10.1007/s10845-010-0395-2

Tavakol, M., & Dennick, R. (2011). Making sense of cronbach's alpha. *International Journal of Medical Education, 2*, 53–55. doi:10.5116/ijme.4dfb.8dfd

Thun, J.-H., & Hoenig, D. (2011). An empirical analysis of supply chain risk management in the German automotive industry. *International Journal of Production Economics, 131*(1), 242–249. doi:10.1016/j.ijpe.2009.10.010

Tidd, J., Bessant, J., & Pavitt, K. (2005). *Managing innovation: integrating technological, market and organizational change*. Chichester: Wiley.

Tiernan, S. D., Flood, P. C., Murphy, E. P., & Carroll, S. J. (2002). Employee Reactions to Flattening Organizational Structures. *European Journal of Work and Organizational Psychology, 11*(1), 47–67. doi:10.1080/13594320143000843

Tortajada Esparza, E., Fernández de Lucio, I., & Gabaldón Estevan, D. (2008a). Competitiveness and Profitability – New Challenges in the Ceramic Frit. *Glaze and Colour Industry, Part 1 Interceram, 57*(6), 436–439.

Tortajada Esparza, E., Fernández de Lucio, I., & Gabaldón Estevan, D. (2009). Competitiveness and Profitability – New Challenges in the Ceramic Frit. *Glaze and Colour Industry Part 2 Interceram*, *58*(1), 36–39.

Tortajada Esparza, E., Gabaldón Estevan, D., & Fernández de Lucio, I. (2008b). La evolución tecnológica del distrito cerámico de Castellón: La contribución de la industria de fritas, colores y esmaltes. *Boletín de la Sociedad Española de Cerámica y Vidrio*, *47*(2), 57–80. doi:10.3989/cyv.2008.v47.i2.198

Tortajada-Esparza, E., Fernández de Lucio, I., & Gabaldón-Estevan, D. (2008c) Competitividad y rentabilidad. Nuevos retos de la industria de fritas, colores y esmaltes cerámicos. At Qualicer 2008, X Congreso Mundial de la Calidad del Azulejo y del Pavimento Cerámico. Cámara Oficial de Comercio, Industria y Navegación de Castellón

Truss, C., Mankin, D., & Kelliher, C. (2012). *Strategic HRM*. Oxford: Oxford University Press.

Ulrich, D., & Brockbank, W. (2005). *The HR Value Proposition*. Harvard Business School Press.

Utterback, J. M. (1994). *Mastering the dynamics of innovation: how companies can seize opportunities in the face of technological change*. Boston, MA: Harvard Business Press.

Uzam, V., Koc, B., Gelen, G., & Haken Aksebzeci, B. (2005). Asynchronous implementation of a Petri net based discrete event control system using a Xilinx FPGA.*Proceedings of the 35th International Conference on Computers and Industrial Engineering* (pp. 2025-2030).

Valckenaers, P., Van Brussel, H., Bongaerts, L., & Wyns, J. (1997). Holonic manufacturing systems. *Integrated Computer Aided Engineering*, *4*, 191–201.

van den Ende, R. (2013). *A guide to an internship in Japan for industrial design engineers*. Lulu Enterprises Inc.

van der Aalst, W. M. P. (1998). The application of Petri nets to workflow management. *Journal of Circuits, Systems, and Computers*, *8*(1), 21–66. doi:10.1142/S0218126698000043

van der Aalst, W. M. P. (2013). Decomposing Petri nets for process mining: A generic approach. *Distributed and Parallel Databases*, *31*(4), 471–507. doi:10.1007/s10619-013-7127-5

Van Der Sluis, L., & Van De Bunt-Kokhuis S. (Eds.), (2009). *Competing for talent*. Assen.

Van Hoek, R. I. (2000). The thesis of leagility revisited. *International Journal of Agile Management Systems*, 2(3), 196–201.

van Hoek, R. I., Harrison, A., & Christopher, M. (2001). Measuring agile capabilities in the supply chain. *International Journal of Operations & Production Management*, *21*(1/2). doi:10.1108/01443570110358495

Varrichio, P., Diogenes, D., Jorge, A., & Garnica, L. (2012). Collaborative networks and sustainable business: A case study in the Brazilian system of innovation. *Procedia: Social and Behavioral Sciences*, *52*, 90–99. doi:10.1016/j.sbspro.2012.09.445

Verchopoulos, A. P. (2004). Mass Customization Challenges in Internet Retailing through Information Management. *International Journal of Information Management*, *24*(1), 59–71. doi:10.1016/j.ijinfomgt.2003.09.013

Veseth, M. (2002). *The Rise of the Global Economy*. Taylor&Francis.

Villani, E., Miyagi, P. E., & Valette, R. (2006). *Modeling and Analysis of Hybrid Supervisory System: A Petri Net Approach*. London, UK: Springer Verlag.

Viswanadham, N., & Srinivasa Raghavan, N. R. (2000). Performance analysis and design of supply chains: A Petri Net approach. *The Journal of the Operational Research Society*, *51*(10), 1158–1169. doi:10.1057/palgrave.jors.2600063

Vollman, T. E., Berry, W. L., Whybark, D. C., & Jacob, F. R. (2004). *Manufacturing Planning and Control Systems for Supply Chain Management*. New York: Irwin/McGraw Hill.

Voss, C., Tsikriktsis, N., & Frohlich, M. (2002). Case research in operations management. *International Journal of Operations & Production Management*, *22*(2), 195–219. doi:10.1108/01443570210414329

Vrijhoef, R., & Koskela, L. (2000). The four roles of supply chain management in construction. *European Journal of Purchasing & Supply Management*, *6*(3-4), 169–178. doi:10.1016/S0969-7012(00)00013-7

Wagner, S. M., & Bode, C. (2009). Dominant risks and risk management practices in supply chains. In G. Zsidisin, & B. Ritchie (Eds.), Supply Chain Risks: A Handbook of Assessment, Management, and Performance (Vol. 129, pp. 271-290). doi:10.1007/978-0-387-79934-6_17

Wagner, C., & Ryan, C. (2009). Innovation Cycle Management of Digital Supply Chains. In E. Sweeney (Ed.), *Supply Chain Management and Logistics in a Volatile Global Environment* (pp. 169–182). Dublin: Blackhall Publishers.

Wagner, S. M., & Bode, C. (2006). An empirical investigation into supply chain vulnerability. *Journal of Purchasing and Supply Management*, *12*(6), 301–312. doi:10.1016/j.pursup.2007.01.004

Walish, P., Michael, E. L. R., & Benyoun, M. D. (2014). *MATLAB for neuroscientists: an introduction to scientific computing in MATLAB*. Academic Press. Retrieved from www.internetworldstats.com

Walker, D., & Hampson, K. (Eds.). (2003). *Procurement Strategies: A Relationship-based Approach*. Oxford: Blackwell Science.

Walter, A., Ritter, T., & Gemunden, H. G. (2001). Value creation in buyer-seller relationships: Theoretical considerations and empirical results from a supplier's perspective. *Industrial Marketing Management*, *30*(4), 365–377. doi:10.1016/S0019-8501(01)00156-0

Walters, C. D. J. (2011). *Supply Chain Risk Management: Vulnerability and Resilience in Logistics* (2nd ed.). UK: Kogan Page Limited.

Wang, H.-L. (2014). Theories for competitive advantage. In H. Hasan (Ed.), Being Practical with Theory: A Window into Business Research (pp. 33–43). Wollongong, Australia: THEORI. Retrieved from http://eurekaconnection.files.wordpress.com/2014/02/p-33-43-theoriesof-competitive-advantage-theori-ebook_finaljan2014-v3.pdf

Wang, E. T. G., Tai, J. C. F., & Wei, H. L. (2006). A Virtual Integration Theory of Improved Supply-Chain Performance. *Journal of Management Information Systems*, *23*(2), 41–64. doi:10.2753/MIS0742-1222230203

Wang, H. L. (2004). A framework to support and understand strategic decision-making in business-to-business electronic commerce.*Proceedings of the International Workshop on Business and Information (BAI2004)*, Taipei.

Wang, X., Tang, J., Gong, J., & Chen, M. (2008). Formation of Multi-Objective Dynamic Cells Using Random Weight Multi-Objective Genetic Algorithm. *Chinese Journal of Management*, *5*(4), 516–521.

Wang, Z., Zhang, D., & Xu, X. (2006). Service-Oriented Infrastructure for Collaborative Product Design in ETO Enterprises. *Proceedings of the 10th International Conference on Computer Supported Cooperative Work in Design,Nanjing, China* (pp. 1–6). doi:10.1007/11686699_1

Wasserman, S., & Faust, K. (1994). *Social Network Analysis.* New York: Cambridge University Press. doi:10.1017/CBO9780511815478

Weick, K. E., & Sutcliffe, K. M. (2007). *Managing the Unexpected: Resilient Performance in an Age of Uncertainty.* San Francisco, USA: Jossey-Bass Publisher.

Wei, J. C., & Gaither, N. (1990). An optimal model for cell formation decisions. *Decision Sciences*, *21*(2), 416–433. doi:10.1111/j.1540-5915.1990.tb01694.x

Whinston, A. O. D., & Choi, S. (1997). *The Economics of Electronic Commerce.* New York: Macmillan Publishing Company.

Wikner, J., & Rudberg, M. (2001). On the customer order decoupling point (Working Paper No. WP-284). *Linkoeping Institute of Technology*, Sweden.

Wikner, J., & Rudberg, M. (2005). Integrating production and engineering perspectives on the customer order decoupling point. *International Journal of Operations & Production Management*, *25*(7), 623–641. doi:10.1108/01443570510605072

Wilburn, D., Salazar, K., & McNutt, M. (2011). Wind energy in the United States and materials required for the land-based wind turbine industry from 2010 through 2030. *U.S. Geological Survey Scientific Investigations Report* (2011–5036).

Williamson, O. E. (1979). Transaction cost economics: The governance of contractual relations. *The Journal of Law & Economics*, *22*(2), 233–261. doi:10.1086/466942

Wind Energy - The facts. The economics of Wind Power. (2009). *EWEA - European Wind Energy Association.*

Winner, R. J., Pennell, J. P., Bertend, H. E., & Slusarczuk, M. M. G. (1988). The Role of Concurrent Engineering in Weapon System Acquisition. *IDA Report R-338*, Institute for Defense Systems Analysis, Alexandra, VA, USA.

Wisniewski, R., Stefanowicz, L., Bukowiec, A., & Lipinski, J. (2014). Theoretical aspects of Petri nets decomposition based on invariants and hypergraphs. Multimedia and Ubiquitous Engineering, LNEE (Vol. 308, pp. 371-376).

Womack, J. P., & Jones, D. T. (1996). *Lean Thinking*. New York, NY: Simon & Schuster.

Womack, J. P., & Jones, D. T. (2003). *Lean Thinking: Banish Waste and Create Wealth in Your Corporation.* New York, NY: Simon & Schuster.

Womack, J. P., Jones, D. T., & Roos, D. (1999). *The Machine that Changed the World.* New York, NY: Rawson Associates.

Won, Y., & Chang Lee, K. (2004). Modified p-median approach for efficient GT cell formation. *Computers & Industrial Engineering*, *46*(3), 495–510. doi:10.1016/j.cie.2004.01.010

Wood, J. T. (2010). *Communication in Our Lives.* MA, USA: Cengage Learning.

Woolcock, M. (2001). The place of social capital in understanding social & economic outcomes'. *Isuma: Canadian Journal of Policy Research*, *2*(1), 11–17.

Xue, X., Shen, Q., & Ren, Z. (2010). Critical Review of Collaborative Working in Construction Projects: Business Environment and Human Behaviors. *Journal of Management Engineering*, *26*(4), 196–208. doi:10.1061/(ASCE)ME.1943-5479.0000025

Yang, J. (2010). The Knowledge Management Strategy and its Effect on Firm Performance: A Contingency Analysis. *International Journal of Production Economics*, *125*(2), 215–223. doi:10.1016/j.ijpe.2010.03.012

Yasuda, G., & Mori, M. (1974, February). Future concepts on grouping robots. *Presented at the Meeting on Neuro-Sciences,* Tokyo, Japan.

Yasuda, G. (2002). Hierarchical and distributed control of discrete event robotic manufacturing processes by extended Petri Nets. *Proceedings of the 4th Asia-Pacific Conference on Industrial Engineering and Management Systems* (pp. 709-712).

Yasuda, G. (2011). Petri net model based design and control of robotic manufacturing cells. In V. Modrak & R. Sudhakara Pandian (Eds.), *Operations Management Research and Cellular Manufacturing Systems: Innovative Methods and Approaches* (pp. 385–400). Hershey, PA, USA: IGI Global.

Yasuda, G., Takai, H., & Tachibana, K. (1996). Petri net based implementation and performance evaluation of parallel control architectures for distributed autonomous mechatronic systems.*Proceedings of the 3rd France-Japan Congress & 1st Europe-Asia Congress on Mechatronics* (pp. 99-104).

Yasuda, G., Takai, H., & Tachibana, K. (1997). Implementation and performance evaluation of parallel processing architectures for distributed autonomous motion control systems.*Proceedings of the 2nd Asian Control Conference* (Vol. 2, pp. 823-826).

Yew Wong, C., Stentoft Arlbjørn, J., & Johansen, J. (2005). Supply chain management practices in toy supply chains. *Supply Chain Management: An International Journal*, *10*(5), 367–378. doi:10.1108/13598540510624197

Yu, W., Yan, C., Ding, Z., Jiang, C., & Zhou, M. (2014). Modeling and validating E-commerce business process based on Petri nets. *IEEE Transactions on Systems, Man, and Cybernetics*, *44*(3), 327–341. doi:10.1109/TSMC.2013.2248358

Zamei, E., Subias, A., Combacau, M., & De Bonneval, A. (1997). A hierarchical structure for control of discrete event systems and monitoring of process failures. *Studies in Information and Control*, 6(1).

Zegordi, S. H., & Davarzani, H. (2012). Developing a supply chain disruption analysis model: Application of colored Petri nets. *Expert Systems with Applications*, *39*(2), 2102–2111. doi:10.1016/j.eswa.2011.07.137

Zeng, Q., Lu, F., Liu, C., Duan, H., & Zhou, C. (2014). Modeling and verification for cross-department collaborative business processes using extended Petri nets. *IEEE Transactions on Systems, Man, and Cybernetics*, *45*(2), 349–362. doi:10.1109/TSMC.2014.2334276

Zhang, J., Chung, J. Y., Chang, C. K., & Kim, S. (2004). WS-Net: a Petri net based specification model for web services. *Proceedings of the IEEE International Conference on Web Services* (pp. 420-427). doi:10.1109/ICWS.2004.1314766

Zhou, H., & Benton, W. C. Jr. (2007). Supply chain practice and information sharing. *Journal of Operations Management*, *25*(6), 1348–1365. doi:10.1016/j.jom.2007.01.009

Zhou, J., Shu, Z., Tiantian, L. I., Dongxue, Y. U., Sheng, Z., & Wang, Y. (2015). Novel fabrication route for non-fired ceramic tiles only using gypsum. *Ceramics International*, *41*(7), 9193–9198.

Zhou, M. C., & Venkatesh, K. (1998). *Modeling, Simulation and Control of Flexible Manufacturing Systems. A Petri Net Approach*. Singapore: World Scientific.

Zimmermann, J. J., & Sorergein, M. O. (1974). Quantitation models for Production Management. PH, New Jersey.

Zorzini, M., Hendry, L., Stevenson, M., & Pozzetti, A. (2008). Customer Inquiry Management and Product Customization: An Empirical multi-case study analysis in the Italian capital goods sector. *International Journal of Operations & Production Management*, *28*(12), 1186–1218. doi:10.1108/01443570810919369

About the Contributors

Richard Addo-Tenkorang is a postdoctoral research fellow (docent) with the Industrial Engineering and Management Unit – Department of Production; University of Vaasa, Finland. He was recently offered an Industrial and Manufacturing Engineering (IME) Lectureship position with the Engineering and Technology College, Botswana International University of Science and Technology (BIUST). His research interests are in the area of ERP – SCM – IT system-solutions and Concurrent Engineering for New/Complex Product Development, as well as Logistics and Supply-Chain Management (L&SCM) projects.

Petri T Helo is a Professor and the Head of Logistics Systems Research Group, Department of Production, University of Vaasa. His research addresses the management of Logistics processes in supply demand networks, which take place in electronics, machine building and food industries.

Jussi I Kantola is a Professor and Head of the Department of Production at University of Vaasa. His research interests include product and service design and development, knowledge management and fuzzy systems.

Ahm Shamsuzzoha has been working as an Assistant Professor, Department of Mechanical and Industrial Engineering, Sultan Qaboos University, Muscat, Sultanate of Oman. He is also working as a Visiting Researcher, Department of Production, University of Vaasa, Finland. He received his PhD in Industrial Management from the University of Vaasa, Finland in 2010. He received his Master of Science (Mechanical Engineering) degree from the University of Strathclyde, Glasgow, UK. His major research and teaching interests lie in the areas of enterprise collaborative networks, operations management, product customization, simulation modelling and supply chain management. He has published several research papers both in reputed international journals and conferences.

***

Clinton Aigbavboa holds a Ph.D. degree in Engineering Management and a certificate in Good Governance in Africa from the Thabo Mbeki Africa Leadership Institute in the University of South Africa (UNISA). His field of expertise and research specialisation is sustainable human development, with the focus on: sustainable housing regeneration (urban renewal and informal housing), Life Cycle Assessment in the Construction Industry, leadership in low-income housing, post occupancy evaluation and green job creation. His research draws concurrently and are applicable to construction industry development via the development of emerging contractors, the economics of infrastructure development, and the adaptation of building information modelling as a solution to the eradication/minimisation of urban and rural housing shortages. During the past five years, he has authored/co-authored more than 20 accredited journal articles, 150 peer-reviewed conference papers and three book chapters in the specialised area as stated above. He is actively involved in postgraduate students' supervision. He is currently the Associate Editor of the Journal of Construction Project Management and Innovation.

Tănăsescu Irina Antoaneta, lecturer in linguistics, Ph.D degrees (2011) from the Academy of Rouen, France and the University of Bucharest, Romania. Research fields: economic communication, linguistics, pragmatics, media discourse analysis and foreign languages teaching methodology. Member of CED-IMES (Centre des Etudes Internationaux et des Mouvements Economiques et Sociaux). Representative publications: Study on the forms and causes of violence in secondary schools (grades 5-8) in Romania, D. Tănăsescu, D. Popescu, I. Ionescu, G. Popa, in Proceedings of 4th WSEAS International Conference of Business Administration, Universitz of Cambridge, Great Britain, 2010, ISBN: 978-960-474-161-8, ISSN: 1790-5109, p.154-158; Directions prioritaires pour l'adaptation du système éducationnel préuniversitaire aux besoins du marché de travail roumain dans le contexte européen (coauteur), Colloque international "Les ruptures dans la société de l'information et l'économie du savoir," 15-17 février 2007, Université de Tunis El Manar, Tunisie. Linguistic diversity and communication, integration means and amortization in the European Union, Proceedings of the 14th International Business Information Management Association Conference, June 23-24, Istanbul, Turkey, ISBN: 978-0-9821489- 3-8, pp. 3125-3130.

Andrea Bikfalvi is teaching and is a research staff member in the Department of Business Administration at the University of Girona. She holds a PhD in Business Administration. She is currently the deputy director of the Chair of Family Business at the same university. During her career she has conducted several research projects for the regional government of Catalonia, Spain, as well as a series of EU projects. She has visited and actively collaborates with researchers in relevant research and Higher Education Institutions, such as the Fraunhofer Institute for Systems and Innovation Research (ISI) in Karlsruhe, Germany or the Tampere University of Technology (TUT) in Pori, Finland. Her main research interest is in holistic approaches of innovation in all types of organizations—mainly, but not limited to—public administration, private enterprises and education. Since 2005, she has been involved in 6 EU financed projects within Erasmus+, Tempus, LLL, CIP, as well as other regional and national financed initiatives.

Jordi Castelló is PhD Student in the Doctoral Programme in Technology at the University of Girona, UdG (Spain). He has a Master in Business Administration (MBA) at the Universitat Politècnica de Catalunya (UPC), a Master in Operations Management at the same university and a Master's degree in Business Innovation and Technology Management at the UdG. He also has more than 15 years of experience in the Wind Business. His current research interest is in the area of Supply Chain Management, Business Process Management and Innovation Management.

Can Celikbilek receives his PhD from Ohio University, Industrial and Systems Engineering Department in 2016. He obtained his Transportation Engineering Master's degree in 2015, and he received his Master's degree in 2011 from the Department of Industrial and Systems Engineering at Ohio University. His research interests focus mainly on supply chain management, manufacturing system design and optimization, transportation and distribution modelling, data envelopment analysis (DEA) and business analytics.

Rodolfo de Castro is an Associate Professor in the Universitat de Girona (Spain). He belongs to the Department of Business Administration. He holds a PhD in Industrial Engineering at the UdG. He was Head of Department and Director of Industrial Engineering Master's Degree. His research area is in operation management, focused on Lean management and in supply chain management. He has several publications in JCR Journals and has collaborated in some projects financed by Spanish government and by private firms. He has supervised PhD Theses related to Supply Chain and Lean Management.

Daniel Gabaldón-Estevan Present Position: PhD Assistant professor at the Department of Sociology and Social Anthropology, Faculty of Social Sciences, University of Valencia. Previously (2004-2011) Doctoral Researcher at the Institute for Innovation and Knowledge Management (INGENIO), joint Institute of the Spanish Council for Scientific Research and the Polytechnic University of Valencia. Training: PhD Awarded Cum Laude, University of Valencia [2010] Graduate in Political Science and Administration, University of Valencia [2008] Masters Diploma of Advanced Studies in Sociology, University of Valencia [2005] Graduate in Sociology, University of Alicante [2000].

Arun Nambiar is an Associate Professor in the Department of Industrial Technology at California State University - Fresno. His research interests include production scheduling, agile manufacturing, lean methodologies, information technology in manufacturing and supply chain management.

Dušan Malindžák is a professor of Industrial Logistics and Industrial Process Control in the Institute of Logistics at the Technical University of Košice. He is a specialist in production process solutions, process control and improvements, SCM and heuristics. He uses his experience in many projects with real industrial companies in order to increase their effectiveness which resulted to reducing of costs and increasing their competitiveness.

Neo Malose Masemeni Construction Site Manager (Aveng Grinaker-LTA) Candidate Pr Construction Project Manager (SACPCMP).

Gürsel A. Süer joined the Industrial and Systems Engineering Department at Ohio University after working 12 years at the University of Puerto Rico-Mayaguez. He is on the editorial board of various journals. Currently, he serves as the Manufacturing Area Editor of the Computers and the Industrial Engineering Journal. He has Co-chaired two Computers and Industrial Engineering Conferences (1997-Puerto Rico, 2005-Istanbul). He also initiated Group Technology/Cellular Manufacturing Conferences held in

2000-Puerto Rico, 2003-Ohio, 2006-Netherlands, 2009-Japan. Most of his research has been motivated by his projects and observations in industrial settings. He has edited seven conference proceedings and four special issues with different journals. He has published over 170 papers in journals, edited books, conference proceedings and made over 100 technical presentations. He also advised around 50 graduate students.

Florea Nicoleta Valentina is a PhD Lecturer in the Management-Marketing Department, Valahia University of Targoviste, Romania, from 2012 and with over 14 years' experience as an economist at Valahia University of Targoviste. She studied Marketing, from 1993-1997, she obtained her Master's degree in Marketing in 2004, and her PhD thesis in Management, in 2011. She has published 39 articles in HRM, Knowledge Management, and Communication, and 4 books. Her main course and seminar activities are: leadership and management communication, PR and protocol, human resource management, marketing, management, simulation in marketing, services management, and international marketing. She was a member of the CEDIMES Institute, Paris, from 2009-2010, and is a member of the Centre for Research in Management and Marketing of Targoviste, since 2012. She received a doctoral scholarship at Paris Est, Creteil, Val-de-Marne, France, during April-July, 2010 and an academic scholarship at ASE Bucharest, between Jan-Feb., 2011.

Claudia-Maria Wagner currently works as a lecturer for the Dublin Institute of Technology. Prior to that, she was Principal Researcher and Programme Director at the National Institute for Transport and Logistics in Dublin, Ireland. She has worked with multinational companies on several industry-based research projects in China, Singapore, Malaysia, Finland, USA and Germany. Her research expertise covers ICT strategies and implementation challenges, with particular emphasis on B2B e-Marketplace diffusion in the transport industry and its impact on competitiveness and performance. Her research interests include global procurement strategies in an SCM context, the role of ICT in the design of sustainable supply chain configurations, as well as transport policy and regulation.

Gen'ichi Yasuda joined Information Science and Control Engineering Course in the Department of Mechanical Engineering, Faculty of Engineering, at Nagasaki Institute of Applied Science, Japan, as a Research Assistant, in 1983. He was a Professor in the Department of Mechanical Engineering during 1996-2005, a Professor in Department of Human and Computer Intelligence, Faculty of Informatics during 2005-2010, and also a Professor in the Division of Instrument and Control, Department of Electronics and Information Technology, Graduate School of Engineering during 1996-2010. He is now a Guest Professor at the Nagasaki Institute of Applied Science. He is a recipient of the CACS International Automatic Control Conference Best Paper Award in 2009. He also served as session chairman at many international conferences. He originated the frontier research on non-centralized systems and grouping robots as a doctoral dissertation at the Tokyo Institute of Technology since 1971. His current research projects include a multi-microcomputer-based development of automated noodle-making machines.

Index

www.ingramcontent.com/pod-product-compliance
Lightning Source LLC
LaVergne TN
LVHW062153300926
849161LV00010B/2657